THERMODYNAMIC CYCLES

CHEMICAL INDUSTRIES

A Series of Reference Books and Textbooks

Founding Editor

HEINZ HEINEMANN

1. *Fluid Catalytic Cracking with Zeolite Catalysts,* Paul B. Venuto and E. Thomas Habib, Jr.
2. *Ethylene: Keystone to the Petrochemical Industry,* Ludwig Kniel, Olaf Winter, and Karl Stork
3. *The Chemistry and Technology of Petroleum,* James G. Speight
4. *The Desulfurization of Heavy Oils and Residua,* James G. Speight
5. *Catalysis of Organic Reactions,* edited by William R. Moser
6. *Acetylene-Based Chemicals from Coal and Other Natural Resources,* Robert J. Tedeschi
7. *Chemically Resistant Masonry,* Walter Lee Sheppard, Jr.
8. *Compressors and Expanders: Selection and Application for the Process Industry,* Heinz P. Bloch, Joseph A. Cameron, Frank M. Danowski, Jr., Ralph James, Jr., Judson S. Swearingen, and Marilyn E. Weightman
9. *Metering Pumps: Selection and Application,* James P. Poynton
10. *Hydrocarbons from Methanol,* Clarence D. Chang
11. *Form Flotation: Theory and Applications,* Ann N. Clarke and David J. Wilson
12. *The Chemistry and Technology of Coal,* James G. Speight
13. *Pneumatic and Hydraulic Conveying of Solids,* O. A. Williams
14. *Catalyst Manufacture: Laboratory and Commercial Preparations,* Alvin B. Stiles
15. *Characterization of Heterogeneous Catalysts,* edited by Francis Delannay
16. *BASIC Programs for Chemical Engineering Design,* James H. Weber
17. *Catalyst Poisoning,* L. Louis Hegedus and Robert W. McCabe
18. *Catalysis of Organic Reactions,* edited by John R. Kosak
19. *Adsorption Technology: A Step-by-Step Approach to Process Evaluation and Application,* edited by Frank L. Slejko
20. *Deactivation and Poisoning of Catalysts,* edited by Jacques Oudar and Henry Wise
21. *Catalysis and Surface Science: Developments in Chemicals from Methanol, Hydrotreating of Hydrocarbons, Catalyst Preparation, Monomers and Polymers, Photocatalysis and Photovoltaics,* edited by Heinz Heinemann and Gabor A. Somorjai
22. *Catalysis of Organic Reactions,* edited by Robert L. Augustine

ADDITIONAL VOLUMES IN PREPARATION

THERMODYNAMIC CYCLES
COMPUTER-AIDED
DESIGN AND OPTIMIZATION

Chih Wu

United States Naval Academy
Annapolis, Maryland, U.S.A.

CRC Press
Taylor & Francis Group
Boca Raton London New York

CRC Press is an imprint of the
Taylor & Francis Group, an **informa** business

CRC Press
Taylor & Francis Group
6000 Broken Sound Parkway NW, Suite 300
Boca Raton, FL 33487-2742

First issued in paperback 2019

© 2004 by Taylor Francis Group, LLC
CRC Press is an imprint of Taylor & Francis Group, an Informa business

No claim to original U.S. Government works

ISBN-13: 978-0-8247-4298-0 (hbk)
ISBN-13: 978-0-367-39491-2 (pbk)

Library of Congress Cataloging-in-Publication Data
A catalog record for this book is available from the Library of Congress.

Visit the Taylor & Francis Web site at
http://www.taylorandfrancis.com

and the CRC Press Web site at
http://www.crcpress.com

To my wife, Hoying Tsai Wu,

and to my children, Anna, Joy, Sheree, and Patricia

Preface

Development in classical thermodynamics is, logically and traditionally, aimed at the analysis of cycles. Computational efforts impose harsh constraints on the kinds and amounts of cycle analyses that can reasonably be attempted. Cycle simulations cannot approach realistic complexity. Even relative sensitivity analyses based on grossly simplified cycle models are computationally taxing compared to their pedagogical benefits.

The desire for more design content in thermodynamic books is a familiar theme emanating from engineering industry and educational oversight bodies. Thermodynamic cycles are fertile ground for engineering design. Niches for innovative power generating systems for example, are being created by deregulation, cogeneration, vacillating fuel costs, and concern over global warming. Computerized look-up tables reduce computational labor somewhat, but modeling cycles with many interactive loops still lies well outside of power-engineering time budgets.

This book is based on intelligent computer software called Cyclepad. Cyclepad was developed by Professor K. Forbus of Northwestern University and evaluated by me. It is a powerful, mature, user-friendly freeware package developed expressly to simulate thermodynamic devices and cycles. It reduces by about 100-fold the computational effort involved in modeling realistically complex systems and cycles. It thus makes it feasible for engineers to run meaningful sensitivity analyses, to consider combinations of design modifications, to make engineering cost–benefit analyses, and to include refinements such as accounting for pressure changes and heat transfers occurring between major cycle components.

A previous book of mine (*Intelligent Computer-based Elementary Applied Thermodynamics*, 2002) covers the basic concept and laws of thermodynamics.

This book, along with Cyclepad, is intended to comprise the cycle portion in thermodynamics or to support cost–benefit analyses and to direct design projects. It is strictly design- and problem-solving oriented.

Released from the drain of repetitive and iterative hand calculations, power engineers can achieve a far wider and deeper analysis of cycles than has previously been possible.

Classical thermodynamics is based on the concept of "equilibrium." Time is not involved in conventional engineering thermodynamic textbooks. Heat transfer deals with rate of energy transfer, but does not cover cycles. There is a gap between thermodynamics and heat transfer. The chapter "Finite-Time Thermodynamics" bridges the gap between thermodynamics and heat transfer.

I have noted some attitudinal benefits during my eight semesters of Cyclepad-assisted teaching of thermodynamics at the U.S. Naval Academy and at Johns Hopkins University. Students tend from the outset to be more positively disposed toward computer-assisted learning of education thermodynamics, with quite a few describing it as "fun". (Not surprisingly, higher education is increasingly embracing computer-assisted course work.) Material that is more positively regarded tends to be better retained. Further, an ability to execute realistically complicated cycle simulations builds confidence and a sense of professionalism.

Both Cyclepad and the text contain pedagogical aids. The intelligent computer software switches to a warning-tutoring mode when users attempt to impose erroneous assumptions or perform inappropriate operations during cycle analyses. Chapter summaries review the more salient points and provide cohesion. Review problems and worked examples appear liberally throughout the text. Both SI and English unit systems are used in the book.

I wish to acknowledge the following individuals who encouraged me and assisted in the text preparation: Dr. Susan Chipman of the Naval Office of Research, Professor Ken Forbus of Northwestern University, Professor Myron Miller of Johns Hopkins University, and Professor Al Adams, Professor Joe Gillerlain, Jr., Associate Professor Karen Flack, Ensign Donald Sherrill, Mr. Doug Richardson, and Ms. Katherine Kindig of the U.S. Naval Academy.

Chih Wu

Contents

1
Thermodynamic Concepts

1.1 INTRODUCTION

Research in cognitive science potentially offers tremendous benefits in education, industry, and training. Psychological studies of human learning can yield insights as to what forms of instruction might be most beneficial for particular purposes, and what mental models and misconceptions students might have. Artificial intelligence research can provide formal representations and reasoning techniques that can be used in new kinds of educational software that itself contains a deep understanding of the domain being taught. These scientific advances, coupled with dramatic changes in computing technology (i.e., increased computing power at decreasing costs), provide new opportunities for exploring how intelligent software can be used in thermodynamic education and industry.

Thermodynamic design, for example, can be relieved of repetitive hand calculation, iterative solution techniques, and two-way interpolations. Beyond this is a pervasive cultural preference of this generation towards interacting with computers.

Throughout, the book is software oriented. The text is tightly integrated with "CyclePad," an intelligent, user-friendly software which performs first and second law analyses for user-defined inputs and processes, solves simultaneous equations, looks up state data, and attractively graphs processes in phase spaces. In reducing computational effort more than 10-fold, this software enables and requires engineers to work many more cycle problems, and just as importantly, to do so in a timely way. Such efficient learning strategy boosts engineer confidence.

1.2 INTELLIGENT COMPUTER-AIDED SOFTWARE

An intelligent computer software called CyclePad is used throughout the book. CyclePad has been codeveloped by Oxford University and

Northwestern University since 1995, and evaluated by Professor C. Wu in the Mechanical Engineering Department at the U.S. Naval Academy.

Normally, calculation in a cycle analysis is lengthy, complex, and tedious for an engineer. The analysis would be more meaningful, time saving, and fun for engineers using CyclePad as a tool to help in the cycle analysis, design, and optimization. Engineers would be able to change any parameter in the cycle and see the effect of the parameter on the performance of the cycle.

The value that CyclePad brings to its users is a reduction in time while moving from concept to finished design. This time saving allows the user to move to his physical prototyping stage faster, thereby eliminating certain prototyping costs and improving cycle design more rapidly. Virtual prototyping with CyclePad follows the same basic steps used in traditional or physical design process. The user builds his cycle design, runs it through analyses, validates results, refines the design, and iterates this process until the cycle's performance is optimized. The difference is that the user manages this process virtually through a computer, much more quickly, easily, and cost-efficiently.

Design is a trying and learning process. Engineers would be able to choose various components in their design of power plants or choose different working fluids in the plants for a specific mission by using CyclePad as a tool. They would be able to gain a tremendous design experience and knowing what they are doing in a relative short time.

CyclePad is made from a design and coaching perspective view. CyclePad is designed to help with the learning and conceptual design of thermodynamic cycles. It works in two phases, build and analysis. There are three modes (build, analysis, and contradiction) in the software.

1.2.1 Build Mode

In the build mode, a user takes components out from either a thermo-dynamic open-system inventory shop or a thermodynamic closed-system inventory shop and connects them to form a state or several states, a process or several processes, or a cycle or several cycles.

1.2.2 Analysis Mode

In the analysis mode, a user chooses working fluid, process assumption for each component, and input numerical property values. All the calculations are then quickly done by the software. There is a sensitivity tool that makes cycle performance parameter effects easy and quick, and generates the effect in graph form.

1.2.3 Contradiction Mode

There is a coach (senior engineer) in the software. If there is a mistake or a contradiction made by the user, the coach will show up and tell the contradiction and suggest ways to solve the contradiction.

CyclePad can be downloaded (free from Northwestern University's web page at http://www.qrg.ils.northwestern.edu) by the following steps:

> Internet Explorer
> http://www.qrg.ils.northwestern.edu
> Software
> Download CyclePad v2.0
> Fill in form—submit form
> License agreement—download software
> Save this program to disk—OK
> Choose a location—save
> Close Internet Explorer
> Open my computer
> Go to the location you save the file to
> Double click on webcpad—#######.exe
> Double click on Setup.exe
> > Install
> > Yes
> > OK
> > OK—CyclePad is downloaded

1.2.4 Installation Into Your Own PC

CyclePad can be installed into your own PC by the following steps:

> Make new folder
> Temp-CyclePad
> Copy files from Zip to new folder Temp-CyclePad
> Go to Temp-CyclePad
> Double click–webcpad–20000504   20000504 (date)
> Double click–cpadinst
> Double click–setup
> Install
> Yes
> Yes
> C:\CyclePad
> OK
> OK

1.3   REVIEW OF THERMODYNAMIC CONCEPTS

Thermodynamics is a science in which the storage, transformation, and transfer of energy E and entropy S are studied. Thermodynamics is governed by four basic laws called (1) the zeroth law of thermodynamics, (2) the first law of thermodynamics, (3) the second law of thermodynamics, and (4) the third law of thermodynamics.

The zeroth law states that if two bodies are in thermal equilibrium with a third body, they are also in thermal equilibrium with each other.

The first law is an expression of the conservation of energy.

The second law states that actual isolated processes occur in the direction of decreasing quality of energy.

The third law states that the entropy of a pure substance at absolute zero temperature is zero.

A thermodynamic system is either a quantity of matter called a control mass or closed system, or a region in space called a control volume or open system chosen for study. The mass or region outside the system is called the surroundings. The real or imaginary surface that separates the system from its surroundings is called the boundary. A closed system or a control mass consists of a fixed amount of mass, and no mass can cross its boundary. It usually encloses a device that involves no mass flow such as gas contained in a cylinder-piston setup. An open system or a control volume is a selected region in space. It usually encloses a device that involves mass flow such as a pump, compressor, turbine, mixing chamber, separator, heat exchanger, boiler, condenser, or evaporator. Flow through these devices is best studied by selecting the region within the device as the open system.

Any characteristic of a system is called a property. The essential feature of a property is that it has a unique value when a system is in a particular state. Properties are considered to be either intensive or extensive. Intensive properties are those that are independent of the size of a system, such as temperature T and pressure p. Extensive properties are those that are dependent on the size of a system, such as volume V, internal energy U, and entropy S. Extensive properties per unit mass are called specific properties such as specific volume v, specific internal energy u, and specific entropy s. Properties can be either measurable such as temperature T, volume V, pressure p, specific heat at constant pressure process c_p, and specific heat at constant volume process c_v, or nonmeasurable such as internal energy U and entropy S. A relatively small number of independent properties suffice to fix all other properties and thus the state of the system. If the system is composed of a single phase, free from magnetic, electrical, chemical, and surface effects, the state is fixed when any two independent intensive properties are fixed.

A system is in equilibrium if it maintains thermal, mechanical, and phase equilibrium. Any change that a thermodynamic system undergoes from one equilibrium state to another is called a process, and the series of states through which a system passes during a process is called the path of the process. A reversible process is a process without irreversibility factors and the process can be reversed without leaving any trace on the surroundings. An irreversible process is a process with irreversibility factors such as friction, heat transfer due to finite temperature difference, free expansion, mixing, etc., and the process cannot be reversed without leaving any trace on the surroundings. An internally reversible process is a process without irreversibility factors within the boundary of the system during the process. An externally reversible process is a process without irreversibility factors outside the boundary of the system during the process. A totally reversible process is a process without irreversibility factors within the boundary of the system and its surroundings during the process. An isochoric or isometric process is a process during which the volume V remains constant. An isobaric process is a process during which the pressure p remains constant. An isothermal process is a process during which the temperature T remains constant. A throttling process is a process during which the enthalpy h remains constant. An adiabatic process is a process during which there is no heat transfer between the system and its surroundings. A polytropic process is a process during which the quantity of pV^n remains constant, where n is a nondimensional polytropic index. An isentropic process is a process during which the entropy S remains constant. An isentropic process is also known as an adiabatic and reversible process. A steady-flow process is a process in which no properties within the control volume change with time. Notice that a large number of engineering systems such as power plants, refrigerators, and heat pumps operate for long periods under the same conditions once the transient start-up period is completed and steady operation (cruise condition) is established, and they are classified as steady-flow systems. Cycles and processes involving such systems are called steady-flow processes. Since the design, analysis, and test of engineering systems are all based on cruise condition, all of our open cycle analyses in this book are assumed to be steady flow.

There are three kinds of thermodynamic substances used in thermodynamic cycles. The three substances are (1) pure substance, (2) ideal gas, and (3) liquid and solid.

A pure substance is a homogeneous simple compressible substance. There are three phases (solid, liquid, and vapor) of a pure substance. A pure substance state may occur in more than one phase, but each phase must have the same chemical composition. A pure substance exists in different phases, depending on its energy level. In the compressed liquid phase, the pressure of the substance is higher than the boiling pressure. During a phase-change process, the substance is a mixture of saturated liquid and saturated vapor, and the boiling pressure and boiling temperature are dependent. In a saturated liquid and saturated vapor mixture, the mass fraction of vapor is called the quality x and is expressed as $x = m_{vapor}/m_{total}$. In the superheated vapor phase, the temperature of the substance is higher than the boiling temperature. Water, ammonia, and freons are pure substances. The properties of a pure substance is usually found in tables such as steam tables and freon tables.

An ideal gas is a relatively low-density gas. The pressure p, temperature T, and specific volume v of an ideal gas are related by an equation of state, $pv = RT$, where R is a constant for a particular gas and is called the gas constant. Air, helium, and carbon dioxide are ideal gases. The properties of an ideal gas can be found in tables such as air tables.

Solids are usually not used in cycles. Liquids can be approximated as incompressible ($dv = 0$) substances since their specific volumes remain near constant during a process.

Energy can cross the boundary of a system without mass transfer in either macroscopic form called work (W) or microscopic form called heat (Q). Boundary work is due to a pressure difference and causes a system volume displacement (dV). The boundary work of a process is given by the expression $W = \int p\,dV$. On a p–V diagram, the boundary work of a process is the area underneath the process path. Heat is due to a temperature difference and causes a system entropy displacement (dS). The heat of a process is given by the expression $Q = \int T\,dS$. On a T–S diagram, the heat of a process is the area underneath the process path.

Elementary thermodynamics is based on the following three fundamental physical relations applied to the system being analyzed:

1. Mass m balance: mass change = mass transfer
2. Energy E balance: energy change = energy transfer
3. Entropy S balance: entropy change = entropy transfer + entropy generation

The relations above are applicable to any kind of system undergoing any kind of process or cycle.

1.4 THERMODYNAMIC CYCLIC SYSTEMS

An energy conversion cyclic system that converts heat to work, and sometimes work to heat, is the foremost objective of thermodynamics.

The former is manifested by heat engines of one sort or another, the latter by heat pumps or refrigeration cycles. The three thermodynamic cyclic systems (heat engine, heat pump, and refrigerator) are described in the following subsections.

1.4.1 Heat Engine

A *heat engine* is a continuous cyclic device that produces positive net work output by adding heat. The energy flow diagram of a heat engine and its thermal reservoirs are shown in Fig. 1.1. A *thermal reservoir* is any object or system that can serve as a heat source or sink for another system. Thermal reservoirs usually have accumulated energy capacities that are very, very large compared with the amounts of heat energy they exchange. Therefore, the thermal reservoirs are considered to operate at constant temperatures. Heat (Q_H) is added to the heat engine from a high-temperature thermal reservoir at T_H, output work (W) is done by the heat engine, and heat (Q_L) is removed from the heat engine to a low-temperature thermal reservoir at T_L.

An example of this is a commercial central power station using a heat engine called a Rankine steam power plant. The Rankine heat engine

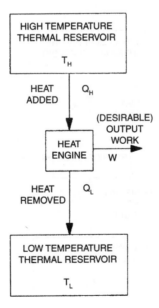

Figure 1.1 Heat engine.

consists of a pump, a boiler, a turbine, and a condenser. Heat (Q_H) is added to the working fluid (water) in the boiler from a high-temperature (T_H) flue gas by burning coal or oil. Output work (W_o) is done by the turbine. Heat (Q_L) is removed from the working fluid in the condenser to low-temperature (T_L) lake cooling water. Input work (W_i) is added to drive the pump. The net work (W) produced by the Rankine heat engine is $W = W_o - W_i$.

The measurement of performance for a heat engine is called the thermal *efficiency,* η. The thermal efficiency of a heat engine is defined as the ratio of the desirable net output work sought to the heat input of the engine: $\eta = W_{net}/Q_{input}$.

1.4.2 Refrigerator

A *refrigerator* is a continuous cyclic device that removes heat from a low-temperature reservoir to a high-temperature reservoir at the expense of work input. The energy flow diagram of a refrigerator and its thermal reservoirs are shown in Fig. 1.2. Input work (W) is added to the refrigerator, desirable heat (Q_L) is removed from the low-temperature thermal reservoir at T_L, and heat (Q_H) is added to the high-temperature thermal reservoir at T_H.

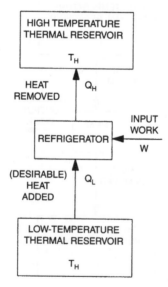

Figure 1.2 Refrigerator.

An example of a refrigerator is a domestic refrigerator, which is made up of a compressor, a condenser, an expansion valve, and an evaporator. Work (W) is added to drive the compressor by an electric motor, heat (Q_L) is added in the evaporator from the low-temperature (T_L) refrigerator inner space by the working fluid (refrigerant), and heat (Q_H) is removed from the condenser from the working fluid to the high-temperature (T_H) reservoir (kitchen).

The measurement of performance for a refrigerator is called the *coefficient of performance* (COP), and is denoted by β_R. The coefficient of performance of a refrigerator is defined as the ratio of the desirable heat removed Q_L to the work input W of the refrigerator: $\beta_R = Q_L/W$.

Q_L is called *refrigerator capacity* and is usually expressed in tons of refrigeration. One *ton of refrigeration* is 3.516 kW or 12,000 Btu/h. The term "ton" is derived from the fact that the rate of heat required to melt one ton of ice is about 12,000 Btu/h.

Notice that the coefficient of performance of a refrigerator may be larger or smaller than unity.

1.4.3 Heat Pumps

A *heat pump* is a continuous cyclic device that pumps heat to a high-temperature reservoir from a low-temperature reservoir at the expense of work input. The energy flow diagram of a heat pump and its thermal reservoirs are shown in Fig. 1.3. Input work (W) is added to the heat pump, desirable heat (Q_H) is pumped to the high-temperature thermal reservoir at T_H, and heat (Q_L) is removed from the low-temperature thermal reservoir at T_L.

An example of a heat pump is a house heat pump, which is made up of a compressor, a condenser, a throttling valve, and an evaporator. Heat (Q_H) is removed to the high-temperature house from the working fluid in the condenser, work is added to the compressor by an electric motor, and heat (Q_L) is added to the evaporator from the low-temperature outside air in the winter season.

Notice that the energy flow diagram and hardware components of a heat pump are exactly the same as those of a refrigerator. The difference between the heat pump and the refrigerator is the function of the cyclic devices. A refrigerator is used to remove heat (Q_L) from a low-temperature (T_L) thermal reservoir by adding work. A heat pump is used to add heat (Q_H) to a high-temperature (T_H) thermal reservoir by adding work.

The measurement of performance for a heat pump is called the coefficient of performance (COP), and is denoted by β_{HP}. The coefficient

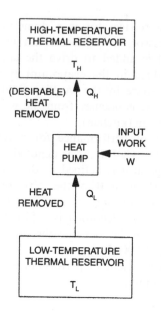

Figure 1.3 Heat pump.

of performance of a heat pump is defined as the ratio of the desirable heat output Q_H to the work input W of the heat pump: $\beta_{HP} = Q_H/W$.

Notice that the coefficient of performance of a heat pump is always larger than unity.

1.5 CYCLES

Thermodynamic cycles can be divided into two general categories: heat engine cycles, which are discussed in Chapters 2–5, and refrigeration and heat pump cycles, which are discussed in Chapter 6.

Thermodynamic cycles can be categorized by phase as vapor cycles and gas cycles, depending on the phase of the working fluid. In vapor cycles, the working fluid exists in the vapor phase during one part of the cycle and in the liquid phase during another part of the cycle. In gas cycles, the working fluid remains in the gaseous phase throughout the entire cycle.

Thermodynamic cycles can also be categorized by system as closed-system cycles and open-system cycles. In closed-system cycles, each component of the cycle is considered as a closed system. In open-system cycles, each component of the cycle is considered as an open system.

1.6 CARNOT CYCLE

The *Carnot cycle* is of historical importance. The reversible cycle was introduced by a French engineer N.S. Carnot in 1824 and led to the development of the second law of thermodynamics. The importance of the Carnot cycle is that it sets up a standard thermal cycle performance for the actual cycles to compare with.

Considering the concepts of reversible processes, a reversible cycle can be carried out for given thermal reservoirs at temperatures T_H and T_L. The Carnot heat engine cycle on a p–V diagram and a T–S diagram, as shown in Fig. 1.4 is composed of the following four reversible processes:

> 1-2 Reversible adiabatic (isentropic) compression
> 2-3 Reversible isothermal heating at T_H
> 3-4 Reversible adiabatic (isentropic) expansion
> 4-1 Reversible isothermal cooling at T_L

Referring to Fig. 1.4, the system undergoes a Carnot heat engine cycle in the following manner:

1. During process 1-2, the system is thermally insulated and the temperature of the working substance is raised from the low temperature T_L to the high temperature T_H. The process is an isentropic process. The amount of heat transfer during the process is $Q_{12} = \int T dS = 0$, because there is no area underneath a constant entropy (vertical) line.

2. During process 2-3, heat is transferred isothermally to the working substance from the high-temperature reservoir at T_H. This process is accomplished reversibly by bringing the system in contact with the high-temperature reservoir whose temperature is equal to or infinitesimally higher than that of the working substance. The amount of heat transfer during the process is $Q_{23} = \int T dS = T_H(S_3 - S_2)$, which can be represented

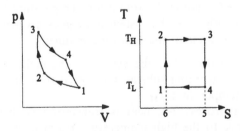

Figure 1.4 Carnot heat engine cycle on p–v and T–s diagram.

by the area 2-3-5-6-2; Q_{23} is the amount of heat added to the Carnot cycle from a high-temperature thermal reservoir.

3. During process 3-4, the system is thermally insulated and the temperature of the working substance is decreased from the high temperature T_H to the low temperature T_L. The process is an isentropic process. The amount of heat transfer during the process is $Q_{34} = \int T dS = 0$, because there is no area underneath a constant entropy (vertical) line.

4. During process 4-1, heat is transferred isothermally from the working substance to the low-temperature reservoir at T_L. This process is accomplished reversibly by bringing the system in contact with the low-temperature reservoir whose temperature is equal to or infinitesimally lower than that of the working substance. The amount of heat transfer during the process is $Q_{41} = \int T dS = T_L(S_1 - S_4)$, which can be represented by the area 1-4-5-6-1; Q_{41} is the amount of heat removed from the Carnot cycle to a low-temperature thermal reservoir.

The net heat added to the cycle is $Q_{net} = Q_{12} + Q_{23} + Q_{34} + Q_{41} = 0 + T_H(S_3 - S_2) + 0 + T_L(S_1 - S_4) = (T_H - T_L)(S_4 - S_1) = $ [area 2-3-5-6-2]-[area 1-4-5-6-1] = area 1-2-3-4-1. Notice that the area 1-2-3-4-1 is the area enclosed by the cycle.

The net work produced to the cycle is $W_{net} = Q_{net} = $ area 1-2-3-4-1.

According to the definition of heat engine efficiency, the *efficiency of the Carnot heat engine* is $\eta_{Carnot} = W_{net\ output}/Q_{input} = $ [area 1-2-3-4-1]/[area 2-3-5-6-2] $= (T_H - T_L)(S_4 - S_1)/[T_H(S_4 - S_1)] = (T_H - T_L)/T_H$, or

$$\eta_{Carnot} = 1 - T_L/T_H \qquad\qquad (1.1)$$

If the Carnot cycle for a heat engine is carried out in the reverse direction, the result will be either a *Carnot heat pump* or a Carnot refrigerator. Such a cycle is shown in Fig. 1.5. Using the same graphical explanation that was used in the Carnot heat engine, the heat added from the low-temperature reservoir at T_L is area 1-4-5-6-1; Q_{41} is the amount of heat added to the Carnot cycle from a low-temperature thermal reservoir.

Referring to Fig. 1.5, the system undergoes a Carnot heat pump or Carnot refrigerator cycle in the following manner:

1. During process 1-2, the system is thermally insulated and the temperature of the working substance is raised from the low temperature T_L to the high temperature T_H.

2. During process 2-3, heat is transferred isothermally from the working substance to the high temperature reservoir at T_H. This process is accomplished reversibly by bringing the system in contact with the

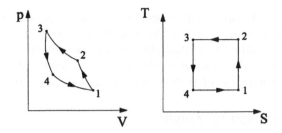

Figure 1.5 Carnot heat pump or Carnot refrigerator cycle on p–V and T–S diagram.

high-temperature reservoir whose temperature is equal to or infinitesimally lower than that of the working substance.

3. During process 3-4, the system is thermally insulated and the temperature of the working substance is decreased from the high temperature T_H to the low temperature T_L.

4. During process 4-1, heat is transferred isothermally to the working substance from the low-temperature reservoir at T_L. This process is accomplished reversibly by bringing the system in contact with the low-temperature reservoir whose temperature is equal to or infinitesimally higher than that of the working substance.

The net heat added to the cycle is $Q_{net} = Q_{12} + Q_{23} + Q_{34} + Q_{41} = 0 + T_H(S_3 - S_2) + 0 + T_L(S_1 - S_4) = (T_H - T_L)(S_4 - S_1) = $ [area 2-3-5-6-2]-[area 1-4-5-6-1] = Area 1-2-3-4-1. Notice that the area 1-2-3-4-1 is the area enclosed by the cycle.

The net work added to the cycle is $W_{net} = Q_{net} = $ area 1-2-3-4-1. According to the COP (β) definition of a refrigerator, the *COP of the Carnot refrigerator* is $\beta_{Carnot,R} = Q_{desirable\ output}/W_{input} = $ [area 4-1-5-6-4]/ [area 1-2-3-4-1] $= (T_L)(S_2 - S_3)/[(T_H - T_L)(S_2 - S_3)] = T_L/(T_H - T_L)$, or

$$\beta_{Carnot,R} = 1/(T_H/T_L - 1) \qquad (1.2)$$

The net work added to the cycle is $W_{net} = Q_{net} = $ area 1-2-3-4-1. According to the COP (β) definition of a heat pump, the *COP of the Carnot heat pump* is $\beta_{Carnot,HP} = Q_{desirable\ output}/W_{input} = $ [area 2-3-5-6-2]/ [area 1-2-3-4-1] $= (T_H)(S_2 - S_3)/[(T_H - T_L)(S_2 - S_3)] = T_H/(T_H - T_L)$, or

$$\beta_{Carnot,HP} = 1/(1 - T_L/T_H) \qquad (1.3)$$

A reversible isothermal heat-transfer process between the Carnot cycle and its surrounding thermal reservoirs is impossible to achieve because it would require either an infinitely large heat exchanger or an infinitely long time to complete the process. Therefore, it is not economic to build a practical cyclic system that would operate closely to the Carnot cycle. However, the Carnot cycle sets up a standard thermal cycle efficiency or coefficient of performance for actual cycles to compare with.

1.7 CARNOT COROLLARIES

Six corollaries deduced from the Carnot cycle are of great use in comparing the performance of cycles. The corollaries are:

1. The efficiency of the Carnot heat engine operating between a fixed high-temperature heat source thermal reservoir at T_H and a fixed low-temperature heat sink thermal reservoir at T_L is irrespective of the working substance.

2. No heat engine operating between a fixed high-temperature heat source thermal reservoir and a fixed low-temperature heat sink thermal reservoir can be more efficient than a Carnot heat engine operating between the same two thermal reservoirs.

3. All reversible heat engines operating between a fixed high-temperature heat source thermal reservoir and a fixed low-temperature heat sink thermal reservoir have the same efficiency.

4. The COP (coefficient of performance) of the Carnot heat pump (or refrigerator) operating between a fixed high-temperature thermal reservoir at T_H and a fixed low-temperature thermal reservoir at T_L is irrespective of the working substance.

5. No heat pump (or refrigerator) operating between a fixed high-temperature thermal reservoir and a fixed low-temperature thermal reservoir can have higher COP (coefficient of performance) than a Carnot heat pump (or refrigerator) operating between the same two thermal reservoirs.

6. All reversible heat pumps (or refrigerators) operating between a fixed high-temperature thermal reservoir and a fixed low-temperature thermal reservoir have the same COP (coefficient of performance).

These corollaries can be proved by demonstrating that the violation of any one of them results in the violation of the second law of thermodynamics.

2
Vapor Cycles

2.1 CARNOT VAPOR CYCLE

Theoretically, the Carnot vapor cycle is the most efficient vapor power cycle operating between two temperature reservoirs.

The *Carnot vapor cycle* as shown in Fig. 2.1 is composed of the following four processes:

1-2 Isentropic compression
2-3 Isothermal heat addition
3-4 Isentropic expansion
4-1 Isothermal heat rejection

In order to achieve the isothermal heat addition and isothermal heat rejection processes, the Carnot simple vapor cycle must operate inside the vapor dome. The *T–S* diagram of a Carnot cycle operating inside the vapor dome is shown in Fig. 2.2. Saturated water at state 2 is evaporated isothermally to state 3, where it is saturated vapor. The steam enters a turbine at state 3 and expands isentropically, producing work, until state 4 is reached. The vapor–liquid mixture would then be partially condensed isothermally until state 1 is reached. At state 1, a pump would isentropically compress the vapor–liquid mixture to state 2.

Applying the first law and second law of thermodynamics of the open system to each of the four processes of the Carnot vapor cycle yields:

$$Q_{12} = 0 \tag{2.1}$$

$$W_{12} = m(h_1 - h_2) \tag{2.2}$$

$$W_{23} = 0 \tag{2.3}$$

$$Q_{23} - 0 = m(h_3 - h_2) \tag{2.4}$$

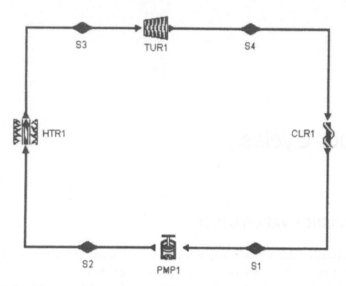

Figure 2.1 Carnot vapor cycle.

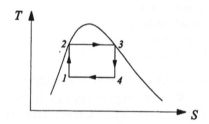

Figure 2.2 Vapor Carnot cycle *T–S* diagram.

$$Q_{34} = 0 \tag{2.5}$$

$$W_{34} = m(h_3 - h_4) \tag{2.6}$$

$$W_{41} = 0 \tag{2.7}$$

and

$$Q_{41} - 0 = m(h_1 - h_4) \tag{2.8}$$

The net work (W_{net}), which is also equal to net heat (Q_{net}), is

$$W_{net} = Q_{net} = Q_{23} + Q_{41} \tag{2.9}$$

The thermal efficiency of the cycle is

$$\eta = W_{net}/Q_{23} = Q_{net}/Q_{23} = 1 - Q_{41}/Q_{23} = 1 - (h_4 - h_1)/(h_3 - h_2) \quad (2.10)$$

Example 2.1

A steam Carnot cycle operates between 250°C and 100°C. Determine the pump work, turbine work, heat added, quality of steam at the exit of the turbine, quality of steam at the inlet of the pump, and cycle efficiency.

To solve this problem by CyclePad, we take the following steps:

1. Build:
   a. Take a pump, a boiler, a turbine, and a condenser from the inventory shop and connect the four devices to form the basic Rankine cycle.
   b. Switch to analysis mode.
2. Analysis:
   a. Assume a process each for the four devices: (1) pump as adiabatic and isentropic, (2) boiler as isothermal (isobaric inside the vapor dome), (3) turbine as adiabatic and isentropic, and (4) condenser as isothermal (isobaric inside the vapor dome).
   b. Input the given information: (1) working fluid is water, (2) inlet temperature and quality of the boiler are 400°C and 0, (3) exit quality of the boiler is 1, and (4) inlet temperature of the condenser is 100°C.
3. Display results:
   a. Display the *T–s* diagram and cycle properties results. The cycle is a heat engine. The answers are pump work = −111.9 kJ/kg, turbine work = 603.7 kJ/kg, and $\eta = 28.67\%$.
   b. Display the sensitivity diagram of cycle efficiency versus boiler temperature. (See Fig. 2.3.)

COMMENTS: The Carnot vapor cycle as illustrated by Example 2.1 is not practical. Difficulties arise in the isentropic processes of the cycle. One difficulty is that the isentropic turbine will have to handle steam of low quality. The impingement of liquid droplets on the turbine blade causes erosion and wear. Another difficulty is the isentropic compression of a liquid–vapor mixture. The two-phase mixture of the steam causes serious cavitation problems during the compression process. Also, since the specific volume of the saturated mixture is high, the pump power required is also very high. Thus, the Carnot vapor cycle is not a realistic model for vapor power cycles.

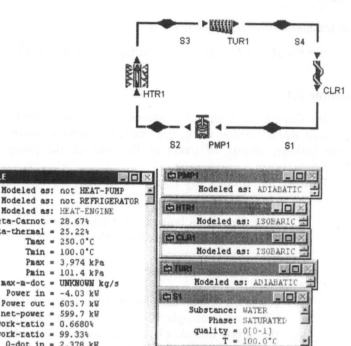

Figure 2.3 Carnot vapor cycle.

Review Problems 2.1 Carnot Vapor Cycle

1. Why is excessive moisture in a steam undesirable in a steam
    turbine? What is the highest moisture content allowed?
2. Why is the Carnot cycle not a realistic model for steam power
    plants?
3. A Carnot engine with a steady flow rate of 1 kg/sec uses water as
    the working fluid. Water changes phase from saturated liquid to
    saturated vapor as heat is added from a heat source at 300°C.
    Heat rejection takes place at a pressure of 10 kPa. Determine
    (1) the quality at the exit of the turbine, (2) the quality at the
    inlet of the pump, (3) the heat transfer added in the boiler, (4)
    the power required for the pump, (5) the power produced by the
    turbine, (6) the heat transfer rejected in the condenser, and (7)
    the cycle efficiency.

2.2 BASIC RANKINE VAPOR CYCLE

The working fluid for vapor cycles is alternately condensed and vaporized. When a working fluid remains in the saturation region at constant pressure, its temperature is also constant. Thus, the condensation or evaporation of a fluid in a heat exchanger is a process that closely approximates the isothermal heat-transfer processes of the Carnot cycle. Owing to this fact, vapor cycles closely approximate the behavior of the Carnot cycle. Thus, in general, they tend to perform efficiently.

The *Rankine cycle* is a modified Carnot cycle for overcoming the difficulties with the latter cycle when the working fluid is a vapor. In the Rankine cycle, the heating and cooling processes occur at constant pressure. Figure 2.4 shows the devices used in a basic Rankine cycle, and Fig. 2.5 is the *T–s* diagram of the basic Rankine cycle.

The *basic Rankine cycle* is composed of the following four processes:

1-2 Isentropic compression
2-3 Isobaric heat addition
3-4 Isentropic expansion
4-1 Isobaric heat removing

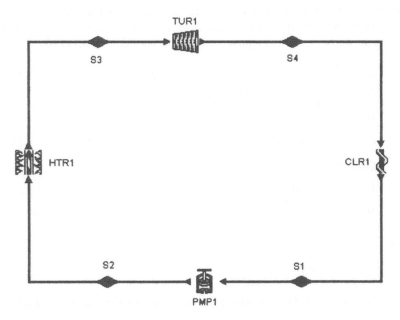

Figure 2.4 Basic Rankine cycle.

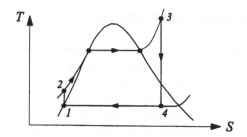

Figure 2.5 Basic Rankine cycle T–S diagram.

Water enters the pump at state 1 as a low-pressure saturated liquid to avoid the cavitation problem and exits at state 2 as a high-pressure compressed liquid. The heat supplied in the boiler raises the water from the compressed liquid at state 2 to saturated liquid to saturated vapor and to a much higher temperature superheated vapor at state 3. The super-heated vapor at state 3 enters the turbine where it expands to state 4. The superheating moves the isentropic expansion process to the right on the T–s diagram as shown in Fig. 2.5, thus preventing a high moisture content of the steam as it exits the turbine at state 4 as a saturated mixture. The exhaust steam from the turbine enters the condenser at state 4 and is condensed at constant pressure to state 1 as saturated liquid.

Applying the first and second laws of thermodynamics of the open system to each of the four processes of the Rankine cycle yields:

$$Q_{12} = 0 \tag{2.11}$$

$$W_{12} = m(h_1 - h_2) = mv_1(p_1 - p_2) \tag{2.12}$$

$$W_{23} = 0 \tag{2.13}$$

$$Q_{23} - 0 = m(h_3 - h_2) \tag{2.14}$$

$$Q_{34} = 0 \tag{2.15}$$

$$W_{34} = m(h_3 - h_4) \tag{2.16}$$

$$W_{41} = 0 \tag{2.17}$$

and

$$Q_{41} - 0 = m(h_1 - h_4) \tag{2.18}$$

The net work (W_{net}), which is also equal to the net heat (Q_{net}), is

$$W_{\text{net}} = Q_{\text{net}} = Q_{23} + Q_{41} \tag{2.19}$$

The thermal efficiency of the cycle is

$$\eta = W_{net}/Q_{23} = Q_{net}/Q_{23} = 1 - Q_{41}/Q_{23} = 1 - (h_4 - h_1)/(h_3 - h_2)$$

(2.20)

Example 2.2

Determine the efficiency and power output of a basic Rankine cycle using steam as the working fluid in which the condenser pressure is 80 kPa. The boiler pressure is 3 MPa. The steam leaves the boiler as saturated vapor. The mass rate of steam flow is 1 kg/sec. Show the cycle on a *T–s* diagram. Plot the sensitivity diagram of cycle efficiency versus boiler pressure.

To solve this problem by CyclePad, we take the following steps:

1. Build:
   a. Take a pump, a boiler, a turbine, and a condenser from the inventory shop and connect the four devices to form the basic Rankine cycle.
   b. Switch to analysis mode.
2. Analysis:
   a. Assume a process each for the four devices: (a) pump as isentropic, (2) boiler as isobaric, (3) turbine as isentropic, and (4) condenser as isobaric.
   b. Input the given information: (1) working fluid is water, (2) inlet pressure and quality of the pump are 80 kPa and 0, (3) inlet pressure and quality of the turbine are 3 MPa and 1, and (4) mass flow rate is 1 kg/sec.
3. Display results:
   a. Display the *T–s* diagram and cycle properties' results. The cycle is a heat engine. The answers are $\eta = 24.61\%$, $Wdot_{pump} = -3.07\,kW$, $Qdot_{boiler} = 2409\,kW$, $Wdot_{turbine} = 595.7\,kW$, $Qdot_{condenser} = -1816\,kW$, and net power output $= 592.7\,kW$.
   b. Display the sensitivity diagram of cycle efficiency versus boiler pressure.
   c. Display the sensitivity diagram of cycle efficiency versus condenser pressure. (See Fig. 2.6.)

COMMENTS: (1) The sensitivity diagram of cycle efficiency versus boiler pressure demonstrates that increasing the boiler pressure increases the boiler temperature. This raises the average temperature at which heat is added to the steam and thus raises the cycle efficiency. Operating pressures of boilers have increased over the years up to 30 MPa (4500 psia) today.

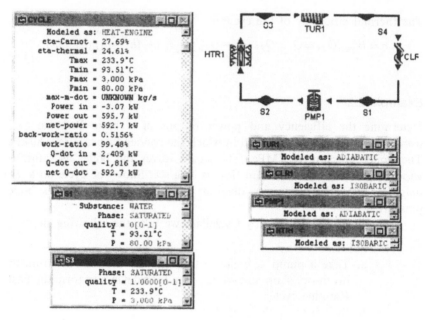

Figure 2.6 Rankine cycle.

(2) The sensitivity diagram of cycle efficiency versus condenser pressure demonstrates that decreasing the condenser pressure decreases the condenser temperature. This drops the average temperature at which heat is removed to the surroundings and thus raises the cycle efficiency. Operating pressures of condensers have decreased over the years to 5 kPa (0.75 psia) today.

The Rankine efficiency could be increased by increasing the boiler pressure, since the area enclosed by the cycle (net work) in the T–s diagram will be increased.

Example 2.3

A simple Rankine cycle using water as the working fluid operates between a boiler pressure of 500 psia and a condenser pressure of 20 psia. The mass flow rate of the water is 3 lbm/sec. Determine (1) the quality of the steam at the exit of the turbine, (2) the net power of the cycle, and (3) the cycle efficiency. Then change the boiler pressure to 600 psia, and determine (4) the quality of the steam at the exit of the turbine and (5) the net power of the cycle.

To solve this problem by CyclePad, we take the following steps:

1. Build:
   a. Take a pump, a boiler, a turbine, and a condenser from the inventory shop and connect the four devices to form the basic Rankine cycle.
   b. Switch to analysis mode.
2. Analysis:
   a. Assume a process each for the four devices: (1) pump as adiabatic and isentropic, (2) boiler as isobaric, (3) turbine as adiabatic and isentropic, and (4) condenser as isobaric.
   b. Input the given information: (1) working fluid is water, (2) the inlet pressure and quality of the pump are 20 psia and 0, (3) the inlet pressure and quality of the turbine are 500 psia and 1, and (4) the mass flow rate is 3 lb/s.
3. Display results:
   a. Display cycle properties' results. The cycle is a heat engine. The answers are $x_{\text{turbine outlet}} = 0.8082$, $W\text{dot}_{\text{net}} = 981.8$ hp, and $\eta = 22.97\%$.
   b. Change the boiler pressure to 600 psia and display cycle properties' results again. The answers are $x_{\text{turbine outlet}} = 0.7953$, $W\text{dot}_{\text{net}} = 1028$ hp, and $\eta = 24.08\%$. (See Fig. 2.7.)

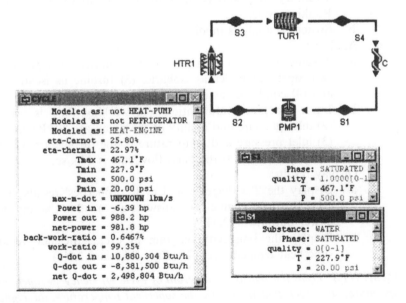

Figure 2.7 Rankine cycle.

COMMENTS: The effect of increasing the boiler pressure on the quality of the steam at the exit of the turbine can be seen by comparing the two cases. The higher the boiler pressure, the higher the moisture content (or the lower the quality) at the exit of the turbine. Steam with qualities less than 90% at the exit of the turbine, cannot be tolerated in the operation of actual Rankine steam power plants. To increase steam quality at the exit of the turbine, superheating and reheating are used.

Example 2.4

Determine the efficiency and power output of a superheated Rankine cycle using steam as the working fluid in which the condenser pressure is 80 kPa. The boiler pressure is 3 MPa. The steam leaves the boiler at 400°C. The mass rate of steam flow is 1 kg/sec. Show the cycle on a *T–s* diagram. Plot the sensitivity diagram of cycle efficiency versus boiler superheat temperature.

Convert the SI units to Imperial units and find the answer.

To solve this problem by CyclePad, we take the following steps:

1. Build:
   a. Take a pump, a boiler, a turbine, and a condenser from the inventory shop and connect the devices to form the basic Rankine cycle.
   b. Switch to analysis mode.
2. Analysis:
   a. Assume a process each for the four devices: (1) pump as isentropic, (2) boiler as isobaric, (3) turbine as isentropic, and (4) condenser as isobaric.
   b. Input the given information: (1) working fluid is water, (2) inlet pressure and quality of the pump are 80 kPa and 0, (3) inlet pressure and temperature of the turbine are 3 MPa and 400°C, and (4) the mass flow rate is 1 kg/sec.
3. Display results:
   a. Display the *T–s* diagram and cycle properties' results. The cycle is a heat engine. The answers are $\eta = 26.11\%$ and net power output = 740.3 kW.
   b. Display the sensitivity diagram of cycle efficiency versus superheat temperature. (See Figs. 2.8a and 2.8b.)

COMMENTS: From the sensitivity diagram of cycle efficiency vs superheat temperature, it is seen that the higher the superheat temperature, the higher the cycle efficiency. The superheat temperature is limited, however, due to

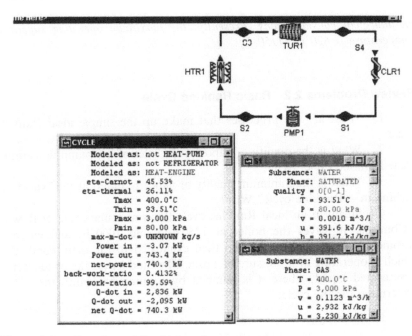

Figure 2.8a Superheated Rankine cycle.

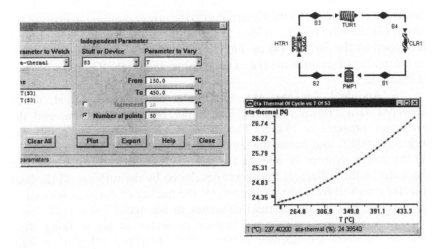

Figure 2.8b Superheated Rankine cycle sensitivity analysis.

metallurgical considerations. Presently, the maximum operating superheat temperature is 620°C (1150°F).

Review Problems 2.2   Basic Rankine Cycle

1.   What are the processes that make up the simple ideal Rankine cycle?

2.   What is the quality of vapor at the inlet of the pump in a simple ideal Rankine cycle? Why?

3.   What is the minimum quality of vapor required at the exit of the turbine in a Rankine cycle? Why?

4.   Steam in an ideal Rankine cycle flows at a mass rate of flow of 14 lbm/sec. It leaves the boiler at 1250 psia and 1000°F, and enters a turbine where it is expanded and then exhausted to the main condenser, which is operating at a pressure of 1 psia. The fluid leaves the condenser as a saturated liquid, where it is pumped by a pump back into the boiler. Determine for the cycle:

1.   Pump power
2.   Rate of heat added by the boiler
3.   Ideal turbine power
4.   Rate of heat rejected by the condenser
5.   Thermal efficiency of the cycle

5.   An ideal Rankine cycle uses water as a working fluid, which circulates at a rate of 80 kg/sec. The boiler pressure is 6 MPa, and the condenser pressure is 10 kPa. Determine (1) the power required to operate the pump, (2) the heat transfer added to the water in the boiler, (3) the power developed by the turbine, (4) the heat transfer removed from the condenser, (5) the quality of steam at the exit of the turbine, and (6) the thermal efficiency of the cycle.

6.   An ideal Rankine cycle uses water as a working fluid, which circulates at a rate of 80 kg/sec. The boiler pressure is 6 MPa, and the condenser pressure is 10 kPa. The steam is superheated and enters the turbine at 600°C and leaves the condenser as a saturated liquid. Determine (1) the power required to operate the pump, (2) the heat transfer added to the water in the boiler, (3) the power developed by the turbine, (4) the heat transfer removed from the condenser, (5) the quality of steam at the exit of the turbine, and (6) the thermal efficiency of the cycle.

7.   A Rankine cycle using 1 kg/sec of water as the working fluid operates between a condenser pressure of 7.5 kPa and a boiler pressure of 17 MPa. The superheater temperature is 550°C. Determine (1) the pump

power, (2) the turbine power, (3) the heat transfer added to the boiler, and (4) the cycle thermal efficiency.

8. In a Rankine power plant, the steam temperature and pressure at the turbine inlet are 1000°F and 2000 psia. The temperature of the condensing steam in the condenser is maintained at 60°F. The power generated by the turbine is 30,000 hp. Assuming all processes to be ideal, determine: (1) the pump power required (hp), (2) the mass flow rate, (3) the heat transfer added in the boiler (Btu/hr), (4) the heat transfer removed from the condenser (Btu/hr), and (5) the cycle thermal efficiency (%).

9. Water circulates at a rate of 80 kg/sec in an ideal Rankine power plant. The boiler pressure is 6 MPa and the condenser pressure is 10 kPa. The steam enters the turbine at 600°C and water leaves the condenser as a saturated liquid. Find: (1) the power required to operate the pump, (2) the heat transfer added to the boiler, (3) the power developed by the turbine, (4) the thermal efficiency of the cycle.

10. For an ideal Rankine cycle, steam enters the turbine at 5 MPa and 400°C, and exhausts to the condenser at 10 kPa. The turbine produces 20 MW of power.

1. Draw a *T–s* diagram for this cycle with respect to the saturation curve.
2. What is the mass flow rate of the steam? (kg/sec)
3. What is the rate of heat rejection from the condenser, and the rate of heat added to the boiler?
4. Find the thermal efficiency for this cycle.

11. Steam is generated in the boiler of a steam power plant operating on an ideal Rankine cycle at 10 MPa and 500°C at a steady rate of 80 kg/sec. The steam expands in the turbine to a pressure of 7.5 kPa. Determine (1) the quality of the steam at the turbine exit, (2) rate of heat rejection in the condenser, (3) the power delivered by the turbine, and (4) the cycle thermal efficiency (%).

12. A steam power plant operating on an ideal Rankine cycle has a boiler pressure of 800 psia and a condenser pressure of 2 psia. The quality at the turbine exit is 95% and the power generated by the turbine is 10,000 hp. Determine (1) the mass flow rate of steam (lbm/s), (2) the turbine inlet temperature (°F), (3) the rate of heat addition to the boiler (Btu/h), and (4) the cycle thermal efficiency (%).

13. Water is the working fluid in an ideal Rankine cycle. The condenser pressure is 8 kPa and saturated vapor enters the turbine at: (1) 15 MPa, (2) 10 MPa, (3) 7 MPa, and (4) 4 MPa. The net power output of the cycle is 100 MW. Determine for each case the mass flow rate of

water, heat transfer in the boiler and the condenser, and the thermal efficiency.

14. A steam power plant operates on the Rankine cycle. The steam enters the turbine at 7 MPa and 550°C. It discharges to the condenser at 20 kPa. Determine the quality of the steam at the exit of the turbine, pump work, turbine work, heat added to the boiler, and thermal cycle efficiency.

15. A steam power plant operates on the Rankine cycle. The steam with a mass rate flow of 10 kg/sec enters the turbine at 6 MPa and 600°C. It discharges to the condenser at 10 kPa. Determine the quality of the steam at the exit of the turbine, pump power, turbine power, rate of heat added to the boiler, and thermal cycle efficiency.

2.3 IMPROVEMENTS TO RANKINE CYCLE

The thermal efficiency of the Rankine cycle can be improved by increasing the average temperature at which heat is transferred to the working fluid in the heating process, or by decreasing the average temperature at which heat is transferred to the surroundings from the working fluid in the cooling process. Several modifications to increase the thermal efficiency of the basic Rankine cycle include increasing boiler pressure, decreasing condenser pressure, and use of a superheater, reheater, regenerator, preheater, etc.

Increasing the average temperature during the heat-addition process increases the boiler pressure. The maximum boiler pressure is limited by the tube metallurgical material problem in the boiler. Increasing the boiler pressure increases the moisture content of the steam at the turbine exit, which is not desirable.

Increasing the average temperature during the heat-addition process without increasing boiler pressure can be done by superheating the steam to high temperature with a superheater. Superheating the steam to a higher temperature also decreases the moisture content of the steam at the turbine exit, which is very desirable.

Increasing the average temperature during the heat-addition process can be accomplished with a superheater. The moisture content of steam at the turbine exhaust can be decreased by reheating the steam between the stages of a multistage turbine.

An increase in the average temperature during the heat-addition process can also be accomplished by regenerating the steam. A portion of the partially expanded steam between the stages of a multistage turbine is drawn off to preheat the condensed liquid before it is returned to the boiler. In this way, the amount of heat added at low temperature is reduced. So the average temperature during the heat-addition process is increased.

Decreasing the average temperature during the heat-rejection process decreases the condenser pressure and increases cycle efficiency. The minimum condenser pressure is limited by the sealing and leakage problem in the condenser.

2.4   ACTUAL RANKINE CYCLE

For actual Rankine cycles, many irreversibilities are present in various components. Fluid friction causes pressure drops in the boiler and condenser. These drops in the boiler and condenser are usually small. The major irreversibilities occur within the turbine and pump. To account for these irreversibility effects, turbine efficiency and pump efficiency must be used in computing the actual work produced or consumed. The *T–s* diagram of the actual Rankine cycle is shown in Fig. 2.9. The effect of irreversibilities on the thermal efficiency of a Rankine cycle is illustrated in the following example.

Example 2.5

Determine the efficiency and power output of an actual Rankine cycle using steam as the working fluid and having a condenser pressure is 80 kPa. The boiler pressure is 3 MPa. The steam leaves the boiler at 400°C. The mass rate of steam flow is 1 kg/sec. The pump efficiency is 85% and the turbine efficiency is 88%. Show the cycle on a *T–s* diagram. Plot the

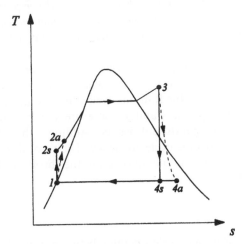

Figure 2.9   Actual Rankine cycle *T–s* diagram.

sensitivity diagram of cycle efficiency versus pump efficiency, and cycle efficiency versus turbine efficiency.

To solve this problem by CyclePad, we take the following steps:

1. Build:
   a. Take a pump, a boiler, a turbine, and a condenser from the inventory shop and connect the devices to form the actual Rankine cycle.
   b. Switch to analysis mode.
2. Analysis:
   a. Assume a process each for the four devices: (1) pump as adiabatic, (2) boiler as isobaric, (3) turbine as adiabatic, and (4) condenser as isobaric.
   b. Input the given information: (1) working fluid is water, (2) inlet pressure and quality of the pump are 80 kPa and 0, (3) inlet pressure and temperature of the turbine are 3 MPa and 400°C, (4) mass flow rate is 1 kg/sec, (5) phase at the exit of turbine is saturated, and (6) pump efficiency is 85% and turbine efficiency is 88%.
3. Display results:
   a. Display the *T–s* diagram (Fig. 2.9a) and cycle properties results (Fig. 2.9b). The cycle is a heat engine. The answers are $\eta = 22.95\%$ and net power output $= 650.5$ kW.
   b. Display cycle efficiency versus pump efficiency (Fig. 2.9c), and cycle efficiency versus turbine efficiency (Fig. 2.9d).

COMMENTS: (1) The pump work is quite small compared to the turbine work. Therefore, it is seen from the sensitivity diagram of cycle efficiency versus pump efficiency that the cycle efficiency is not sensitive to the pump efficiency.

(2) The sensitivity diagram of cycle efficiency versus turbine efficiency demonstrates that the cycle efficiency is sensitive to the turbine efficiency.

The power output of the Rankine cycle can be controlled by a throttling valve. The inlet steam pressure and temperature may be throttled down to a lower pressure and temperature if desired. Adding a throttling valve to the Rankine cycle decreases the cycle efficiency. The throttling Rankine cycle is shown in Fig. 2.10. An example illustrating the throttling Rankine cycle is given in Example 2.6.

Example 2.6

An actual steam Rankine cycle operates between a condenser pressure of 1 psia and a boiler pressure of 600 psia. The outlet temperature of the

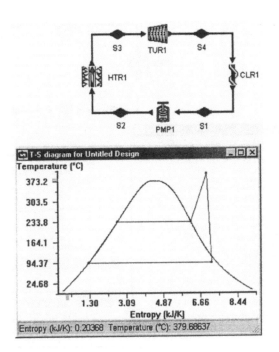

Figure 2.9a Rankine cycle *T–s* diagram.

superheater is 600°F and the turbine efficiency is 80%. The rate of mass flow in the cycle is 1 lbm/sec.

1. Find the pump power required, turbine power produced, rate of heat added to the boiler, and the cycle efficiency.
2. If the pressure is throttled down to 400 psia at the inlet of the turbine, find the pump power required, turbine power produced, rate of heat added to the boiler, rate of heat removed in the condenser, and the cycle efficiency. Draw the *T–s* diagram.

To solve part (1) of this problem, let us make use of Fig. 2.10.

(1) Assume that the pump is isentropic, the boiler is isobaric, the turbine is adiabatic with 80% efficiency, and the condenser is isobaric; (2) input $p_1 = 1$ psia, $x_1 = 0$, $p_2 = 600$ psia, $T_3 = 600°$F, $mdot = 1$ lbm/s, and $p_4 = 600$ psia.

The results are: $Wdot_{pump} = -2.55$ hp, $Wdot_{turbine} = 491.0$ hp, $Qdot_{boiler} = 1218$ Btu/sec, and $\eta = 28.36\%$ as shown in Fig. 2.10a.

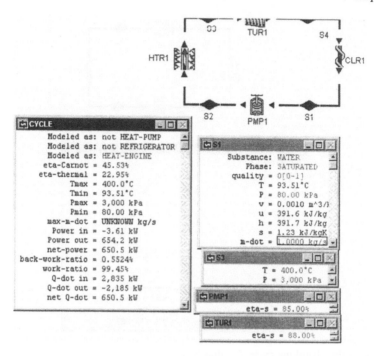

Figure 2.9b   Rankine cycle results.

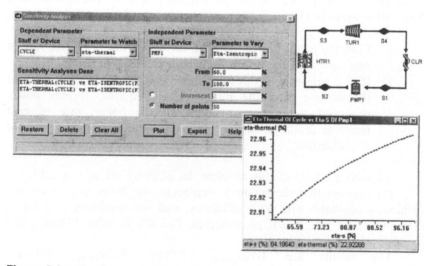

Figure 2.9c   Rankine cycle efficiency versus pump efficiency sensitivity analysis.

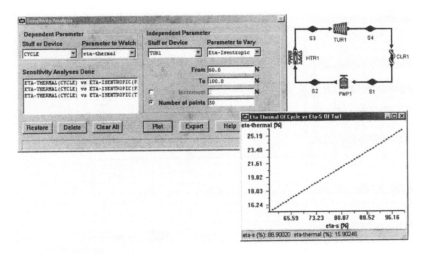

Figure 2.9d Rankine cycle efficiency versus turbine efficiency sensitivity analysis.

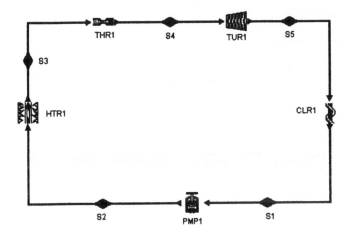

Figure 2.10 Throttling Rankine cycle.

To solve part (2) of this problem, let us make use of Fig. 2.10.

(1) Assume that the pump is isentropic, the boiler is isobaric, the turbine is adiabatic with 80% efficiency, and the condenser is isobaric; (2) input $p_1 = 1$ psia, $x_1 = 0$, $p_2 = 600$ psia, $T_3 = 600°F$, mdot $= 1$ lbm/sec, and $p_4 = 400$ psia;

The results are: $W\text{dot}_{pump} = -2.55$ hp, $W\text{dot}_{turbine} = 461.2$ hp, $Q\text{dot}_{boiler} = 1218$ Btu/sec, and $\eta = 26.62\%$ as shown in Fig. 2.10b. The T–s diagram is shown in Fig. 2.10c.

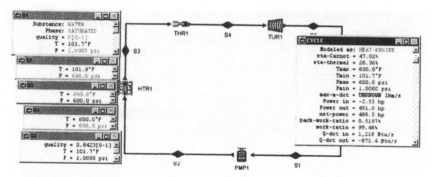

Figure 2.10a Rankine cycle with throttling valve off.

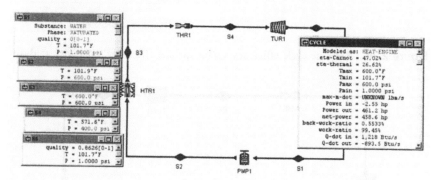

Figure 2.10b Rankine cycle with throttling valve on.

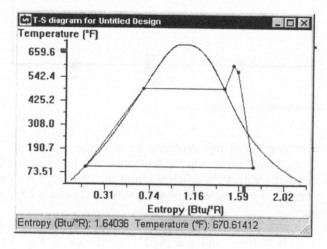

Figure 2.10c Rankine throttling cycle *T–s* diagram.

Review Problems 2.4   Actual Rankine Cycle

1. Is Rankine cycle efficiency sensitive to the pump inefficiency? Why?
2. Is Rankine cycle efficiency sensitive to the turbine inefficiency? Why?
3. What is the purpose of the throttling valve in the Rankine throttling cycle? Does it improve the cycle efficiency?
4. Steam enters a turbine of a Rankine power plant at 5 MPa and 400°C, and exhausts to the condenser at 10 kPa. The turbine produces a power output of 20 MW. Given a turbine isentropic efficiency of 90% and a pump isentropic efficiency of 100%:
    a. What is the mass flow rate of the steam around the cycle?
    b. What is the rate of heat rejection from the condenser?
    c. Find the thermal efficiency of the power plant.
5. 12.7 kg/sec of superheated steam flow at 2 MPa and 320°C enters the turbine of a Rankine power plant and expands to a condenser pressure of 10 kPa. Assuming that the isentropic efficiencies of the turbine and pump are 85 and 100%, respectively, find:
    a. The actual pump power required.
    b. The quality of steam at the exit of the turbine.
    c. The actual turbine power produced.
    d. The rate of heat supplied in the boiler.
    e. The rate of heat removed in the condenser.
    f. The cycle efficiency.
6. Water circulates at a rate of 80 kg/sec in a Rankine power plant. The boiler pressure is 6 MPa and the condenser pressure is 10 kPa. The steam enters the turbine at 700°C and water leaves the condenser as a saturated liquid. The actual turbine efficiency is 90%. Find:
    a. The power required to operate the pump.
    b. The heat transfer added in the boiler.
    c. The power developed by the turbine.
    d. The thermal efficiency of the cycle.
7. A Rankine cycle using water as the working fluid operates between a condenser pressure of 7.5 kPa and a boiler pressure of 17 MPa. The superheater temperature is 550°C. The rate of mass flow in the cycle is 2.3 lbm/sec. The turbine efficiency is 85%. Determine:
    a. The pump power
    b. The turbine power
    c. The rate of heat added in the boiler
    d. The cycle thermal efficiency

8. A throttling Rankine cycle using water as the working fluid operates between a condenser pressure of 7.5 kPa and a boiler pressure of 17 MPa. The superheater temperature is 550°C. The rate of mass flow in the cycle is 2.3 lbm/sec. The turbine efficiency is 85%. If the pressure at the exit of the throttling valve is 12 MPa, determine:
   a. The pump power
   b. The turbine power
   c. The rate of heat added to the boiler
   d. The cycle thermal efficiency

2.5   REHEAT RANKINE CYCLE

The thermal efficiency of the Rankine cycle can be significantly increased by using higher boiler pressure, but this requires ever-increasing superheats. Since the maximum temperature in the superheater is limited by the temperature the boiler tubes can stand, superheater temperatures are usually restricted. Since the major fraction of the heat supplied to the Rankine cycle is supplied in the boiler, the boiler temperatures (and hence pressures) must be increased if cycle efficiency improvements are to be obtained.

The problem of excessive superheater temperatures may be solved while avoiding two-phase saturated mixtures in the expansion, by reheating the expanding steam part way through the expansion as shown in Fig. 2.11. The steam leaving the boiler section as saturated vapor is superheated to

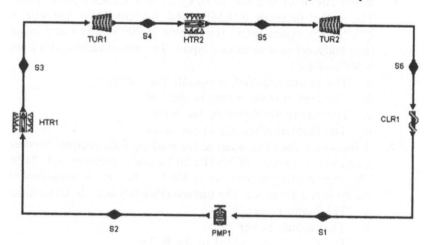

Figure 2.11   Reheat Rankine cycle.

an acceptable temperature and then expanded (while producing work) until it intersects with the maximum moisture (complement of quality, or minimum quality) curve. The steam is then reheated in a second superheater section and expanded in a second turbine (while producing more work) until it intersects with the maximum moisture (complement of quality, or minimum quality) curve again. The steam is then condensed and pumped back into the boiler.

The one *reheat Rankine basic cycle* is composed of the following six processes:

1-2 Isentropic compression
2-3 Isobaric heat addition
3-4 Isentropic expansion
4-5 Isobaric heat addition
5-6 Isentropic expansion
6-1 Isobaric heat removing

Applying the first law of thermodynamics of the open system to each of the six processes of the reheat Rankine cycle yields:

$$Q_{12} = 0 \tag{2.21}$$

$$W_{12} = m(h_1 - h_2) = mv_1(p_1 - p_2) \tag{2.22}$$

$$W_{23} = 0 \tag{2.23}$$

$$Q_{23} - 0 = m(h_3 - h_2) \tag{2.24}$$

$$Q_{34} = 0 \tag{2.25}$$

$$W_{34} = m(h_3 - h_4) \tag{2.26}$$

$$W_{45} = 0 \tag{2.27}$$

$$Q_{45} - 0 = m(h_5 - h_4) \tag{2.28}$$

$$Q_{56} = 0 \tag{2.29}$$

$$W_{56} = m(h_5 - h_6) \tag{2.30}$$

$$W_{61} = 0 \tag{2.31}$$

and

$$Q_{61} - 0 = m(h_1 - h_6) \tag{2.32}$$

The net work (W_{net}), which is also equal to net heat (Q_{net}), is

$$W_{net} = Q_{net} = Q_{23} + Q_{45} + Q_{61} \tag{2.33}$$

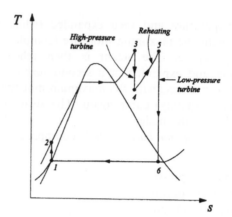

Figure 2.12   Reheat Rankine T-S diagram.

The thermal efficiency of the cycle is

$$\eta = W_{net}/(Q_{23} + Q_{45}) = Q_{net}/(Q_{23} + Q_{45}) = 1 - Q_{61}/(Q_{23} + Q_{45})$$
$$= 1 - (h_6 - h_1)/[(h_3 - h_2) + (h_5 - h_4)] \qquad (2.34)$$

The T–S diagram of the reheat Rankine cycle is shown in Fig. 2.12.

Example 2.7

A steam reheat Rankine cycle operates between the pressure limits of 5 and 1600 psia. Steam is superheated to 600°F before it is expanded to the reheat pressure of 500 psia. Steam is reheated to 600°F. The steam flow rate is 800 lbm/hr. Determine the quality of steam at the exit of the turbine, the cycle efficiency, and the power produced by the cycle.

To solve this problem by CyclePad, we take the following steps:

1. Build:
    a. Take a pump, a boiler, a turbine, a reheater, another turbine, and a condenser from the inventory shop and connect the devices to form the reheat Rankine cycle.
    b. Switch to analysis mode.
2. Analysis:
    a. Assume a process each for the six devices: (1) pump as adiabatic, (2) boiler and reheater as isobaric, (3) turbines as adiabatic, and (4) condenser as isobaric.
    b. Input the given information: (1) working fluid is water, (2) inlet pressure and quality of the pump are 5 psia and 0,

(3) inlet pressure and temperature of the first turbine are 1600 psia and 600°F, (4) mass flow rate is 800 lbm/hr, and (5) inlet pressure and temperature of the first turbine are 500 psia and 600°F.

3. Display results: display the *T–s* diagram and cycle properties results. The cycle is a heat engine. The answers are $x = 82.52\%$, $\eta = 30.04\%$, and net power output $= 111.4$ hp. (See Fig. 2.13.)

COMMENTS: The sole purpose of the reheat cycle is to reduce the moisture content of the steam at the final stages of the turbine expansion process. The more reheating processes, the higher the quality of the steam at the exit of the last-stage turbine. The reheat temperature is often very close or equal to the turbine inlet temperature. The optimum reheat pressure is about one-fourth of the maximum cycle pressure.

Example 2.8

Determine the efficiency and power output of a reheat Rankine cycle, using steam as the working fluid, in which the condenser pressure is 80 kPa. The boiler pressure is 3 MPa. The steam leaves the boiler at 400°C. The mass flow rate of steam is 1 kg/sec. The pump efficiency is 85% and the turbine efficiency is 88%. After expansion in the high-pressure turbine to 800 kPa, the steam is reheated to 400°C and then expanded in the low-pressure turbine to the condenser.

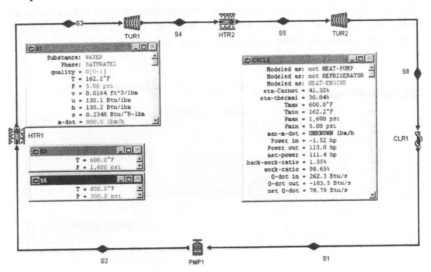

Figure 2.13 Reheat Rankine cycle.

To solve this problem by CyclePad, we take the following steps:

1. Build:
   a. Take a pump, a boiler, a turbine, a reheater, another turbine, and a condenser from the inventory shop and connect the devices to form the reheat Rankine cycle.
   b. Switch to analysis mode.
2. Analysis:
   a. Assume a process each for the six devices: (1) pump as adiabatic, (2) boiler and reheater as isobaric, (3) turbines as adiabatic, and (4) condenser as isobaric.
   b. Input the given information: (1) working fluid is water, (2) inlet pressure and quality of the pump are 80 kPa and 0, (3) inlet pressure and temperature of the turbine are 3 MPa and 400°C, (4) mass flow rate is 1 kg/sec, and (5) pump efficiency is 85% and turbine efficiency is 88%.
3. Display results: display the cycle properties' results. The cycle is a heat engine.

The answers are $\eta = 24.57\%$ and net power output = 779.9 kW. (See Fig. 2.14.)

COMMENTS: The advantage of using reheat is to reduce the moisture content at the exit of the low-pressure turbine and increase the net power of the Rankine power plant. The one reheat Rankine basic cycle shown in Fig. 2.13 can be expanded into more than one reheat if desired. In this fashion it is possible to use higher boiler pressure without having to increase the maximum superheater temperature above the limit of the superheater tubes.

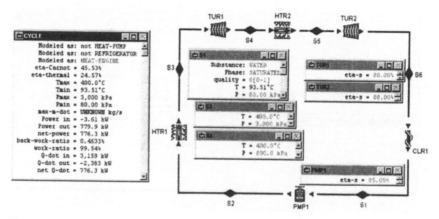

Figure 2.14 Reheat Rankine cycle.

Review Problems 2.5 Reheat Rankine Cycle

1. What is the purpose of reheat in a reheat Rankine cycle?
2. Is the efficiency of a reheat Rankine cycle always higher than the efficiency of a simple Rankine cycle operating between the same boiler pressure and condenser pressure?
3. Consider a steam power plant operating on the ideal reheat Rankine cycle; 1 kg/sec of steam flow enters the high-pressure turbine at 15 MPa and 600°C and leaves at 5 MPa. Steam is reheated to 600°C and enters the low-pressure turbine. Exhaust steam from the turbine is condensed in the condenser at 10 kPa. Determine:
   a. The pump power required.
   b. Rate of heat added to the boiler.
   c. Rate of heat added in the reheater.
   d. Power produced by the high-pressure turbine.
   e. Power produced by the low-pressure turbine.
   f. Rate of heat removed from the condenser.
   g. Quality of steam at the exit of the low-pressure turbine.
   h. Thermal cycle efficiency.
4. Consider a steam power plant operating on the ideal reheat Rankine cycle; 1 kg/sec of steam flow enters the high-pressure turbine at 15 MPa and 600°C and leaves at 5 MPa. Steam is reheated to 600°C and enters the low-pressure turbine. Exhaust steam from the turbine is condensed in the condenser at 10 kPa. Both turbines have 90% efficiency. Determine:
   a. Pump power required.
   b. Rate of heat added in the boiler.
   c. Rate of heat added in the reheater.
   d. Power produced by the high-pressure turbine.
   e. Power produced by the low-pressure turbine.
   f. Rate of heat removed from the condenser.
   g. Quality of steam at the exit of the low-pressure turbine.
   h. Thermal cycle efficiency.

2.6 REGENERATIVE RANKINE CYCLE

The thermal efficiency of the Rankine cycle can be increased by the use of regenerative heat exchange as shown in Fig. 2.15. In the regenerative cycle, a portion of the partially expanded steam is drawn off between the high- and low-pressure turbines. The steam is used to preheat the condensed

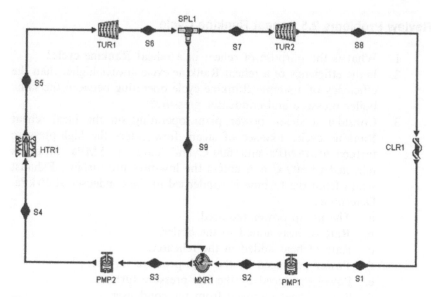

Figure 2.15   Regenerative Rankine cycle.

liquid before it returns to the boiler. In this way, the amount of heat added at low temperatures is reduced. Therefore, the mean effective temperature of heat addition is increased, and cycle efficiency is increased. In the case of an ideal regenerative Rankine cycle, the best result is obtained by heating the feed-water to a temperature equal to the saturation temperature corresponding to the boiler pressure. To carry out the ideal regenerative process, the regenerative heat exchanger is called the feed-water heater. The *T–s* diagram of the ideal *regenerative Rankine cycle* is shown in Fig. 2.16.

The one regenerative Rankine basic cycle is composed of the following processes:

> 1-2 Isentropic compression
> 2-3 Isobaric heat addition
> 3-4 Isentropic compression
> 4-5 Isobaric heat addition
> 5-6 Isentropic expansion
> 6-7 Isentropic expansion
> 7-1 Isobaric heat removing

Applying the mass balance and the first law of thermodynamics of the open system to each of the seven processes of the regenerative Rankine

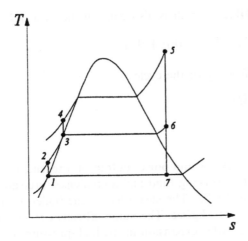

Figure 2.16   Regenerative Rankine cycle T–S diagram.

cycle yields:

$$m_1 = m_2 = m_7 \tag{2.35}$$

$$m_4 = m_5 \tag{2.36}$$

$$m_4 = m_2 + m_6 \tag{2.37}$$

$$Q_{12} = 0 \tag{2.38}$$

$$W_{12} = m_1(h_1 - h_2) = m_1 v_1 \ (p_1 - p_2) \tag{2.39}$$

$$m_4(h_4 - h_3) + m_2(h_2 - h_3) + m_6(h_6 - h_3) = 0 \tag{2.40}$$

$$Q_{34} = 0 \tag{2.41}$$

$$W_{34} = m_4(h_3 - h_4) = m_4 v_3 \ (p_3 - p_4) \tag{2.42}$$

$$W_{45} = 0 \tag{2.43}$$

$$Q_{45} - 0 = m_4(h_5 - h_4) \tag{2.44}$$

$$Q_{56} = 0 \tag{2.45}$$

$$W_{56} = m_4(h_3 - h_4) \tag{2.46}$$

$$Q_{67} = 0, \tag{2.47}$$

$$W_{67} = m_1(h_6 - h_7) \tag{2.48}$$

$$W_{71} = 0 \tag{2.49}$$

and

$$Q_{71} - 0 = m_1(h_1 - h_6) \tag{2.50}$$

The net work (W_{net}), which is also equal to net heat (Q_{net}), is

$$W_{net} = W_{56} + W_{67} + W_{12} + W_{34} \qquad (2.51)$$

The thermal efficiency of the cycle is

$$\eta = W_{net}/Q_{45} \qquad (2.52)$$

Example 2.9

Determine the efficiency and power output of a regenerative Rankine cycle using steam as the working fluid and a condenser pressure of 80 kPa. The boiler pressure is 3 MPa. The steam leaves the boiler at 400°C. The mass rate of steam flow is 1 kg/sec. The pump efficiency is 85% and the turbine efficiency is 88%. After expansion in the high-pressure turbine to 400 kPa, some of the steam is extracted from the turbine exit for the purpose of heating the feed-water in an open feed-water heater, the rest of the steam is reheated to 400°C and then expanded in the low-pressure turbine to the condenser. The water leaves the open feed-water heater at 400 kPa as saturated liquid. Determine the steam fraction extracted from the turbine exit, cycle efficiency, and net power output of the cycle.

To solve this problem by CyclePad, we take the following steps:

1. Build:
   a. Take two pumps, a boiler, a turbine, a reheater, another turbine, a splitter, a mixing chamber (open feed-water heater), and a condenser from the inventory shop and connect the devices to form the regenerating Rankine cycle.
   b. Switch to analysis mode.
2. Analysis:
   a. Assume a process for each of the devices: (1) pumps as adiabatic, (2) boiler and reheater as isobaric, (3) turbines as adiabatic, (4) splitter as isoparametric, and (5) condenser as isobaric.
   b. Input the given information: (1) working fluid is water, (2) inlet pressure and quality of the pump are 80 kPa and 0, (3) inlet pressure and temperature of the turbine are 4 MPa and 400°C, (4) mass flow rate is 1 kg/sec through the boiler, and (5) pump efficiency is 85% and turbine efficiency is 88%.
3. Display results:
   a. Display the cycle properties results. The cycle is a heat engine. The answers are fraction extraction = 0.0877, $\eta = 24.28\%$ and net power output = 636.8 kW.

b. Display the sensitivity diagram of cycle efficiency versus reheater pressure.

c. Locate the optimum regeneration pressure (about 730 kPa) from the sensitivity diagram.

d. Redo the problem with regenerator pressure at 730 kPa. The answers are: fraction extraction $= 0.1233$, $\eta = 24.42\%$, and net power output $= 616.1$ kW. (See Figs. 2.17a and 2.17b.)

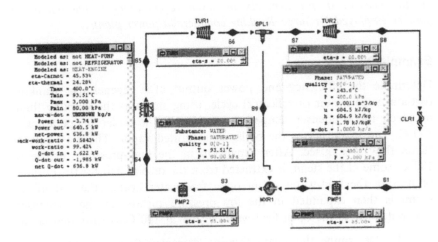

Figure 2.17a Regenerative Rankine cycle.

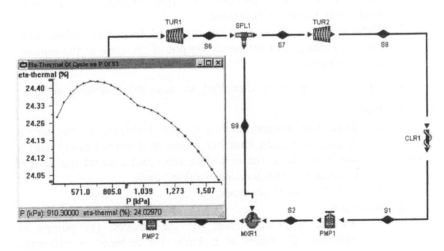

Figure 2.17b Regenerative Rankine cycle.

COMMENTS: (1) From the example, it is seen that the efficiency of the regenerative Rankine cycle is better than that of the Rankine cycle without regenerator.

(2) Suppose an infinite number of regenerators are used; the regenerative cycle would have the same cycle efficiency as that of a Carnot cycle operating between the same temperature limits. This is physically not practical. Large number of regenerators may not be economically justified. Consequently, the finite number of regenerators is a design decision. In practice, six or seven regenerators is the maximum number employed for large Rankine commercial power plants.

Example 2.10

Determine the efficiency and power output of a regenerative Rankine (without superheater or reheater) cycle, using steam as the working fluid, in which the condenser temperature is 50°C. The boiler temperature is 350°C. The steam leaves the boiler as saturated vapor. The mass rate of steam flow is 1 kg/sec. After expansion in the high-pressure turbine to 100°C, some of the steam is extracted from the turbine exit for the purpose of heating the feed-water in an open feed-water heater; the rest of the steam is then expanded in the low-pressure turbine to the condenser. The water leaves the open feed-water heater at 100°C as saturated liquid.

1. Determine the steam fraction extracted from the turbine exit, cycle efficiency, and net power output of the cycle.
2. Plot the sensitivity diagram of cycle efficiency versus open feed-water heater temperature.
3. Determine the steam fraction extracted from the turbine exit, cycle efficiency, and net power output of the cycle at the optimal cycle efficiency.

To solve this problem by CyclePad, we take the following steps:

1. Build:
   a. Take two pumps, a boiler, two turbines, a reheater, a splitter, a mixing chamber (open feed-water heater), and a condenser from the inventory shop and connect the devices to form the regenerating Rankine cycle.
   b. Switch to analysis mode.
2. Analysis:
   a. Assume a process each for the devices: (1) pumps as adiabatic, (2) boiler as isobaric, (3) turbines as adiabatic, (4) splitter as isoparametric, and (5) condenser as isobaric.

b. Input the given information: (1) working fluid is water, (2) inlet temperature and quality of the pump are 50°C and 0, (3) turbine inlet steam quality and temperature are 1 and 350°C, (4) mass flow rate is 1 kg/sec through the boiler, and (5) regenerator exit steam quality and temperature are 0 and 100°C.

3. Display results:
   a. Display the cycle properties' results. The cycle is a heat engine. The answers are fraction extraction = 0.1258, $\eta = 40.16\%$, and net power output = 871.4 kW. (See Fig. 2.18a.)
   b. The sensitivity diagram of cycle efficiency versus open feedwater heater temperature (η vs. T_3) is shown in Fig. 2.18b.

COMMENT: The temperature difference between the condenser and regenerator is approximately equal to the temperature difference between the boiler and regenerator. From the diagram, the temperature is approximately 187°C.

c. Change the temperature of water, leaving the open feedwater heater as saturated liquid, to 187°C. The answers given by the cycle properties are: fraction extraction = 0.2970, $\eta = 41.55\%$, and net power output = 727.9 kW. (See Fig. 2.18c.)

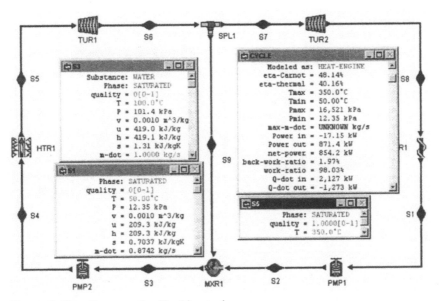

Figure 2.18a Regenerative Rankine cycle.

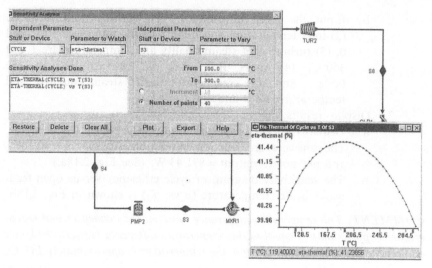

Figure 2.18b   Regenerative Rankine cycle sensitivity diagram.

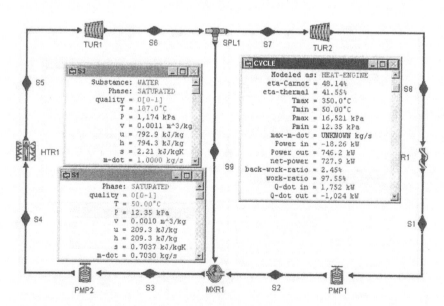

Figure 2.18c   Regenerative Rankine cycle optimization.

Example 2.11

Determine the efficiency and power output of a four-stage steam regenerative Rankine cycle (Fig. 2.19a). The following information is provided:

$p_1 = p_{24} = 10 \, \text{kPa}$, $\quad p_2 = p_3 = p_{21} = p_{22} = p_{23} = p_{28} = 3 \, \text{MPa}$, $\quad p_4 = p_5 = p_{18} = p_{19} = p_{20} = p_{27} = 7 \, \text{MPa}$, $\quad p_6 = p_7 = p_{15} = p_{16} = p_{17} = p_{26} = 10 \, \text{MPa}$, $\quad p_8 = p_9 = p_{12} = p_{13} = p_{14} = p_{25} = 13 \, \text{MPa}$, $p_{10} = p_{11} = 16 \, \text{MPa}$, $T_{11} = T_{14} = T_{17} = T_{20} = T_{23} = 400°\text{C}$, $\quad x_1 = x_3 = x_5 = x_7 = x_9 = 0$, $\quad mdot_9 = 1 \, \text{kg/sec}$, $\quad \eta_{Tur1} = \eta_{Tur2} = \eta_{Tur3} = \eta_{Tur4} = \eta_{Tur5} = 85\%$, and $\eta_{pmp1} = \eta_{pmp2} = \eta_{pmp3} = \eta_{pmp4} = \eta_{pmp5} = 85\%$.

To solve this problem by CyclePad, we take the following steps:

1. Build:
   a. Take five pumps, five heaters (one boiler and four reheaters), five turbines, four splitters, four mixing chambers (open feed-water heaters), and a condenser from the inventory shop and connect the devices to form the regenerating Rankine cycle.
   b. Switch to analysis mode.
2. Analysis:
   a. Assume a process each for the devices: (1) pumps as adiabatic, (2) boiler, reheater, and regenerators as isobaric, (3) turbines as adiabatic, (4) splitters as isoparametric, and (5) condenser as isobaric.

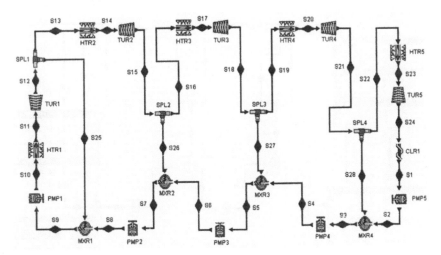

Figure 2.19a Four-stage steam regenerative Rankine cycle.

b. Input the given information: working fluid is water, $p_1 = p_{24} = 10$ kPa, $p_2 = p_3 = p_{21} = p_{22} = p_{23} = p_{28} = 3$ MPa, $p_4 = p_5 = p_{18} = p_{19} = p_{20} = p_{27} = 7$ MPa, $p_6 = p_7 = p_{15} = p_{16} = p_{17} = p_{26} = 10$ MPa, $p_8 = p_9 = p_{12} = p_{13} = p_{14} = p_{25} = 13$ MPa, $p_{10} = p_{11} = 16$ MPa, $T_{11} = T_{14} = T_{17} = T_{20} = T_{23} = 400°$C, $x_1 = x_3 = x_5 = x_7 = x_9 = 0$, mdot$_9 = 1$ kg/sec, $\eta_{\text{tur1}} = \eta_{\text{tur2}} = \eta_{\text{tur3}} = \eta_{\text{tur4}} = \eta_{\text{tur5}} = 85\%$, and $\eta_{\text{pmp1}} = \eta_{\text{pmp2}} = \eta_{\text{pmp3}} = \eta_{\text{pmp4}} = \eta_{\text{pmp5}} = 85\%$, as shown in Fig. 2.19b.

3. Display results: display the cycle properties results. The cycle is a heat engine.

The results are: $\eta_{\text{cycle}} = 36.92\%$, $W\text{dot}_{\text{input}} = -97.09$ kW, $W\text{dot}_{\text{output}} = -777.3$ kW, $W\text{dot}_{\text{net output}} = 680.2$ kW, $Q\text{dot}_{\text{add}} = 1842$ kW, $Q\text{dot}_{\text{remove}} = -1162$ kW, $W\text{dot}_{\text{pmp1}} = -44.23$ kW, $W\text{dot}_{\text{pmp2}} = -42.45$ kW, $W\text{dot}_{\text{pmp3}} = -4.14$ kW, $W\text{dot}_{\text{pmp4}} = -4.33$ kW, $W\text{dot}_{\text{pmp5}} = -1.93$ kW, $W\text{dot}_{\text{tur1}} = 39.74$ kW, $W\text{dot}_{\text{tur2}} = 50.45$ kW, $W\text{dot}_{\text{tur3}} = 69.28$ kW, $W\text{dot}_{\text{tur4}} = 147.7$ kW, $W\text{dot}_{\text{tur5}} = 470.1$ kW, $Q\text{dot}_{\text{htr1}} = 1372$ kW, $Q\text{dot}_{\text{htr2}} = 112.8$ kW, $Q\text{dot}_{\text{htr3}} = 107.2$ kW, $Q\text{dot}_{\text{htr4}} = 107.2$ kW, $Q\text{dot}_{\text{htr5}} = 143.5$ kW, $Q\text{dot}_{\text{clr1}} = -1162$ kW, mdot$_2 =$ mdot$_{22} = 0.5372$ kg/sec, mdot$_3 =$ mdot$_{19} = 0.7605$ kg/sec, mdot$_5 =$ mdot$_{16} = 0.8705$ kg/sec, mdot$_8 =$ mdot$_{12} = 0.9459$ kg/sec, mdot$_{25} = 0.0541$ kg/sec, mdot$_{26} = 0.0754$ kg/sec, mdot$_{27} = 0.1100$ kg/sec, and mdot$_{28} = 0.2234$ kg/sec. (See Fig. 2.19c.)

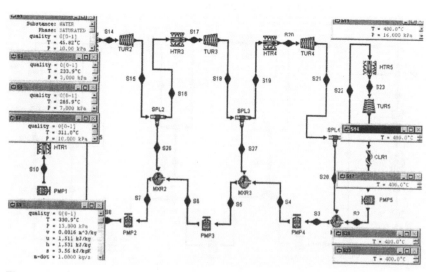

Figure 2.19b Four-stage steam regenerative Rankine cycle input.

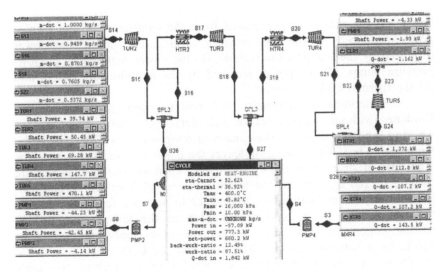

Figure 2.19c Four-stage steam regenerative Rankine cycle output.

Review Problems 2.6 Regenerative Rankine Cycle

1. What is the purpose of regeneration in a regenerative Rankine cycle?

2. Consider a steam power plant operating on the ideal regenerating Rankine cycle; 1 kg/sec of steam flow enters the turbine at 15 MPa and 600°C and is condensed in the condenser at 10 kPa. Some steam leaves the high-pressure turbine at 1.2 MPa and enters the open feed-water heater. If the steam at the exit of the open feed-water heater is saturated liquid, determine (1) the fraction of steam not extracted from the high-pressure turbine, (2) the rate of heat added to the boiler, (3) the rate of heat removed from the condenser, (4) the turbine power produced by the high-pressure turbine, (5) the turbine power produced by the low-pressure turbine, (6) the power required by the low-pressure pump, (7) the power required by the high-pressure pump, and (8) the thermal cycle efficiency.

3. Consider a steam power plant operating on the ideal regenerating Rankine cycle; 1 kg/sec of steam flow enters the turbine at 15 MPa and 600°C and is condensed in the condenser at 10 kPa. Some steam leaves the high-pressure turbine at 1.2 MPa and enters the open feed-water heater. The turbine efficiency is 90%. If the steam at the exit of the open feed-water heater is a saturated liquid, determine (1) the fraction of steam not extracted from the high-pressure turbine, (2) the rate of heat added in the

boiler, (3) the rate of heat removed from the condenser, (4) the turbine power produced by the high-pressure turbine, (5) the turbine power produced by the low-pressure turbine, (6) the power required by the low-pressure pump, (7) the power required by the high-pressure pump, and (8) the thermal cycle efficiency.

4. Determine the efficiency and power output of a regenerative Rankine (without superheater or reheater) cycle using steam as the working fluid and in which the condenser temperature is 50°C. The boiler temperature is 350°C. The steam leaves the boiler as saturated vapor. The mass rate of steam flow is 1 kg/sec. After expansion in the high-pressure turbine to 100°C, some of the steam is extracted from the turbine exit for the purpose of heating the feed-water in an open feed-water heater; the rest of the steam is then expanded in the low-pressure turbine to the condenser. The water leaves the open feed-water heater at 120°C as a saturated liquid. The efficiency of both turbines is 85%. (a) Determine the steam fraction extracted from the turbine exit, cycle efficiency, and net power output of the cycle, (b) Plot the sensitivity diagram of cycle efficiency versus open feed-water heater temperature, and (c) Determine the steam fraction extracted from the turbine exit, cycle efficiency, and net power output of the cycle at the optimal cycle efficiency.

ANSWERS: (a) $y = 0.1605$, $\eta = 35.17\%$, $W\text{dot}_{net} = 718.4\,\text{kW}$; (b) $T = 187°C$; (c) $y = 0.2884$, $\eta = 35.97\%$, $W\text{dot}_{net} = 630.2\,\text{kW}$.

2.7 LOW-TEMPERATURE RANKINE CYCLES

Much of this chapter has been concerned with various modifications to the simple Rankine cycle at high temperature. In the following five sections, the Rankine cycle that makes possible use of energy sources at low temperature, such as solar, geothermal, ocean thermal, solar pond, and waste heat, will be discussed. Because of the small temperature range available, only a simple Rankine cycle can be used and the cycle efficiency will be low. This is not critical economically, because the fuel is free.

The working fluids, such as ammonia and freons, used in refrigerators and heat pumps are more desirable than steam for the very low-temperature Rankine cycles. The reason is that the specific volume of such working fluids at low temperature is much less than that of steam, so the resulting turbine sizes can be much smaller and less expensive.

The sun provides a direct flux of thermal radiation of about $1350\,\text{kW/m}^2$ outside the atmosphere, and about $630\,\text{kW/m}^2$ on average at the earth's surface. If a flat-plate collector is used, the solar thermal power system is a low-temperature heat engine. If a concentrate collector is used to focus the power of the sun, the *solar thermal heat engine* can be operated at a higher temperature.

All the geothermal heat sources of interest have been created by the intrusion of hot magma from deep in the earth up into rock strata close to the surface. The average temperature gradient is small. The regions of interest for geothermal power production are those in which the temperature gradient exceeds 20°C/km. There are regions where geothermal energy in dry-steam form, hot-water form, or hot-rock form could be tapped and used for power generation. More commonly than it puts out dry steam, a geothermal well puts out a mixture of steam and water. A separator is needed to separate the flashing steam from the hot water. The steam is then used to drive the turbine. In a hot-rock (no steam or hot water) geothermal well, water will need to be injected into the well to tap the thermal energy. As with the hot-water geothermal energy, a secondary closed, simple Rankine cycle will be required to produce *geothermal power*.

Incident solar energy is absorbed by the surface water of the oceans. Ocean surface temperatures in excess of 26°C occur near the equator. Pure water has a maximum density at a temperature of 4°C. The chilled water tends to settle to the depths of the ocean. The combination of the warmed ocean surface water and cold deep ocean water provides the thermodynamic condition needed to operate a heat engine called *ocean thermal energy conversion* (OTEC). A typical closed-cycle OTEC Rankine cycle using a working fluid such as ammonia or a freon is suggested.

Temperature differences have been found in natural ponds having high concentration gradients of dissolved salt. Solar radiation is absorbed in the lower water levels and at the bottom of the pond. The water near the bottom (70–80°C) is at a higher temperature than that of the top surface (30°C), with the density of the hot concentrated lower level water higher than the density of the more dilute and cooler top levels. A typical closed-cycle *solar pond* Rankine cycle using a working fluid such as ammonia or a freon is also suggested.

Waste heat from farming, animal manure, crop production, and municipal solid residues could also be used for power generation.

There is no question that power can be produced from these various natural and waste energy sources. The question is the cost. Significant efforts have been underway for many years to produce power from these free energy sources. However, there are only a few commercial power plants presently utilizing these energy sources.

Review Problems 2.7 Low-Temperature Rankine Cycles

1. Is the cycle efficiency of the low-temperature heat engine higher than that of the high-temperature heat engine?
2. What is the fuel cost of the low-temperature heat engine?
3. List at least three low-temperature energy resources.
4. Why are working fluids such as ammonia and freons, used in refrigerators and heat pumps, more desirable than steam for the low-temperature Rankine cycles?

2.8 SOLAR HEAT ENGINES

The abundance of incident solar energy, particularly in large desert regions with few interruptions due to cloud cover, lends to the appeal of solar heat engines. Electrical power produced via thermal conversion of solar energy by means of a conventional Rankine cycle is technically achievable.

If a flat-plate collector is used, the solar thermal power system is a low-temperature heat engine. The conversion efficiency of a thermo-dynamic cycle depends on the collector temperature achieved, i.e., the higher the collector temperature the higher the heat engine efficiency. On the other hand, the collector efficiency also depends on the collector temperature achieved, i.e., the higher the collector temperature the lower the collector efficiency. Therefore, for a solar-collector Rankine cycle to operate at high collector efficiency and high heat engine efficiency, the heat input to the solar low-temperature heat engine can be derived from several collectors, as shown in Fig. 2.20. The heat exchangers (collectors) in Fig. 2.20 are operated at different temperatures.

Another solar heat engine, suitable for a moderate sized power-generating facility, utilizes a concentrate collector (receiver) mounted on top of a high tower. Radiation from the sun is reflected by a field of mirrors on to the receiver to achieve a high concentration ratio. The orientation of each mirror depends on its location relative to the central receiver. Each mirror is controlled to keep the sun's reflected radiation concentrated on the central receiver. The central receiver can serve as the boiler for the steam Rankine solar heat engine. Thermal storage will be required for a commercial facility to mitigate interruptions caused by clouds and to provide for an evening and night-time output.

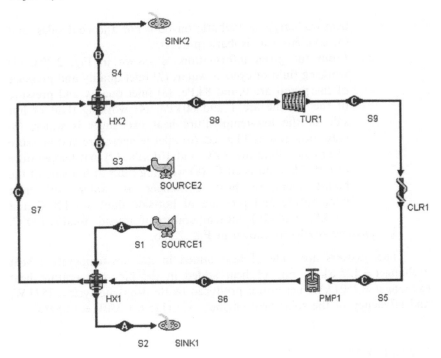

Figure 2.20 Solar heat engine.

Example 2.12

A solar heat engine with two collectors as shown in Figure 2.20 is proposed. Water enters the low-temperature heat exchanger from a low-temperature collector at 1 kg/sec and 80°C and 101 kPa. Water enters the high-temperature heat exchanger from a high-temperature collector at 120°C and leaves the heat exchanger at 100°C and 101 kPa. Cycle water enters the pump of the Rankine heat engine at 8 kPa and 0% quality. Cycle water enters the high-temperature heat exchanger at 80°C. Saturated steam enters the turbine at 80 kPa. Find the power produced by the solar heat engine.

To solve this problem with CyclePad, we take the following steps:

1. Build as shown in Fig. 2.20.
2. Analysis:
   a. Assume a process each for the five devices: (1) turbine as adiabatic with 100% efficiency, (2) pump as adiabatic with 100% efficiency, (3) low-temperature heat exchanger as isobaric on both hot and cold sides, (4) high-temperature

heat exchanger as isobaric on both hot and cold sides, and (5) condenser as isobaric process.

    b. Input the given information as shown in Fig. 2.20a: (1) working fluid of cycle is water, (2) inlet quality and pressure of the pump are 0 and 8 kPa, (3) inlet quality and pressure of the turbine are 1 and 80 kPa, (4) working fluid of hot side of the low-temperature heat exchanger is water, (5) mass flow rate is 1 kg/sec, (6) inlet temperature and pressure of hot-side fluid are 80°C and 101 kPa, (7) exit temperature of hot-side fluid is 50°C, (8) working fluid of hot side of the high-temperature heat exchanger is water, (9) inlet temperature and pressure of hot-side fluid are 120°C and 101 kPa, and (10) exit temperature of hot-side fluid is 100°C.

3. Display results as shown in Fig. 2.20b.

The answers are: rate of heat added in the low-temperature heat exchanger = 125.6 kW, rate of heat added in the high-temperature heat exchanger = 2501 kW, net power produced by the Rankine cycle = 357 kW, and efficiency of the solar heat engine = 357/(125.6 + 2501) = 13.59%.

Example 2.13

A solar Rankine heat engine with one concentrated collector used as the boiler is proposed. Saturated water at 1 kg/sec enters the pump of the

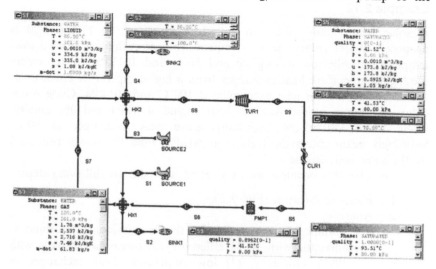

Figure 2.20a Solar heat engine input information.

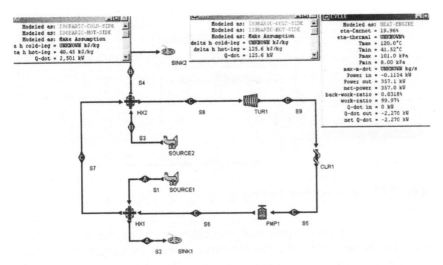

Figure 2.20b Solar heat engine output information.

Rankine heat engine at 10 kPa. Steam enters the turbine at 1 MPa and 250°C. Find the cycle efficiency and power produced by the solar heat engine.

To solve this problem with CyclePad, we take the following steps:

1. Build a Rankine cycle.
2. Analysis:
    a. Assume a process each for the four devices: (1) turbine as adiabatic with 100% efficiency, (2) pump as adiabatic with 100% efficiency, (3) boiler as isobaric process, and (4) condenser as isobaric process.
    b. Input the given information: (1) working fluid of cycle is water, (2) inlet quality and pressure of the pump are 0 and 10 kPa, (3) inlet temperature and pressure of the turbine are 250°C and 1 MPa, and (4) mass flow rate is 1 kg/sec.
3. Display result as shown in Fig. 2.21.

The answers are: rate of heat added in the boiler = 2749 kW, net power produced by the Rankine cycle = 747.7 kW, and efficiency of the solar heat engine = 27.2%.

Review Problems 2.8   Solar Heat Engine

1. Why are several solar collectors at different temperatures desirable for a solar-collector Rankine heat engine?

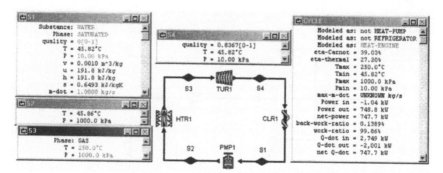

Figure 2.21   Solar Rankine heat engine.

2. A solar heat engine with two collectors as shown in Fig. 2.20 is proposed. Water enters the low-temperature heat exchanger from a low-temperature collector at 100°C and 101 kPa. Water enters the high-temperature heat exchanger from a high-temperature collector at 120°C and leaves the heat exchanger at 100°C and 101 kPa. Cycle water enters the pump of the Rankine heat engine at 1 kg/sec and 8 kPa. Cycle water enters the high-temperature heat exchanger at 100°C. Saturated steam enters the turbine at 80 kPa. Find the water mass flow rates of the low-temperature collector and the high-temperature collector.

ANSWERS: mass flow rate of the low-temperature collector = 57.61 kg/sec, mass flow rate of the high-temperature collector = 57.61 kg/ sec.

3. A solar heat engine with two collectors as shown in Fig. 2.20 is proposed. Water enters the low-temperature heat exchanger from a low-temperature collector at 100°C and 101 kPa. Water enters the high-temperature heat exchanger from a high-temperature collector at 140°C and leaves the heat exchanger at 100°C and 101 kPa. Cycle water enters the pump of the Rankine heat engine at 1 kg/sec and 8 kPa. Cycle water enters the high-temperature heat exchanger at 100°C. Saturated steam enters the turbine at 80 kPa. Find the water mass flow rates of the low-temperature collector and the high-temperature collector.

ANSWERS: mass flow rate of the low-temperature collector = 29 kg/ sec; mass flow rate of the high-temperature collector = 29 kg/sec.

2.9   GEOTHERMAL HEAT ENGINES

Since interior regions of the earth have temperatures higher than those at the surface, an outward flux of heat is observed. The earth's interior is a vast thermal reservoir, which can be used as a source of energy if favorable geological conditions exist. There are many areas that have a high heat flow rate. Also, there are many geological formations resulting in thermal reservoirs located within a short distance of the earth's surface. By drilling into these high-temperature reservoirs, useful quantities of geothermal energy are obtained.

In a few regions, porous rock is overlain by a low-permeability stratum and above that an aquifer, which allows water to trickle into the hot porous rock at a rate such that a steady flow of dry steam is generated. Dry steam is the most desirable form of geothermal energy. For these dry-steam wells, the obvious course is to use the dry steam directly in the turbine after filtering it to remove mineral particulates. The dry steam can be expanded in a turbine and exhausted directly to the atmosphere. While this is the simplest and least costly type of geothermal power installation, its efficiency in converting geothermal energy to electrical energy is low. Leaving the turbine would be a mixture of vapor and liquid, the pressure of which must be above atmospheric pressure. The exhaust geothermal fluid temperature must, therefore, be above 100°C. A considerable improvement on the efficient use of the geothermal power plant can be achieved by reducing the turbine exhaust temperature to 50°C. A condenser with an internal pressure less than atmospheric pressure is required. Figure 2.22a shows the basic dry steam geothermal power plant for such an arrangement.

Example 2.14

At a geothermal energy source, dry steam at 700 kPa and 170°C is available at a mass flow rate of 100 kg/sec. A barometric condenser at 10 kPa is used to decrease the turbine exhaust temperature. Find (a) the power produced by the geothermal power plant as shown in Fig. 2.22a. (b) What is the power produced without the barometric condenser?

To solve this problem with CyclePad, we take the following steps:

1. Build as shown in Fig. 2.22a.

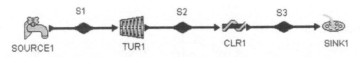

Figure 2.22a Dry-steam geothermal power plant.

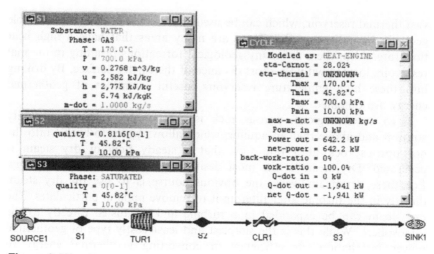

Figure 2.22b Dry-steam geothermal power plant without condenser.

    2.   Analysis:

        a.   Assume a process each for the two devices: (1) turbine as adiabatic with 100% efficiency, and (2) condenser as isobaric process.

        b.   Input the given information: (1) working fluid of cycle is water, (2) inlet temperature and pressure of the turbine are 170°C and 700 kPa, (3) the inlet pressure of the condenser is 10 kPa, and (4) mass flow rate is 100 kg/sec.

    3.   Display result:

The answers are: (a) With condenser, power = 642.2 kW as shown in Fig. 2.22a, and (b) without condenser, changing the inlet pressure of the condenser to 100 kPa gives: power = 332.7 kW as shown in Fig. 2.22b.

    More commonly than it puts out dry steam, a geothermal well puts out a mixture of steam and water above 130°C, or just hot water. A separator is needed in a hot water–steam mixture geothermal power plant to separate the flashing steam from the hot water, as shown in Figure 2.23. An additional throttling valve is required to generate saturated steam in a hot-water geothermal power plant.

Example 2.15

At a geothermal energy source, a mixture of 80% steam and 20% water at 140°C is available at a mass flow rate of 1 kg/sec. A barometric condenser

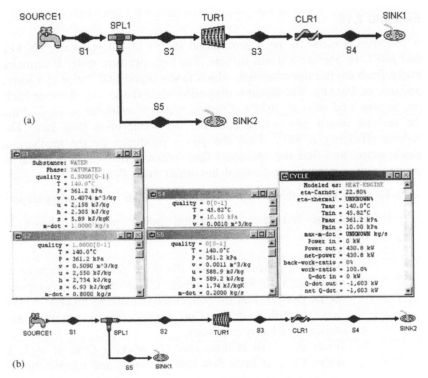

Figure 2.23 Hot water–steam mixture geothermal power plant.

at 10 kPa is used to decrease the turbine exhaust temperature. Find the power produced by the geothermal power plant.

To solve this problem with CyclePad, we take the following steps:

1. Build as shown in Fig. 2.23.
2. Analysis:
   a. Assume a process each for the three devices: (1) turbine as adiabatic with 100% efficiency, (2) splitters as nonisoparametric devices, and (3) condenser as isobaric process.
   b. Input the given information: (1) working fluid of cycle is water, (2) inlet mass flow rate, quality and temperature of the separator (splitter) are 1 kg/sec, 0.8, and 140°C, (3) inlet quality of the turbine is 1, (4) inlet quality of the sink1 is 0, and (5) pressure of the condenser is 10 kPa.
3. Display result.

The answer is power = 430.8 kW as shown in Fig. 2.23a.

Example 2.16

A proposal is made to use a geothermal supply of hot water at 1500 kPa and 180°C to operate a steam turbine. The high-pressure water is throttled into a flash evaporator chamber, which forms liquid and vapor at a lower pressure of 400 kPa. The liquid is discarded while the saturated vapor feeds the turbine and exits at 10 kPa. Cooling water is available at 15°C. Find the turbine power per unit geothermal hot-water mass flow rate. The turbine efficiency is 88%. Find the power produced by the geothermal power plant, and find the optimized flash pressure that will give the most turbine power per unit geothermal hot water mass flow rate.

To solve this problem with CyclePad, we take the following steps:

1. Build the hot-water geothermal power plant as shown in Fig. 2.23.
2. Analysis:
   a. Assume a process each for the three devices: (1) turbine as adiabatic with 88% efficiency, (2) splitters as nonisoparametric devices, and (3) condenser as isobaric process. Notice that throttling devices in CyclePad are automatically constant-enthalpy processes.
   b. Input the given information: (1) working fluid of cycle is water, (2) inlet mass flow rate, pressure, and temperature of the separator (splitter) are 1 kg/sec, 1500 kPa, and 180°C, (3) inlet quality and pressure of the turbine are 1 and 400 kPa, (4) inlet quality of the sink1 is 0, and (5) inlet pressure of the condenser is 10 kPa.
3. Display result.

The answer is power = 32.95 kW as shown in Fig. 2.24a.

To find the optimized flash pressure that will give the most turbine power per unit geothermal hot-water mass flow rate, we use sensitivity analysis. A plot of power versus turbine inlet pressure is made as shown in Fig. 2.24b. The maximum power is found to be about 42 kW at a pressure of about 140 kPa.

Example 2.17

A proposal is made to use a geothermal supply of hot water at 1500 kPa and 180°C to operate a two-flash evaporator and two geothermal steam turbines system as shown in Fig. 2.25a. The high-pressure water is throttled into a flash evaporator chamber, which forms liquid and vapor at a lower pressure of 400 kPa. The saturated vapor at 400 kPa

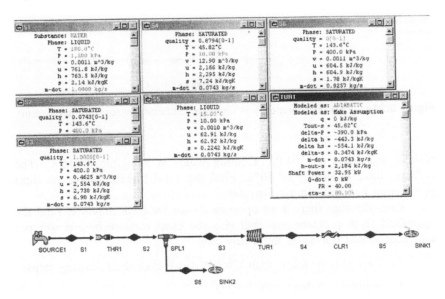

Figure 2.24a Geothermal hot-water power plant.

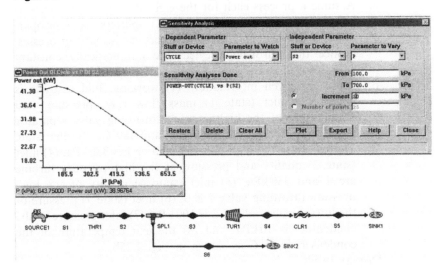

Figure 2.24b Optimization of geothermal hot-water power plant.

feeds the high-pressure turbine and exits at 10 kPa. The saturated liquid at 400 kPa is then throttled into a flash evaporator chamber, which forms liquid and vapor at a lower pressure of 100 kPa. The liquid at 100 kPa is discarded while the saturated vapor at 100 kPa feeds the low-pressure

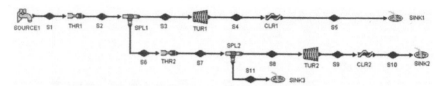

Figure 2.25a Two-flash evaporators and two geothermal steam turbines systems.

turbine and exits at 10 kPa. Cooling water is available at 15°C. The turbines have efficiency of 80%. Find the total turbine power per unit geothermal hot water mass flow rate.

Consider that there is an optional choice for flash pressure. Find the optimized flash pressure that will give the most total turbine power per unit geothermal hot-water mass flow rate.

To solve this problem with CyclePad, we take the following steps:

1. Build as shown in Fig. 2.25a.
2. Analysis:
   a. Assume a process each for the eight devices: (1) turbines as adiabatic with 80% efficiency, (2) splitters as nonisoparametric devices, and (3) condensers as isobaric processes. Notice that throttling devices are automatically constant enthalpy processes.
   b. Input the given information: (1) working fluid of cycle is water, (2) inlet (state 1) mass flow rate, pressure, and temperature of the high-pressure throttling valve separator (splitter 1) are 1 kg/sec, 1500 kPa, and 180°C, (3) inlet (state 2) pressure of the high-pressure splitter 1 is 350 kPa, (4) inlet (state 3) quality and pressure of the high-pressure turbine are 1 and 350 kPa, (5) inlet (state 6) quality of the low-pressure throttling valve 2 is 0, (6) inlet (state 7) pressure of the low-pressure splitter 2 is 100 kPa, (7) inlet pressure of the condensers is 10 kPa, and (8) the exit temperature of the condensers is 15°C as shown in Fig. 2.25b.
3. Display result.

The answers are power of turbine 1 = 35.59 kW, power of turbine 2 = 18.61 kW, and total turbine power = 35.59 + 18.61 = 54.2 kW as shown in Fig. 2.25c.

To find the optimized flash pressure that will give the most turbine power per unit geothermal hot-water mass flow rate, we use the sensitivity analysis. A plot of total turbine power versus high-pressure turbine 1 inlet

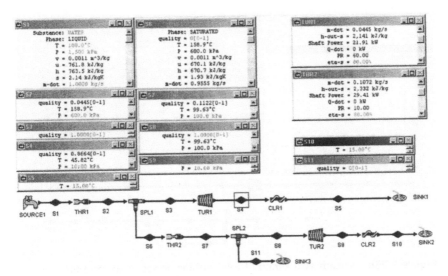

Figure 2.25b   Input information.

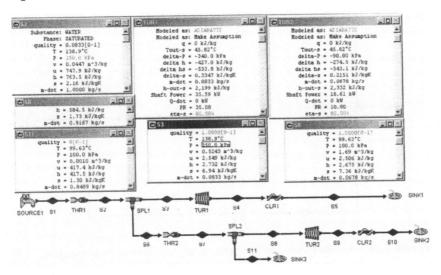

Figure 2.25c   Output information

pressure (p_2) is made as shown in Fig. 2.25d. The maximum total turbine power is found about 54.3 kW at a pressure of 400 kPa.

Electrical power can be produced by geothermal fields, in which there is either hot water or steam below 130°C, by using a secondary closed Rankine cycle as shown in Fig. 2.26.

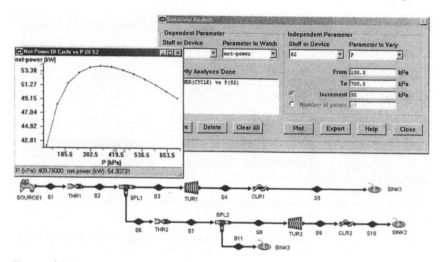

Figure 2.25d Sensitivity diagram of total turbine power versus pressure.

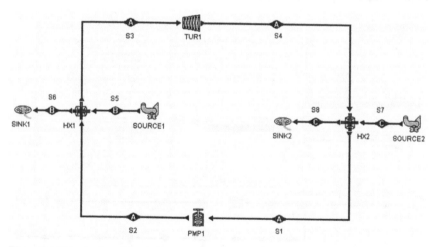

Figure 2.26 A closed-cycle low-temperature dry geothermal Rankine cycle.

Dry geothermal fields (high-temperature rocks) in which no water is present are another potential source of geothermal energy. Water will need to be injected into the field. After drilling, fracturing of the high-temperature rocks will be required to improve heat transfer areas with water. A secondary closed Rankine cycle as shown in Fig. 2.26 will be required for producing power.

Example 2.18

At a dry geothermal energy source, hot rock is available. Water is injected into the field. Geothermal energy is transferred from the hot rock to a proposed Rankine heat engine by a heat exchanger. The information for the proposed geothermal steam–Rankine cycle are: fluid mass flow rate, quality, and pressure at the inlet of the pump are 1 kg/sec, 0, and 8 kPa; fluid quality and pressure at the inlet of the turbine are 1 and 140 kPa; and hot-water temperature at the inlet and exit of the heat exchanger are 120 and 70°C. Find the power produced by the geothermal power plant.

To solve this problem with CyclePad, we take the following steps:

1. Build as shown in Fig. 2.26.
2. Analysis:
   a. Assume a process each for the four devices: (1) turbine as adiabatic with 100% efficiency, (2) pump as adiabatic with 100% efficiency, (3) condenser as isobaric process, and (4) both hot and cold sides of the heat exchanger are isobaric.
   b. Input the given information: (1) working fluid of cycle is water, (2) inlet mass flow rate, quality, and pressure of the pump are 1 kg/sec, 0, and 8 kPa, (3) inlet quality and pressure at the inlet of the turbine are 1 and 140 kPa, (4) inlet and exit of the heat exchanger are 120 and 70°C, and (5) inlet pressure of the hot water to the heat exchanger is 100 kPa.
3. Display result.

The answer is power = 422.5 kW as shown in Fig. 2.27.

Review Problems 2.9   Geothermal Heat Engines

1. A proposal is made to use a geothermal supply of hot water at 1 MPa and 170°C to operate a steam turbine as shown in Fig. 2.24a. The high-pressure water is throttled into a flash evaporator chamber, which forms liquid and vapor at a lower pressure of 400 kPa. The liquid is discarded while the saturated vapor feeds the turbine and exits at 10 kPa. Cooling water is available at 15°C. The turbine has an isentropic efficiency of 88%. Find the turbine power per unit geothermal hot-water mass flow rate. Find the optimized flash pressure that will give the most turbine power per unit geothermal hot-water mass flow rate.

ANSWERS: turbine power = 26.51 kW; maximum turbine power = 40.8 kW at throttling pressure = 106 kPa.

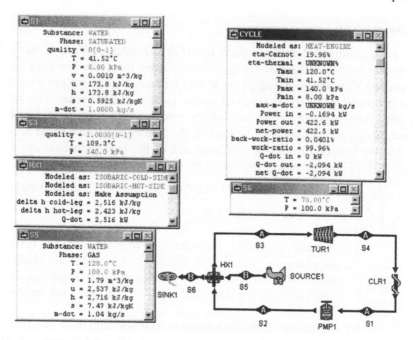

Figure 2.27 A closed-cycle low-temperature dry geothermal Rankine cycle.

2. A proposal is made to use a geothermal supply of hot water at 800 kPa and 170°C to operate a steam turbine as shown. The high-pressure water is throttled into a flash evaporator chamber, which forms liquid and vapor at a lower pressure of 300 kPa. The liquid is discarded while the saturated vapor feeds the turbine and exits at 10 kPa. Cooling water is available at 15°C. The turbine has an isentropic efficiency of 85%. Find the turbine power per unit geothermal hot-water mass flow rate. Find the optimized flash pressure that will give the most turbine power per unit geothermal hot-water mass flow rate.

ANSWERS: turbine power = 31.62 kW; maximum turbine power = 39.3 kW at throttling pressure = 107 kPa.

3. A proposal is made to use a geothermal supply of hot water at 1 MPa and 150°C to operate a two flash evaporators and two geothermal steam turbine systems as shown. The high-pressure water is throttled into a flash evaporator chamber, which forms liquid and vapor at a lower pressure of 400 kPa. The saturated vapor at 400 kPa feeds the high-pressure turbine and exits at 10 kPa. The high-pressure turbine has an isentropic efficiency of 88%. The saturated liquid at 400 kPa is then throttled into a flash evaporator chamber, which forms liquid and vapor

at a lower pressure of 100 kPa. The liquid at 100 kPa is discarded while the saturated vapor at 100 kPa feeds the low-pressure turbine and exits at 10 kPa. The low-pressure turbine has an isentropic efficiency of 87%. Cooling water is available at 15°C. Find the turbine power per unit geothermal hot-water mass flow rate. Consider that there is an optional choice for flash pressure. Find the optimized flash pressure that will give the most total turbine power per unit geothermal hot-water mass flow rate.

ANSWERS: high-pressure turbine power = 6.34 kW; low-pressure turbine power = 24.45 kW; maximum turbine power = 33.5 kW at throttling pressure = 230 kPa.

4. A proposal is made to use a geothermal supply of hot water at 1200 kPa and 170°C to operate a two-flash evaporator and two geothermal steam turbine systems as shown. The high-pressure water is throttled into a flash evaporator chamber, which forms liquid and vapor at a lower pressure of 400 kPa. The saturated vapor at 400 kPa feeds the high-pressure turbine and exits at 10 kPa. The high-pressure turbine has an isentropic efficiency of 85%. The saturated liquid at 400 kPa is then throttled into a flash evaporator chamber, which forms liquid and vapor at a lower pressure of 100 kPa. The liquid at 100 kPa is discarded while the saturated vapor at 100 kPa feeds the low-pressure turbine and exits at 10 kPa. The low-pressure turbine has an isentropic efficiency of 85%. Cooling water is available at 15°C. Find the turbine power per unit geothermal hot-water mass flow rate. Consider that there is an optional choice for flash pressure. Find the optimized flash pressure that will give the most total turbine power per unit geothermal hot-water mass flow rate.

ANSWERS: high-pressure turbine power = 25.3 kW; low-pressure turbine power = 22.9 kW; maximum turbine power = 48.8 kW at throttling pressure = 295 kPa.

5. A proposal is made to use a geothermal supply of hot water at 1200 kPa and 180°C to operate a two-flash evaporator and two geothermal steam turbine systems as shown. The high-pressure water is throttled into a flash evaporator chamber, which forms liquid and vapor at a lower pressure of 400 kPa. The saturated vapor at 600 kPa feeds the high-pressure turbine and exits at 10 kPa. The high-pressure turbine has an isentropic efficiency of 85%. The saturated liquid at 600 kPa is then throttled into a flash evaporator chamber, which forms liquid and vapor at a lower pressure of 100 kPa. The liquid at 100 kPa is discarded while the saturated vapor at 100 kPa feeds the low-pressure turbine and exits at 10 kPa. The low-pressure turbine has an isentropic efficiency of 85%. Cooling water is available at 15°C. Find the turbine power per unit

geothermal hot-water mass flow rate. Consider that there is an optional choice for flash pressure. Find the optimized flash pressure that will give the most total turbine power per unit geothermal hot-water mass flow rate.

ANSWERS: high-pressure turbine power = 23.24 kW; low pressure turbine power = 31.25 kW; maximum turbine power = 57.5 kW at throttling pressure = 340 kPa.

6. A proposal is made to use a geothermal supply of hot water at 1200 kPa and 170°C to operate a two-flash evaporator and two geothermal steam turbine systems as shown. The high-pressure water is throttled into a flash evaporator chamber, which forms liquid and vapor at a lower pressure of 400 kPa. The saturated vapor at 400 kPa feeds the high-pressure turbine and exits at 10 kPa. The high-pressure turbine has an isentropic efficiency of 88%. The saturated liquid at 400 kPa is then throttled into a flash evaporator chamber, which forms liquid and vapor at a lower pressure of 100 kPa. The liquid at 100 kPa is discarded while the saturated vapor at 100 kPa feeds the low-pressure turbine and exits at 10 kPa. The low-pressure turbine has an isentropic efficiency of 87%. Cooling water is available at 15°C. Find the turbine power per unit geothermal hot-water mass flow rate. Consider that there is an optional choice for flash pressure. Find the optimized flash pressure that will give the most total turbine power per unit geothermal hot-water mass flow rate.

ANSWERS: high-pressure turbine power = 26.19 kW; low-pressure turbine power = 23.44 kW; maximum turbine power = 49.9 kW at throttling pressure = 300 kPa.

7. A geothermal supply of hot water at 800 kPa and 150°C is fed to the throttling valve of a geothermal power plant. A stream of saturated vapor at 300 kPa is drawn from the separator and fed to the turbine. The turbine has an isentropic efficiency of 85% and an exit pressure of 10 kPa. Cooling water is available at 15°C. Find the turbine power per unit geothermal hot-water mass flow rate.

ANSWERS: turbine power = 14.21 kW; maximum turbine power = 27.9 kW at throttling pressure = 80 kPa.

8. A geothermal supply of hot water at 800 kPa and 160°C is fed to the throttling valve of a geothermal power plant. A stream of saturated vapor at 300 kPa is drawn from the separator and fed to the turbine. The turbine has an isentropic efficiency of 88% and an exit pressure of 10 kPa. Cooling water is available at 15°C. Find the turbine power per unit geothermal hot-water mass flow rate.

ANSWERS: turbine power = 23.96 kW; maximum turbine power = 27.8 kW at throttling pressure = 79 kPa.

9.   A proposal is made to use a geothermal supply of hot water at 1200 kPa and 180°C to operate a two-flash evaporator and two geothermal steam turbine systems as shown. The high-pressure water is throttled into a flash evaporator chamber, which forms liquid and vapor at a lower pressure of 400 kPa. The saturated vapor at 400 kPa feeds the high-pressure turbine and exits at 10 kPa. The saturated liquid at 400 kPa is then throttled into a flash evaporator chamber, which forms liquid and vapor at a lower pressure of 100 kPa. The liquid at 100 kPa is discarded while the saturated vapor at 100 kPa feeds the low-pressure turbine and exits at 10 kPa. Cooling water is available at 15°C. The turbines have efficiency of 80%. Find the turbine power per unit geothermal hot-water mass flow rate.

ANSWER: total turbine power = 54.49 kW.

10.   A proposal is made to use a geothermal supply of hot water at 1200 kPa and 170°C to operate a two-flash evaporator and two geothermal steam turbine systems as shown. The high-pressure water is throttled into a flash evaporator chamber, which forms liquid and vapor at a lower pressure of 400 kPa. The saturated vapor at 400 kPa feeds the high-pressure turbine and exits at 10 kPa. The saturated liquid at 400 kPa is then throttled into a flash evaporator chamber, which forms liquid and vapor at a lower pressure of 100 kPa. The liquid at 100 kPa is discarded while the saturated vapor at 100 kPa feeds the low-pressure turbine and exits at 10 kPa. Cooling water is available at 15°C. The turbines have efficiency of 80%. Find the turbine power per unit geothermal hot-water mass flow rate.

ANSWER: total turbine power = 44.18 kW.

2.10   OCEAN THERMAL ENERGY CONVERSION

Since the oceans comprise over 70% of the earth's surface area, the absorbed solar energy that is stored as latent heat of the oceans represents a very large potential source of energy. As a result of variation in the density of ocean water with temperature, the ocean water temperature is not uniform with depth. Warm surface ocean water with low density tends to stay on the surface and cold water with high density within a few degree of 4°C tends to settle to the depths of the ocean. In the tropics, ocean surface temperatures in excess of 25°C occur. The combination of the warmed surface water and cold deep water provides two different temperature thermal reservoirs needed to operate a heat engine called OTEC (ocean thermal energy conversion). Since the temperature difference of the OTEC between the heat source and the heat sink is small, the OTEC power plant cycle efficiency

---

is small. It means that enormous quantities of ocean water must be handled and the heat exchangers and turbine must be very large.

There are two principal approaches to build OTEC power plants. The first approach called the open OTEC cycle involves a flash boiler to obtain steam directly from the warm surface ocean water. The open OTEC cycle requires a very large turbine. The second approach is called the closed OTEC cycle, which involves heat exchangers and a secondary thermodynamic working fluid such as ammonia or freon to reduce the size of the plant.

Ocean water is the working fluid of the open OTEC cycle, as shown in Fig. 2.28. For conditions typical of an open OTEC plant, the vapor pressure of the boiler at 26°C is 3.37 kPa and the vapor pressure of the condenser at 5°C is 0.874 kPa. Boiling of the warm water occurs at a pressure of only 3% of atmospheric pressure. The steam is expanded in a low-pressure, low-temperature, high-volume turbine before being condensed by the cold water. An advantage of this cycle is that heat exchangers with their attendant temperature differentials are unnecessary. The disadvantage is the very small pressure drop and the large specific volumes that must be utilized by the turbine.

The closed OTEC cycle as shown in Fig. 2.29 uses a secondary thermodynamic working fluid such as ammonia or freon to reduce the size of the plant. For a boiler temperature of 25°C, the vapor pressure

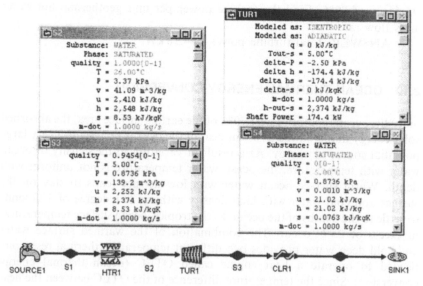

Figure 2.28 Open OTEC cycle.

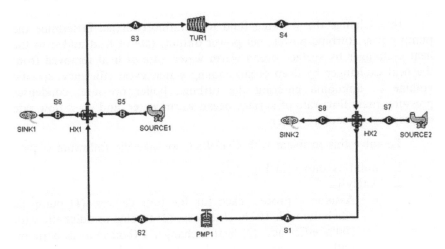

Figure 2.29   Closed OTEC cycle.

of ammonia is nearly 10 times the atmospheric pressure and the specific volume is comparable to that of a conventional steam power plant. While the size of a turbine is smaller than that of a comparable low-pressure steam turbine, large heat exchangers are required.

Example 2.19

A typical closed-cycle OTEC Rankine cycle using ammonia is suggested, as illustrated in Fig. 2.29, with the following information:

Condenser temperature	12°C
Boiler temperature	24°C
Mass flow rate of ammonia	1 kg/sec
Surface ocean warm water entering heat exchanger	28°C
Surface ocean warm water leaving heat exchanger	26°C
Deep ocean cooling water entering heat exchanger	5°C
Deep ocean cooling water leaving heat exchanger	9°C
Turbine efficiency	100%
Pump efficiency	100%

(a)   Determine the pump power, turbine power, net power output, rate of heat added to the heat exchanger by surface ocean warm water, rate of heat removed from the heat exchanger by deep ocean cooling water, cycle efficiency, specific volume of ammonia entering the turbine, boiler pressure, condenser pressure, mass flow rate of surface ocean warm water, and mass flow rate of deep ocean cooling water.

(b) Change the working fluid to tetrafluoroethane. Determine the pump power, turbine power, net power output, rate of heat added to the heat exchanger by surface ocean warm water, rate of heat removed from the heat exchanger by deep ocean cooling water, cycle efficiency, specific volume of ammonia entering the turbine, boiler pressure, condenser pressure, mass flow rate of surface ocean warm water, and mass flow rate of deep ocean cooling water.

To solve this problem with CyclePad, we take the following steps:

1. Build as shown in Fig. 2.29.
2. Analysis:
    a. Assume a process each for the four devices: (1) pump as adiabatic with 100% efficiency, (2) turbine as adiabatic with 100% efficiency, (3) heat exchanger 1 (boiler) as isobaric on both cold and hot sides, and (4) heat exchanger 2 (condenser) as isobaric on both cold and hot sides.
    b. Input the given information: (1) working fluid of cycle A is ammonia, working fluid of cycle B is water, and working fluid of cycle C is water, (2) inlet temperature and quality of the pump are 12°C and 0, (3) inlet temperature and quality of the turbine are 24°C and 1, and (4) mass flow rate is 1 kg/sec.
3. Display result.

The answers are: (a) $Wdot_{pump} = -6.58\,kW$, $Wdot_{turbine} = 48.91\,kW$, $Wdot_{net} = 42.33\,kW$, $Qdot_{boiler} = 1220\,kW$, $Qdot_{condenser} = -1178\,kW$, $\eta = 3.47\%$, $v = 0.1321\,m^3/kg$, $p_{boiler} = 972.4\,kPa$, $p_{condenser} = 658.5\,kPa$, $mdot_{warm\ water} = 145.8\,kg/sec$, and $mdot_{cold\ water} = 70.23\,kg/sec$.

4. (a) Retract the working fluid: working fluid of cycle A is R-134a, and (b) Display result as shown in Fig. 2.30.

The answers are: (b) $Wdot_{pump} = -1.38\,kW$, $Wdot_{turbine} = 7.74\,kW$, $Wdot_{net} = 6.36\,kW$, $Qdot_{boiler} = 194.2\,kW$, $Qdot_{condenser} = -187.9\,kW$, $\eta = 3.27\%$, $v = 0.0320\,m^3/kg$, $p_{boiler} = 445.3\,kPa$, $p_{condenser} = 647.6\,kPa$, $mdot_{warm\ water} = 23.21\,kg/sec$, and $mdot_{cold\ water} = 11.21\,kg/sec$.

Review Problems 2.10 Ocean Thermal Energy Conversion

1. A typical closed-cycle OTEC Rankine cycle using ammonia is suggested, as illustrated in Fig. 2.30, with the following information:

Condenser temperature	10°C
Boiler temperature	22°C
Mass flow rate of ammonia	1 kg/sec

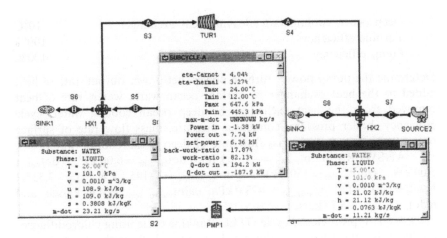

Figure 2.30 OTEC Rankine cycle.

Surface ocean warm water entering heat exchanger	28°C
Surface ocean warm water leaving heat exchanger	22°C
Deep ocean cooling water entering heat exchanger	4°C
Deep ocean cooling water leaving heat exchanger	10°C
Turbine efficiency	100%
Pump efficiency	100%

Determine the pump power, turbine power, net power output, rate of heat added to the heat exchanger by surface ocean warm water, rate of heat removed from the heat exchanger by deep ocean cooling water, cycle efficiency, boiler pressure, condenser pressure, mass flow rate of surface ocean warm water, and mass flow rate of deep ocean cooling water.

ANSWERS: $Wdot_{pump} = -6.36\,kW$, $Wdot_{turbine} = 49.55\,kW$, $Wdot_{net} = 43.18\,kW$, $Qdot_{boiler} = 1228\,kW$, $Qdot_{condenser} = -1185\,kW$, $\eta = 3.52\%$, $p_{boiler} = 913.4\,kPa$, $p_{condenser} = 614.9\,kPa$, $mdot_{warm\ water} = 47.08\,kg/sec$, and $mdot_{cold\ water} = 36.71\,kg/sec$.

2. A typical closed-cycle OTEC Rankine cycle using dichloro-difluoromethane is suggested, as illustrated in Fig. 2.30, with the following information:

Condenser temperature	10°C
Boiler temperature	22°C
Mass flow rate of ammonia	1 kg/sec
Surface ocean warm water entering heat exchanger	28°C
Surface ocean warm water leaving heat exchanger	22°C
Deep ocean cooling water entering heat exchanger	4°C

Deep ocean cooling water leaving heat exchanger   10°C
Turbine efficiency   100%
Pump efficiency   100%

Determine the pump power, turbine power, net power output, rate of heat added to the heat exchanger by surface ocean warm water, rate of heat removed from the heat exchanger by deep ocean cooling water, cycle efficiency, boiler pressure, condenser pressure, mass flow rate of surface ocean warm water, and mass flow rate of deep ocean cooling water.

ANSWERS: $Wdot_{pump} = -1.06 \, kW$, $Wdot_{turbine} = 6.06 \, kW$, $Wdot_{net} = 5.00 \, kW$, $Qdot_{boiler} = 150.1 \, kW$, $Qdot_{condenser} = -145.1 \, kW$, $\eta = 3.33\%$, $p_{boiler} = 600.7 \, kPa$, $p_{condenser} = 423.0 \, kPa$, $mdot_{warm \ water} = 4.49 \, kg/sec$, and $mdot_{cold \ water} = 5.77 \, kg/sec$.

3. A typical closed-cycle OTEC Rankine cycle using chlorodifluoromethane is suggested, as illustrated in Fig. 2.30, with the following information:

Condenser temperature   10°C
Boiler temperature   22°C
Mass flow rate of ammonia   1 kg/sec
Surface ocean warm water entering heat exchanger   28°C
Surface ocean warm water leaving heat exchanger   22°C
Deep ocean cooling water entering heat exchanger   4°C
Deep ocean cooling water leaving heat exchanger   10°C
Turbine efficiency   100%
Pump efficiency   100%

Determine the pump power, turbine power, net power output, rate of heat added to the heat exchanger by surface ocean warm water, rate of heat removed from the heat exchanger by deep ocean cooling water, cycle efficiency, boiler pressure, condenser pressure, mass flow rate of surface ocean warm water, and mass flow rate of deep ocean cooling water.

ANSWERS: $Wdot_{pump} = -2.23 \, kW$, $Wdot_{turbine} = 8.09 \, kW$, $Wdot_{net} = 5.86 \, kW$, $Qdot_{boiler} = 198.4 \, kW$, $Qdot_{condenser} = -192.5 \, kW$, $\eta = 2.95\%$, $p_{boiler} = 963.5 \, kPa$, $p_{condenser} = 680.7 \, kPa$, $mdot_{warm \ water} = 5.93 \, kg/sec$, and $mdot_{cold \ water} = 7.65 \, kg/sec$.

2.11 SOLAR POND HEAT ENGINES

A solar pond heat engine is a small-scale, inverse OTEC system. In this system, a shallow (1–2 m deep) pond saturated with a salt is used as the primary solar collector. As the surface waters of the pond are heated by the solar radiation, the solubility of this warm water increases and the

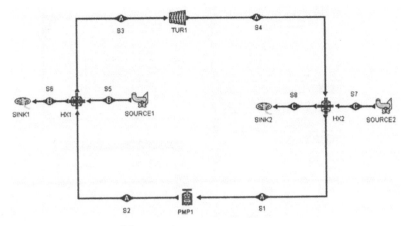

Figure 2.31   Solar pond heat engine.

solution becomes heavier as it absorbs more salt. This causes the hot water
to sink to the bottom of the pond. Consequently, the bottom water of
the pond becomes very hot (65°–82°C) while the surface water remains
at a temperature below 32°C.

The hot water from the bottom of the pond is pumped through a
boiler, where it boils a working fluid in a Rankine power cycle, as shown
in Fig. 2.31. The cooler water from the surface of the pond is used to cool
the turbine exhaust vapor in the condenser. This is the same concept that
is employed in the OTEC system, except that in the OTEC system the
surface waters are warmer than that of the deep ocean water.

Example 2.20

A proposal is made to use a solar pond supply of bottom pond hot water
at 100 kPa and 80°C to operate a steam turbine. The 100 kPa-pressure
bottom pond water is throttled into a flash evaporator chamber, which
forms liquid and vapor at a lower pressure of 20 kPa. The liquid is
discarded while the saturated vapor feeds the turbine and exits at 10 kPa.
Cooling water is available at 15°C. Find the turbine power per unit
geothermal hot-water mass flow rate. The turbine efficiency is 80%. Find
the power produced by the solar pond power plant.

To solve this problem with CyclePad, we take the following steps:

1. Build as shown in Fig. 2.32.
2. Analysis:
    a. Assume a process each for the four devices: (1) turbine as adia-
        batic with 80% efficiency, (2) splitter as nonisoparametric

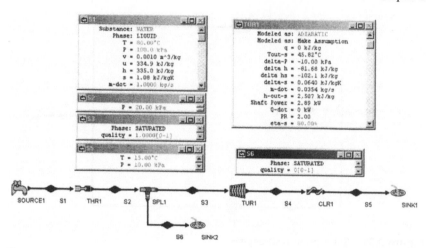

Figure 2.32   Solar pond heat engine.

devices, and (3) condenser as isobaric process. Notice that throttling devices are automatically constant enthalpy processes.

b. Input the given information: (1) working fluid of cycle is water, (2) inlet mass flow rate, pressure, and temperature of the separator (splitter) are 1 kg/sec, 100 kPa, and 80°C, (3) inlet quality and pressure of the turbine are 1 and 20 kPa, (4) inlet quality of sink2 is 0, and (5) exit pressure and temperature of the condenser are 10 kPa and 15°C.

3. Display result.

The answer is power = 2.89 kW as shown in Fig. 2.32.

Review Problems 2.11   Solar Pond Heat Engines

1. What is a solar pond heat engine?
2. A proposal is made to use a solar pond supply of bottom pond hot water at 100 kPa and 80°C to operate a steam turbine. The 100 kPa-pressure bottom pond water is throttled into a flash evaporator chamber, which forms liquid and vapor at a lower pressure of 20 kPa. The liquid is discarded while the saturated vapor feeds the turbine and exits at 10 kPa. Cooling water is available at 15°C. Find the turbine power per unit geothermal hot-water mass flow rate. The turbine efficiency is 85%. Find the power produced by the solar pond power plant.

ANSWER: power produced by the solar pond power plant =
3.07 kW.

2.12 WASTE HEAT ENGINES

Waste sources are variable in both type and availability, depending on
season, location, and socioeconomic factors. Municipal solid residues
generated by large metropolitan areas are large. Large amounts of waste
materials are also generated from farming, animal manure, and crop
production. The quantity and heating values of solid residue are large.
These wastes are often flushed, buried, or burned. Disposal practices
of these wastes are wasteful of resources and create pollution of water
and land. Conversion of waste material to usable thermal energy on
a large scale has been found to be cost-effective and results in a net energy
gain.

Biomass energy created by waste and residues left after food
processing operations, and landfill gas mainly produced during anerobic
decomposition of organic waste material seem to offer the most promising
source of waste heat engines. The material is already concentrated at the
processing site and it creates a disposal pollution problem.

There are three major types of processes for direct combustion of
waste biomass: water-wall incineration, supplementary fuel co-firing with
coal or oil, and fluidized bed combustion.

In water-wall incineration, unprocessed municipal solid residues is
loaded into the incinerator and burned on traveling grates. Low pressure
and temperature (4 MPa, 260°C) steam is produced.

Burning biomass as a supplementary fuel in combination with
steam–electric power production is a proved and established technology.

Fluidized-bed combustion uses air-classified municipal solid residues
to provide heat for a conventional gas turbine to produce power. Several
stages of cyclone separators are also used to remove particulates from the
gas prior to its expansion through the turbines. An advantage of the
process is reduction of noxious gas emission.

Example 2.21

At a solid-waste energy source, steam at 4 MPa and 260°C is available at a
mass flow rate of 1 kg/sec. A barometric condenser at 10 kPa is used to
decrease the turbine exhaust temperature. The turbine efficiency is 85%,
and cooling water is available at 25°C. Find the power produced by the
solid-waste power plant.

To solve this problem with CyclePad, we take the following steps:

1. Build as shown in Fig. 2.33.
2. Analysis:
   a. Assume a process each for the two devices: (1) turbine as adiabatic with 85% efficiency, and (2) condenser as isobaric process.
   b. Input the given information: (1) working fluid of cycle is water, (2) inlet temperature and pressure of the turbine are 260°C and 4000 kPa, (3) exit pressure and temperature of the condenser are 10 kPa and 25°C, and (d) the mass flow rate is 1 kg/sec.
3. Display result.

The answers are: power = 759.8 kW as shown in Fig. 2.33.

Review Problems 2.12   Waste Heat Engines

At a solid-waste energy source, steam at 3 MPa and 250°C is available at a mass flow rate of 1 kg/sec. A barometric condenser at 10 kPa is used to decrease the turbine exhaust temperature. The turbine efficiency is 85%, and cooling water is available at 30°C. Find the power produced by the solid-waste power plant.

ANSWER: power = 753.0 kW

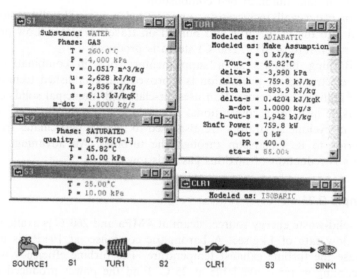

Figure 2.33 Waste heat engine.

2.13 VAPOR CYCLE WORKING FLUIDS

Water has been used mainly as the working fluid in the vapor power examples of this chapter. In fact, water is the most common fluid in large central power plants, though by no means is it the only one used in vapor power cycles. The desirable properties of the vapor cycle working fluid include the following important characteristics.

1. High critical temperature—to permit evaporation at high temperature.
2. Low saturation (boiling) pressure at high temperature—to minimize the pressure vessel and piping costs.
3. Pressure around ambient at condenser temperature—to eliminate serious air leakage and sealing problems.
4. Rapidly diverging pressure lines on the *h–s* diagram—to minimize the back-work ratio and to make reheat modification most effective.
5. Large enthalpy of evaporation—to minimize the mass flow rate for given power output.
6. No degrading aspects—noncorrosive, nonclogging, etc.
7. No hazardous features—nontoxic, nonflammable, etc.
8. Low cost readily available.

There are six vapor working fluids listed on the menu of CyclePad. The fluids are ammonia, methane, refrigerants 12, 22, and 134a, and water. Water has the characteristics of items 4, 5, 7, and 8 above and it remains a top choice for industrial central vapor power plants. Hence, steam power engineering remains the most important area of applied thermodynamics.

Review Problems 2.13 Vapor Cycle Working Fluids

1. Why is water the most popular working fluid choice in central vapor power plants?
2. The ocean surface water is warm (27°C at equator) and deep ocean water is cold (5°C at 2000 m depth). If a vapor cycle operates between these two thermal reservoirs, is water or refrigerant a better choice as the working fluid for this power plant?

2.14 KALINA CYCLE

Thermal reservoirs are not infinitely large in the real world. Therefore, the temperature of a thermal reservoir is not constant when heat is added to or removed from the reservoir.

Kalina and his associates (Kalina, A.I., Combined-cycle system with novel bottoming cycle. *ASME Transaction Journal of Engineering for Gas Turbines and Power*, vol. 106, no. 4, pp. 737–742, 1984) have proposed the use of a mixture of ammonia and water as the working fluid for a vapor Rankine power plant. Since ammonia is more volatile than water, boiling of an ammonia–water mixture starts at a lower temperature and the vapor phase has a higher concentration of ammonia than the liquid phase. Moreover, the mixture temperature increases as the vaporization process progresses. Thus, the constant-pressure heat-transfer process temperature curve of the working fluid more closely matches that of the temperature distribution of its surrounding finite capacity thermal reservoir. The two isobaric processes lead to a higher (better) degree of heat transfer. These differences result in higher efficiency and specific work output. An additional advantage is a condenser pressure near atmospheric pressure.

Review Problems 2.14   Kalina Cycle

1. What is a Kalina cycle?
2. What are the advantages of a Kalina cycle?

2.15   NONAZEOTROPIC MIXTURE RANKINE CYCLE

The thermodynamic performance of a vapor Rankine cycle may be improved potentially by using a nonazeotropic mixture working fluid such as ammonia–water (Wu, C. Nonazeotropic mixture energy conversion. *Energy Conversion and Management*, vol. 25, no. 2, pp. 199–206, 1985). The *nonazeotropic mixture Rankine cycle* is a generalized Kalina cycle. A mixture of two or more different fluids is classified as an azeotrope when such a mixture possesses its own thermodynamic properties, quite unlike the thermal and chemical characteristics of its components. A distinguishing feature of this type of fluid is its ability to maintain a permanent composition and uniform boiling point during evaporation, much the same as a pure simple fluid in that its transition from liquid to vapor phase (or vice versa) occurs at a constant pressure and temperature without any change in the composition. Otherwise, the mixture is called a nonazeotrope. A nonazeotropic mixture has a temperature distribution parallel to that of the thermal reservoir. Note that one of the requirements for the nonazeotropic mixture energy conversion improvement is to have a nonconstant temperature heat source and heat sink. The proper choice of the best combination for a nonazeotropic mixture is still not entirely understood. Uncertainties in modeling the thermodynamic and heat-transfer aspects of the nonazeotropic mixture cycle are such that the probability of realizing significant net benefits in actual application is also not fully known.

An ideal nonazeotropic mixture Rankine cycle and an ideal Carnot cycle operating between a nonconstant temperature heat source and a nonconstant temperature heat sink are shown in the *T–s* diagram of Fig. 2.34. The ideal Carnot cycle consists of an isentropic compression process from state 1 to state 2, an isobaric heat-addition process from state 2 to state 3, an isentropic expansion process from state 3 to state 4, and an isobaric heat-removing process from state 3 to state 4. The ideal nonazeotropic mixture Rankine cycle consists of an isentropic compression process from state 5 to state 2, an isobaric heat-addition process from state 2 to state 6, an isentropic expansion process from state 6 to state 4, and an isobaric heat-removing process from state 4 to state 5. The inlet and exit temperature of the heating fluid (finite-heat-capacity heat source) in the hot-side heat exchanger are T_7 and T_8, and the inlet and exit temperature of the cooling fluid (finite-heat-capacity heat sink) in the hot-side heat exchanger are T_9 and T_{10}, respectively. It is clearly demonstrated that the temperature distribution curves of the ideal nonazeotropic mixture Rankine cycle (curves 2–6 and 4–5) are more closely matched to the temperature distribution curves of the heat source and heat sink (curves 7–8 and 9–10) than the temperature distribution curves of the Carnot cycle (curves 2–3 and 4–1).

Referring to Fig. 2.34, the net work and heat added to the ideal nonazeotropic mixture Rankine cycle are $W_{net,nonaze}$ = area 52645 and

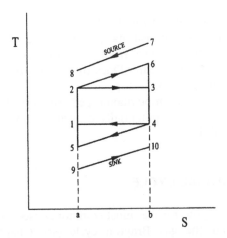

Figure 2.34 *T–s* diagram of ideal nonazeotropic cycle and Carnot cycle.

Q_{nonaze} = area 26ba2, and the net work and heat added to the Carnot cycle are $W_{net,Carnot}$ = area 12341 and Q_{Carnot} = area 23ba2, respectively. The cycle efficiency of the ideal nonazeotropic mixture Rankine cycle is $\eta_{nonaze} = W_{net,nonaze}/Q_{nonaze}$ = area 52645/area 26ba2. Similarly, the cycle efficiency of the Carnot cycle is $\eta_{Carnot} = W_{net,Carnot}/Q_{Carnot}$ = area 12341/ area 23ba2. Rearranging the expression of the cycle efficiency of the ideal nonazeotropic mixture Rankine cycle gives

$$\eta_{nonaze} = W_{net,nonaze}/Q_{nonaze} = \text{area } 52645/\text{area } 26ba2$$

$$= [\text{area } 12341 + \text{area } 2632 + \text{area } 4514]/[\text{area } 23ba2 + \text{area } 2632]$$

$$= \{\text{area } 12341[1 + (\text{area } 2632/\text{area } 12341)$$

$$+ (\text{area } 4514/\text{area } 12341)]\}$$

$$/\{\text{area } 23ba2 \times [1 + (\text{area } 2632/\text{area } 23ba2)]\}$$

$$= \eta_{Carnot}[1 + (\text{area } 2632/\text{area } 12341)$$

$$+ (\text{area } 4514/\text{area } 12341)]/[1 + (\text{area } 2632/\text{area } 23ba2)]$$

Since (area 2632/area 12341) is larger than (area 2632/area 23ba2), the factor $[1 + (\text{area } 2632/\text{area } 12341)+(\text{area } 4514/\text{area } 12341)]/[1 + (\text{area } 2632/\text{area } 23ba2)]$ is larger than 1, and, therefore, η_{nonaze} is larger than η_{Carnot}.

Review Problems 2.15 Nonazeotropic Mixture Rankine Cycle

1. What is a nonazeotropic mixture?
2. Draw an isobaric heating process on a *T–s* diagram for a nonazeotropic mixture from a compressed liquid state to a superheated vapor state. Does the temperature remain the same in the boiling region?
3. Why may the thermodynamic performance of a vapor Rankine cycle be improved potentially by using a nonazeotropic mixture working fluid?

2.16 SUPERCRITICAL CYCLE

The thermodynamic power cycles most commonly used today are the vapor Rankine cycle and the gas Brayton cycle (see Chapter 4). Both are characterized by two isobaric and two isentropic processes. The vapor

Rankine cycle operates mainly in the saturated region of its working fluid whereas the gas Brayton cycle processes are located entirely in the superheat or gas region.

The simple Rankine cycle is inherently efficient. Heat is added and rejected isothermally and, therefore, the ideal Rankine cycle can achieve a high percentage of Carnot cycle efficiency between the same temperatures. Pressure rise in the cycle is accomplished by pumping a liquid, which is an efficient process requiring small work input. The back-work ratio is large.

However, the temperature range of the Rankine cycle is severely limited by the nature of the working fluid—water. Adding superheat in an attempt to circumvent this will remove the cycle from isothermal heat addition. Increasing the temperature range without superheating leads to excessive moisture content in the vapor turbines, resulting in blade erosion.

The gas Brayton cycle adds heat in a isobaric process over a large temperature range. The temperature level is independent of the pressure level. No blade erosion occurs in the gas turbine. However, the compression process of the gas Brayton cycle requires large work input. The back-work ratio is small.

A cycle that retains the advantages and avoids the problems of the two cycles has been devised. This cycle operates entirely above the critical pressure of its working fluid. The cycle is called a *supercritical cycle*.

The supercritical cycle is shown on the *T–s* diagram of a pure substance (Fig. 2.35). The cycle is composed of the following four processes:

1-2 Isentropic compression
2-3 Isobaric heat addition

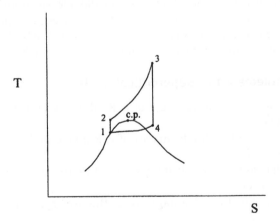

Figure 2.35   *T–s* diagram of supercritical cycle.

3-4 Isentropic expansion
4-1 Isobaric heat removing

Pure substance enters the pump at state 1 as a low-pressure saturated liquid to avoid the cavitation problem and exits at state 2 as a
high-pressure (over critical pressure) compressed liquid. The heat supplied
in the boiler raises the liquid from the compressed liquid at state 2 to a
much higher temperature superheated vapor at state 3. The superheated
vapor at state 3 enters the turbine where it expands to state 4. The exhaust
vapor from the turbine enters the condenser at state 4 and is condensed at
constant pressure to state 1 as saturated liquid.

Analysis of the supercritical cycle is done in the same way as that
of the Rankine cycle. However, it requires special thermodynamic
property tables in the high-temperature and high-pressure range (over
critical temperature and critical pressure) of the working fluid. There is no
such table in the CyclePad working fluid menu. In principle, the
supercritical cycle can be operated with any pure substance. In practice,
the choice of working fluid controls the range of cycle operating pressures
and temperatures. For example, the critical pressure and critical
temperature of ammonia, carbon dioxide, and water are 11.28 MPa
and 405.5 K, 7.39 MPa and 304.2 K, and 22.09 MPa and 647.3 K,
respectively.

Carbon dioxide is a good potential working fluid for the supercritical
cycle for several good reasons. The critical pressure of carbon dioxide is
one-third that of water. Carbon dioxide is known to be a stable and inert
material through the temperature range of industrial power generation. It
is also abundant, nontoxic, and relatively inexpensive.

A numerical example of the carbon dioxide supercritical cycle has
been made by Feher (Feher, E.G., The super-critical thermodynamic
power cycle. *Energy Conversion*, vol. 8, pp. 85–90, 1968). The reasons for
the neglect of the supercritical cycle until now are not known.

Review Problems 2.16 Supercritical Cycle

1. What are the advantages of a Rankine cycle in the compression
   process?
2. What are the disadvantages of a Rankine cycle in the expansion
   process?
3. What are the disadvantages of a Brayton cycle in the compression process?
4. What are the advantages of a Brayton cycle in the expansion
   process?
5. What is the concept of a supercritical cycle?

6. What are the processes of a supercritical cycle?
7. Why is carbon dioxide a better working fluid than water for a supercritical cycle?

2.17 DESIGN EXAMPLES

CyclePad is to a power engineer what a word processor is to a journalist. The benefits of using this software are numerous. The first is that significantly less time is spent doing numerical analysis. As an engineer, this is much appreciated because design computation work that would have taken hours before can now be done in seconds. Second, CyclePad is capable of analyzing cycles with various working fluids. Third, due to its computer-assisted modeling capabilities, the software allows users to view immediately the effects of varying input parameters, either through calculated results or in the form of graphs and diagrams, giving the user a greater appreciation of how a system actually works. More specifically, there is the feature that provides the designer the opportunity to optimize a specific power cycle parameter. The designer is able to gain extensive design experience in a short time. Last, and most important, is the built-in coaching facility that provides definitions of terms and descriptions of calculations. CyclePad goes a step further by informing the inexperienced designer if a contradiction or an incompatability exists within a cycle and why.

When viewing the applicability of CyclePad, users can have benefits at all stages of an engineering career. For the young engineer, just beginning the design learning process, less time is spent doing iterations, resulting in more time dedicated to reinforcing the fundamentals and gaining valuable experience. In the case of the seasoned engineer, who is well indoctrinated in the principles and has gained an engineer's intuition, they can augment their abilities by becoming more computer literate. The following examples illustrate the design of vapor power cycles using CyclePad.

Example 2.22

A four-stage turbine with reheat and three-stage regenerative steam Rankine cycle as shown in Fig. 2.36a was designed by a junior engineer. The following design information is provided:

$p_1 = 103\,\text{kPa}$, $\quad T_1 = 15°\text{C}$, $\quad T_2 = 25°\text{C}$, $\quad p_4 = 16\,\text{MPa}$, $\quad T_4 = 570°\text{C}$, $m\text{dot}_4 = 1000\,\text{kg/sec}$, $\quad p_5 = 8\,\text{MPa}$, $\quad T_6 = 540°\text{C}$, $\quad p_8 = 4\,\text{MPa}$, $\quad p_{10} = 2\,\text{MPa}$,

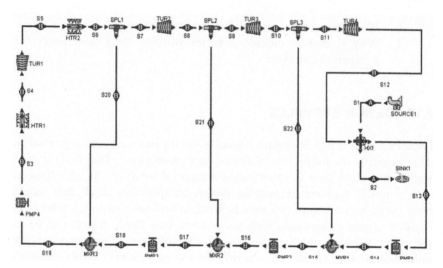

Figure 2.36a Four-stage turbine with reheat and three-stage regenerative Rankine cycle.

$p_{12} = 15\,\text{kPa}$, $x_{13} = 0$, $x_{15} = 0$, $x_{19} = 0$, $\eta_{\text{turbine\#1}} = 0.85$, $\eta_{\text{turbine\#2}} = 0.85$, $\eta_{\text{turbine\#3}} = 0.85$, $\eta_{\text{turbine\#4}} = 0.85$, $\eta_{\text{pump\#1}} = 0.9$, $\eta_{\text{pump\#2}} = 0.9$, $\eta_{\text{pump\#3}} = 0.9$, and $\eta_{\text{pump\#4}} = 0.9$.

Find η_{cycle}, $W\text{dot}_{\text{input}}$, $W\text{dot}_{\text{output}}$, $W\text{dot}_{\text{net output}}$, $Q\text{dot}_{\text{add}}$, $Q\text{dot}_{\text{remove}}$, $W\text{dot}_{\text{turbine\#1}}$, $W\text{dot}_{\text{turbine\#2}}$, $W\text{dot}_{\text{turbine\#3}}$, $W\text{dot}_{\text{turbine\#4}}$, $W\text{dot}_{\text{pump\#1}}$, $W\text{dot}_{\text{pump\#2}}$, $W\text{dot}_{\text{pump\#3}}$, $W\text{dot}_{\text{pump\#4}}$, $Q\text{dot}_{\text{htr\#1}}$, $Q\text{dot}_{\text{htr\#2}}$, $Q\text{dot}_{\text{HX1}}$, $m\text{dot}_{20}$, $m\text{dot}_{21}$, $m\text{dot}_{22}$, $m\text{dot}_{12}$, $m\text{dot}_{15}$, $m\text{dot}_{17}$, and $m\text{dot}_{19}$.

Based on the preliminary design results, try to improve his design. Use η_{cycle} as the objective function and p_5, p_8, and p_{10} as design parameters.

Draw the η_{cycle} versus p_5 sensitivity diagram, the η_{cycle} versus p_8 sensitivity diagram, and the η_{cycle} versus p_{10} sensitivity diagram.

To solve this problem by CyclePad, we take the following steps:

1. Build:
    (a) Take a source, a sink, four pumps, a boiler (HTR1), four turbines, a reheater (HTR2), three splitters, three mixing chambers (open feed-water heaters), and a heat exchanger (condenser) from the inventory shop and connect the devices to form a four-stage turbine with reheat and three-stage regenerative Rankine cycle.
    (b) Switch to analysis mode.

2. Analysis:
   (a) Assume a process for each of the devices: (1) pumps as adiabatic, (2) boiler and reheater as isobaric, (3) turbines as adiabatic, (4) splitters as isoparametric, (5) mixing chambers, and (6) heat exchanger as isobaric.
   (b) Input the given information: working fluid is water in cycles A and B, $p_1 = 103$ kPa, $T_1 = 15°C$, $T_2 = 25°C$, $p_4 = 16$ MPa, $T_4 = 570°C$, $mdot_4 = 1000$ kg/sec, $p_5 = 8$ MPa, $T_6 = 540°C$, $p_8 = 4$ MPa, $p_{10} = 2$ MPa, $p_{12} = 15$ kPa, $x_{13} = 0$, $x_{15} = 0$, $x_{19} = 0$, $\eta_{turbine\#1} = 0.85$, $\eta_{turbine\#2} = 0.85$, $\eta_{turbine\#3} = 0.85$, $\eta_{turbine\#4} = 0.85$, $\eta_{pump\#1} = 0.9$, $\eta_{pump\#2} = 0.9$, $\eta_{pump\#3} = 0.9$, and $\eta_{pump\#4} = 0.9$ as shown in Fig. 2.36b.

3. Display the preliminary design results as shown in Figs. 2.36c and 2.36d.

The results are: $\eta_{cycle} = 41.24\%$, $Wdot_{input} = -20,903$ kW, $Wdot_{output} = 994,618$ kW, $Wdot_{net\ output} = 973,716$ kW, $Qdot_{add} = 2,361,193$ kW, $Qdot_{remove} = -1,387,477$ kW, $Wdot_{turbine\#1} = 193,305$ kW, $Wdot_{turbine\#2} = 1,76,531$ kW, $Wdot_{turbine\#3} = 144,004$ kW, $Wdot_{turbine\#4} = 480,778$ kW, $Wdot_{pump\#1} = -1464$ kW, $Wdot_{pump\#2} = -2201$ kW, $Wdot_{pump\#3} = -5034$ kW, $Wdot_{pump\#4} = -12,203$ kW, $Qdot_{htr\#1} = 2,163,587$ kW, $Qdot_{htr\#2} = 197,605$ kW, $Qdot_{HX1} = -1,387,477$ kW, $mdot_{20} = 93.11$ kg/sec, $mdot_{21} = 66.74$ kg/sec, $mdot_{22} = 197.0$ kg/sec, $mdot_{12} = 643.2$ kg/sec, $mdot_{15} = 840.2$ kg/sec, $mdot_{17} = 906.9$ kg/sec, and $mdot_{19} = 1000$ kg/sec.

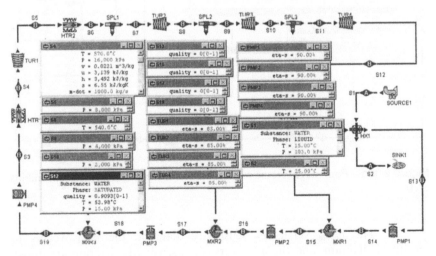

Figure 2.36b Four-stage turbine with reheat and three-stage regenerative Rankine cycle

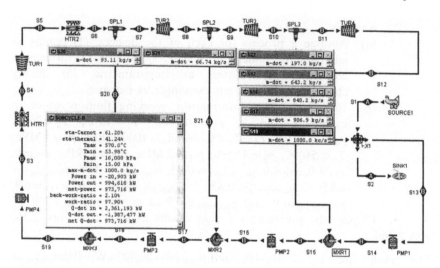

Figure 2.36c Four-stage turbine with reheat and three-stage regenerative Rankine cycle.

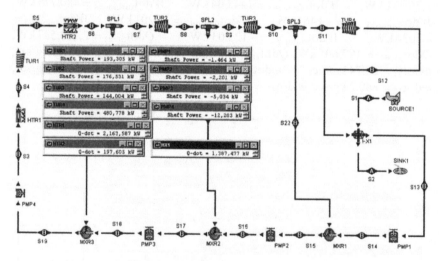

Figure 2.36d Four-stage turbine with reheat and three-stage regenerative Rankine cycle.

4. The η_{cycle} versus p_5 sensitivity diagram, the η_{cycle} versus p_8 sensitivity diagram, and the η_{cycle} versus p_{10} sensitivity diagram are drawn as shown in Figs. 2.36e, f, and g, respectively. Based on these sensitivity diagrams, the η_{cycle} can be optimized.

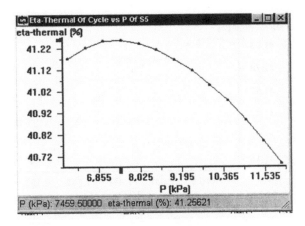

Figure 2.36e Rankine cycle sensitivity diagram.

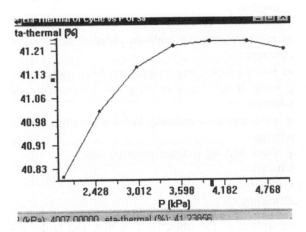

Figure 2.36f Four-stage turbine with reheat and three-stage regenerative Rankine cycle sensitivity diagram.

Example 2.23

A closed-cycle steam Rankine cycle without superheating has been designed by a junior engineer as illustrated in Fig. 2.37a with the following preliminary design information:

Condenser pressure	5 psia
Boiler pressure	3000 psia
Mass flow rate of steam	1 lbm/sec
Flue gas temperature entering high-temperature side heat exchanger	3500°F

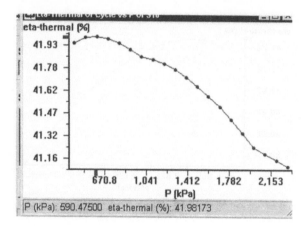

Figure 2.36g Four-stage turbine with reheat and three-stage regenerative Rankine cycle sensitivity diagram.

Flue gas pressure entering high-temperature side heat exchanger	14.7 psia
Flue gas leaving high-temperature side heat exchanger	1500°F
Cooling water temperature entering low-temperature side heat exchanger	60°F
Cooling water pressure entering low-temperature side heat exchanger	14.7 psia
Cooling water leaving low-temperature side heat exchanger	80°F
Turbine efficiency	88%
Pump efficiency	88%

Use net power output as the objective function and boiler pressure as the independent design parameter. Try to improve the preliminary design.

(a) To improve the design with CyclePad, we take the following steps:

1. Build as shown in Fig. 2.37a.
2. Analysis.

Assume a process for each of the four devices: (1) pump as adiabatic with 88% efficiency, (2) turbine as adiabatic with 88% efficiency, (3) heat exchanger 1 (boiler) as isobaric on both cold and hot sides, and (4) heat exchanger 2 (condenser) as isobaric on both cold and hot-sides.

Input the given information: (1) working fluid of heat source is air (flue gas), $p_5 = 14.7$ psia, $T_5 = 3500°F$ and $T_6 = 1500°F$, (2) working fluid of Rankine cycle is water, $p_1 = 5$ psia, $x_1 = 0$, $p_3 = 3000$ psia, and $x_3 = 1$,

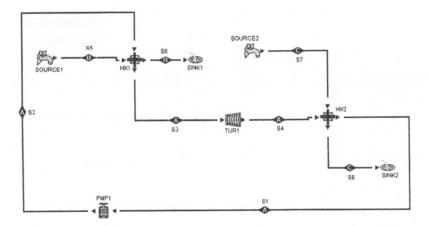

Figure 2.37a Rankine cycle preliminary design.

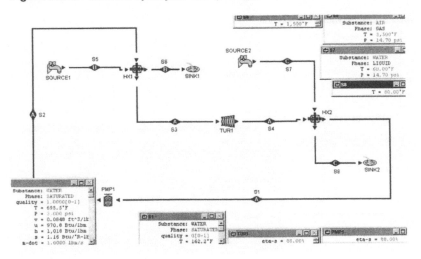

Figure 2.37b Rankine cycle preliminary design input data.

and (3) working fluid of heat sink is water, $p_7 = 14.7$ psia, $T_7 = 60°$F, and $T_6 = 80°$F as shown in Fig. 2.37b.

(b) Determine the preliminary design results.

Display result: The preliminary design results are given in Fig. 2.37c as follows:

$$W\text{dot}_{pump} = -14.58 \text{ hp}, \quad W\text{dot}_{turbine} = 389.3 \text{ hp}, \quad W\text{dot}_{net} = 374.7 \text{ hp},$$
$$Q\text{dot}_{boiler} = 877.5 \text{ Btu/sec}, \quad Q\text{dot}_{condenser} = -612.6 \text{ Btu/sec}, \quad \eta = 30.18\%,$$
$$m\text{dot}_{flue \ gas} = 1.83 \text{ lbm/sec, and } m\text{dot}_{cold \ water} = 30.66 \text{ lbm/sec.}$$

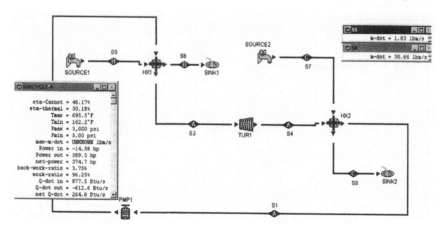

Figure 2.37c   Rankine cycle preliminary design output data

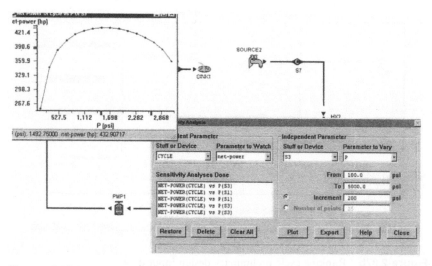

Figure 2.37d   Rankine cycle sensitivity diagram; (e) Rankine cycle optimized design output data.

(c)   Draw the net power versus p_3 sensitivity diagram as shown in Fig. 2.37d. It is shown that the maximum net power is about 430 hp at about 1500 psia.

(d)   Change design input information:

Change p_3 from 3000 psia to 1500 psia and displace results. The results are shown in Fig. 2.37e. Display result: The optimized design results are:

$$Wdot_{pump} = -7.31\,hp, \quad Wdot_{turbine} = 440.5\,hp, \quad Wdot_{net} = 433.2\,hp,$$
$$Qdot_{boiler} = 1033\,Btu/sec, \quad Qdot_{condenser} = -727.7\,Btu/sec, \quad \eta = 29.63\%,$$
$$mdot_{flue\ gas} = 2.16\,lbm/sec, \text{ and } mdot_{cold\ water} = 36.39\,lbm/sec.$$

Review Problems 2.17   Design

1. A four-stage turbine with reheat and three-stage regenerative Rankine cycle as shown in Fig. 2.37a using steam as the working fluid. The following information is provided:

$p_1 = 103\,kPa,$ $T_1 = 15°C,$ $T_2 = 25°C,$ $p_4 = 16\,MPa,$ $T_4 = 600°C,$ $mdot_4 = 1000\,kg/sec,$ $p_5 = 8\,MPa,$ $T_6 = 540°C,$ $p_8 = 4\,MPa,$ $p_{10} = 2\,MPa,$ $p_{12} = 15\,kPa,$ $x_{13} = 0,$ $x_{15} = 0,$ $x_{19} = 0,$ $\eta_{turbine\#1} = 0.85,$ $\eta_{turbine\#2} = 0.85,$ $\eta_{turbine\#3} = 0.85,$ $\eta_{turbine\#4} = 0.85,$ $\eta_{pump\#1} = 0.85,$ $\eta_{pump\#2} = 0.85,$ $\eta_{pump\#3} =$ 0.85, and $\eta_{pump\#4} = 0.85.$

Find $\eta_{cycle},$ $Wdot_{input},$ $Wdot_{output},$ $Wdot_{net\ output},$ $Qdot_{add},$ $Qdot_{remove},$ $Wdot_{turbine\#1},$ $Wdot_{turbine\#2},$ $Wdot_{turbine\#3},$ $Wdot_{turbine\#4},$ $Qdot_{htr\#1},$ $Qdot_{htr\#2},$ $Qdot_{HX1},$ $mdot_{20},$ $mdot_{21},$ $mdot_{22},$ $mdot_{12},$ $mdot_{15},$ $mdot_{17},$ and $mdot_{19}.$

Draw the η_{cycle} versus p_5 sensitivity diagram, the η_{cycle} versus p_8 sensitivity diagram, and the η_{cycle} versus p_{10} sensitivity diagram.

ANSWERS: $\eta_{cycle} = 41.75\%,$ $Wdot_{input} = -20,985\,kW,$ $Wdot_{output}$ $= 1,067,259\,kW,$ $Wdot_{net\ output} = 1,046,274\,kW,$ $Qdot_{add} = 2,505,993\,kW,$ $Qdot_{remove} = -1,459,718\,kW,$ $Wdot_{turbine\#1} = 193,305\,kW,$ $Wdot_{turbine\#2}$ $= 193,484\,kW,$ $Wdot_{turbine\#3} = 158,701\,kW,$ $Wdot_{turbine\#4} = 521,769\,kW,$ $Qdot_{htr\#1} = 2,163,587\,kW,$ $Qdot_{htr\#2} = 342,405\,kW,$ $Qdot_{HX1} =$ $-1,459,718\,kW,$ $mdot_{20} = 87.82\,kg/sec,$ $mdot_{21} = 63.73\,kg/sec,$ $mdot_{22} =$ $191.5\,kg/sec,$ $mdot_{12} = 656.9\,kg/sec,$ $mdot_{15} = 848.5\,kg/sec,$ $mdot_{17} = 912.2$ kg/sec, and $mdot_{19} = 1000\,kg/sec.$

2. A four-stage turbine with reheat and three-stage regenerative Rankine cycle as shown in Fig. 2.37a using steam as the working fluid. The following information is provided:

$p_1 = 103\,kPa,$ $T_1 = 15°C,$ $T_2 = 25°C,$ $p_4 = 16\,MPa,$ $T_4 = 600°C,$ $mdot_4 = 1000\,kg/sec,$ $p_5 = 7\,MPa,$ $T_6 = 540°C,$ $p_8 = 4\,MPa,$ $p_{10} = 2\,MPa,$ $p_{12} = 15\,kPa,$ $x_{13} = 0,$ $x_{15} = 0,$ $x_{19} = 0,$ $\eta_{turbine\#1} = 0.85,$ $\eta_{turbine\#2} = 0.85,$ $\eta_{turbine\#3} = 0.85,$ $\eta_{turbine\#4} = 0.85,$ $\eta_{pump\#1} = 0.9,$ $\eta_{pump\#2} = 0.9,$ $\eta_{pump\#3} = 0.9,$ and $\eta_{pump\#4} = 0.9.$

Find $\eta_{cycle},$ $Wdot_{input},$ $Wdot_{output},$ $Wdot_{net\ output},$ $Qdot_{add},$ $Qdot_{remove},$ $Wdot_{turbine\#1},$ $Wdot_{turbine\#2},$ $Wdot_{turbine\#3},$ $Wdot_{turbine\#4},$ $Qdot_{htr\#1},$ $Qdot_{htr\#2},$ $Qdot_{HX1},$ $mdot_{20},$ $mdot_{21},$ $mdot_{22},$ $mdot_{12},$ $mdot_{15},$ $mdot_{17},$ and $mdot_{19}.$

Draw the η_{cycle} versus p_5 sensitivity diagram, the η_{cycle} versus p_8 sensitivity diagram, and the η_{cycle} versus p_{10} sensitivity diagram.

3. A closed-cycle steam Rankine cycle without superheating was designed by a junior engineer as illustrated in Fig. 2.37a with the following preliminary design information:

Condenser pressure	5 psia
Boiler pressure	2000 psia
Mass flow rate of steam	1 lbm/sec
Flue gas temperature entering high-temperature side heat exchanger	3000°F
Flue gas pressure entering high-temperature side heat exchanger	14.7 psia
Flue gas leaving high-temperature side heat exchanger	1000°F
Cooling water temperature entering low-temperature side heat exchanger	50°F
Cooling water pressure entering low-temperature side heat exchanger	14.7 psia
Cooling water leaving low-temperature side heat exchanger	70°F
Turbine efficiency	85%
Pump efficiency	85%

Use net power output as the objective function and boiler pressure as the independent design parameter. Try to improve the preliminary design.

4. A closed-cycle steam Rankine cycle without superheating was designed by a junior engineer as illustrated in Fig. 2.37a with the following preliminary design information:

Condenser pressure	10 kPa
Boiler pressure	16000 kPa
Mass flow rate of steam	1 kg/sec
Flue gas temperature entering high-temperature side heat exchanger	2000°C
Flue gas pressure entering high-temperature side heat exchanger	101 psia
Flue gas leaving high-temperature side heat exchanger	1000°C
Cooling water temperature entering low-temperature side heat exchanger	14°C
Cooling water pressure entering low-temperature side heat exchanger	101 kPa

Cooling water leaving low-temperature side	
heat exchanger	20°C
Turbine efficiency	85%
Pump efficiency	85%

Use net power output as the objective function and boiler pressure as the independent design parameter. Try to improve the preliminary design.

2.18 SUMMARY

The Carnot cycle is not a practical model for vapor power cycles because of cavitation and corrosion problems. The modified Carnot model for vapor power cycles is the basic Rankine cycle, which consists of two isobaric and two isentropic processes. The basic elements of the basic Rankine cycle are pump, boiler, turbine, and condenser. The Rankine cycle is the most popular heat engine to produce commercial power. The thermal cycle efficiency of the basic Rankine cycle can be improved by adding a superheater, regenerating, and reheater, among other means.

3

Gas Closed-System Cycles

3.1 OTTO CYCLE

A four-stroke internal combustion engine was built by a German engineer, Nicholas Otto, in 1876. The cycle patterned after his design is called the *Otto cycle*. It is the most widely used internal combustion heat engine in automobiles.

The piston in a four-stroke internal combustion engine executes four complete strokes as the crankshaft completes two revolutions per cycle, as shown in Fig. 3.1. On the intake stroke, the intake valve is open and the piston moves downward in the cylinder, drawing in a premixed charge of gasoline and air until the piston reaches its lowest point of the stroke called *bottom dead center* (BDC). During the compression stroke the intake valve closes and the piston moves toward the top of the cylinder, compressing the fuel–air mixture. As the piston approaches the top of the cylinder called *top dead center* (TDC), the spark plug is energized and the mixture ignites, creating an increase in the temperature and pressure of the gas. During the expansion stroke the piston is forced down by the high-pressure gas, producing a useful work output. The cycle is then completed when the exhaust valve opens and the piston moves toward the top of the cylinder, expelling the products of combustion.

The thermodynamic analysis of an actual Otto cycle is complicated. To simplify the analysis, we consider an ideal Otto cycle composed entirely of internally reversible processes. In the Otto cycle analysis, a closed piston–cylinder assembly is used as a control mass system.

The cycle consists of the following four processes:

1-2 Isentropic compression
2-3 Constant-volume heat addition
3-4 Isentropic expansion
4-1 Constant-volume heat removal

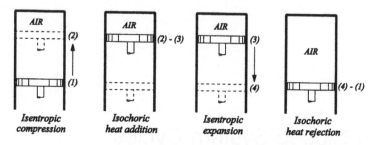

Figure 3.1 Otto cycle.

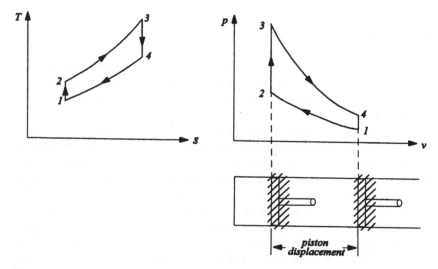

Figure 3.2 Otto cycle p–v and T–s diagrams.

The p–v and T–s process diagrams for the ideal Otto cycle are illustrated in Fig. 3.2.

Applying the first and second laws of thermodynamics of the closed system to each of the four processes of the cycle yields:

$$W_{12} = \int p\,dV \tag{3.1}$$

$$Q_{12} - W_{12} = m(u_2 - u_1) \qquad Q_{12} = 0 \tag{3.2}$$

$$W_{23} = \int p\,dV = 0 \tag{3.3}$$

$$Q_{23} - 0 = m(u_3 - u_2) \tag{3.4}$$

$$W_{34} = \int p\,dV \tag{3.5}$$

$$Q_{34} - W_{34} = m(u_4 - u_3) \qquad Q_{34} = 0 \tag{3.6}$$

$$W_{41} = \int p\,dV = 0 \tag{3.7}$$

and

$$Q_{41} - 0 = m(u_1 - u_4) \tag{3.8}$$

The net work (W_{net}), which is also equal to net heat (Q_{net}), is

$$W_{net} = W_{12} + W_{34} = Q_{net} = Q_{23} + Q_{41} \tag{3.9}$$

The thermal efficiency of the cycle is

$$\eta = W_{net}/Q_{23} = Q_{net}/Q_{23} = 1 - Q_{41}/Q_{23} = 1 - (u_4 - u_1)/(u_3 - u_2) \tag{3.10}$$

This expression for the thermal efficiency of an ideal Otto cycle can be simplified if air is assumed to be the working fluid with constant specific heat. Equation (3.10) is reduced to

$$\eta = 1 - (T_4 - T_1)/(T_3 - T_2) = 1 - (r)^{1-k} \tag{3.11}$$

where r is the *compression ratio* for the engine defined by the equation:

$$r = V_1/V_2 \tag{3.12}$$

The compression ratio is the ratio of the cylinder volume at the beginning of the compression process (BDC) to the cylinder volume at the end of the compression process (TDC).

Equation (3.11) shows that the thermal efficiency of the Otto cycle is only a function of the compression ratio of the engine. Therefore, any engine design that increases the compression ratio should result in an increased engine efficiency. The compression ratio cannot be increased indefinitely. As the compression ratio increases, the temperature of the working fluid also increases during the compression process. Eventually, a temperature is reached that is sufficiently high to ignite the air–fuel mixture prematurely without the presence of a spark. This condition causes the engine to produce a noise called knock. The presence of *engine knock* places a barrier on the upper limit of Otto engine compression ratios. To reduce the engine knock problem of a high compression ratio Otto cycle, one must use gasoline with higher octane rating. In general, the higher the octane rating number of gasoline, the higher the resistance of engine knock.

One way to simplify the calculation of the net work of the cycle and to provide a comparative measure of the performance of an Otto heat engine is to introduce the concept of the mean effective pressure. The *mean effective pressure* (MEP) is the average pressure of the cycle. The net work of the cycle is equal to the mean effective pressure multiplied by the displacement volume of the cylinder. That is,

MEP = (cycle net work)/(cylinder displacement volume)
$$= W_{net}/(V_1 - V_2)$$

The engine with the larger MEP value of two engines of equal cylinder displacement volume would be the better one, because it would produce a greater net work output.

Example 3.1

An engine operates on the Otto cycle and has a compression ratio of 8. Fresh air enters the engine at 27°C and 100 kPa. The amount of heat addition is 700 kJ/kg. The amount of air mass in the cylinder is 0.01 kg. Determine the pressure and temperature at the end of the combustion, the pressure and temperature at the end of the expansion, MEP, efficiency, and work output per kilogram of air. Show the cycle on a *T–s* diagram. Plot the sensitivity diagram of cycle efficiency versus compression ratio.

To solve this problem by CyclePad, we take the following steps:

1. Build:
   a. Take a compression device, a combustion chamber, an expander, and a cooler from the closed-system inventory shop and connect the four devices to form the Otto cycle as shown in Fig. 3.2.
   b. Switch to analysis mode.
2. Analysis:
   a. Assume a process for each of the four devices: (1) compression device as adiabatic and isentropic, (2) combustion chamber as isochoric, (3) expander as adiabatic and isentropic, and (4) cooler as isochoric.
   b. Input the given information: (1) working fluid is air, (2) inlet pressure and temperature of the compression device are 100 kPa and 27°C, (3) compression ratio of the compression device is 8, (4) heat addition is 700 kJ/kg in the combustion chamber, and (5) $m = 0.01$ kg.

3. Display results:
   a. Display the *T–s* diagram and cycle properties' results. The cycle is a heat engine. The answers are: $p = 4441\,\text{kPa}$ and $T = 1393°\text{C}$ (the pressure and temperature at the end of the combustion), $p = 241.6\,\text{kPa}$ and $T = 452.1°\text{C}$ (the pressure and temperature at the end of the expansion), $\text{MEP} = 525.0\,\text{kPa}$, $\eta = 56.47\%$ and $W_{\text{net}} = 3.95\,\text{kJ}$.
   b. Display the sensitivity diagram of cycle efficiency versus compression ratio. (See Figs. 3.3a–3.3c.)

COMMENT: Efficiency increases as compression ratio increases.

Example 3.2

The compression ratio in an Otto cycle is 8. If the air before compression (state 1) is at 60°F and 14.7 psia, and 800 Btu/lbm is added to the cycle and the mass of air contained in the cylinder is 0.025 lbm, calculate (1) temperature and pressure at each point of the cycle, (2) the heat that must be removed, (3) the thermal efficiency, and (4) the MEP of the cycle.

To solve this problem by CyclePad, we take the following steps:

1. Build:
   a. Take a compression device, a combustion chamber, an expander, and a cooler from the closed-system inventory shop and connect the four devices to form the Otto cycle.
   b. Switch to analysis mode.

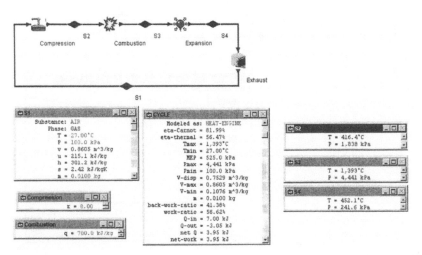

Figure 3.3a Otto cycle.

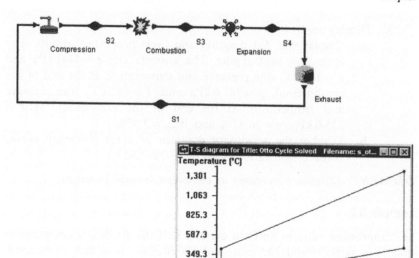

Figure 3.3b Otto cycle *T–s* diagram.

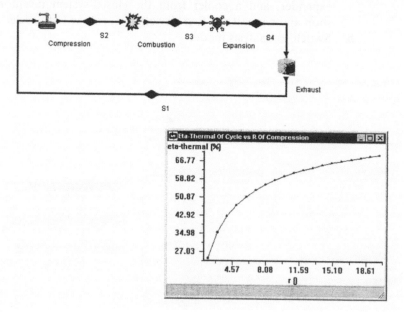

Figure 3.3c Otto cycle sensitivity analysis.

2. Analysis:
    a. Assume a process for each of the four devices: (1) compression device as isentropic, (2) combustion chamber as isochoric, (3) expander as isentropic, and (4) cooler as isochoric.
    b. Input the given information: (1) working fluid is air, (2) inlet pressure and temperature of the compression device are 60°F and 14.7 psia, (3) compression ratio of the compression device is 8, (4) heat addition is 800 Btu/lbm in the combustion chamber, and (5) mass of air is 0.025 lbm.
3. Display results:
    a. Display the T–s diagram and cycle properties' results. The cycle is a heat engine. The answers are $T_2 = 734.2°F$, $p_2 = 270.2$ psia, $T_3 = 5407°F$, $p_3 = 1328$ psia, $T_4 = 2094°F$, $p_4 = 72.24$ psia, $\eta = 56.47\%$, $Q_{41} = -8.71$ Btu, and MEP = 213.3 psia.
    b. Display the T–s diagram. (See Figs. 3.4a and 3.4b.)

The power output of the Otto cycle can be increased by *turbocharging* the air before it enters the cylinder in the Otto engine. Since the inlet air density is increased due to higher inlet air pressure, the mass of air in the cylinder is increased. Turbocharging raises the inlet air pressure of the engine above atmospheric and raises the power output of the engine, but it may not improve the efficiency of the cycle. A schematic

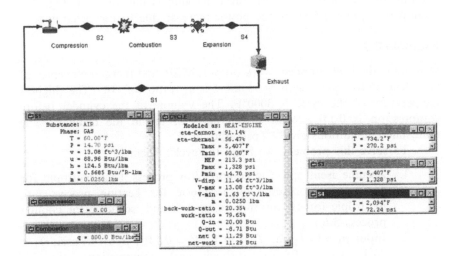

Figure 3.4a Otto cycle.

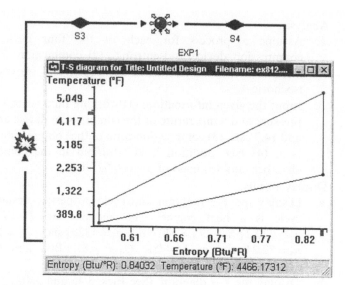

Figure 3.4b   Otto cycle *T–s* diagram plot.

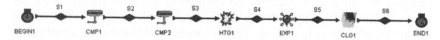

Figure 3.5   Otto engine with turbocharging.

diagram of the Otto cycle with turbocharging is illustrated in Fig. 3.5. Examples 3.3 and 3.4 show the power increase due to turbocharging.

Example 3.3

Determine the heat supplied, work output, MEP, and thermal efficiency of an ideal Otto cycle with a compression ratio of 10. The highest temperature of the cycle is 3000°F. The volume of the cylinder before compression is $0.1\,\text{ft}^3$. What is the mass of air in the cylinder? The atmosphere conditions are 14.7 psia and 70°F.

To solve this problem, we build the cycle. Then:

1. Assume isentropic for compression process 1-2, isentropic for compression process 2-3, isochoric for heating process 3-4, isentropic for expansion process 4-5, and isochoric for cooling process 5-6.

2. Input $p_1 = 14.7\,\text{psia}$, $T_1 = 70°\text{F}$, $V_1 = 0.1\,\text{ft}^3$, $p_2 = 14.7\,\text{psia}$, $T_2 = 70°\text{F}$, $V_2 = 0.1\,\text{ft}^3$ (no turbocharger), compression ratio $= 10$, $T_4 = 3000°\text{F}$, $p_6 = 14.7\,\text{psia}$, and $T_6 = 70°\text{F}$.

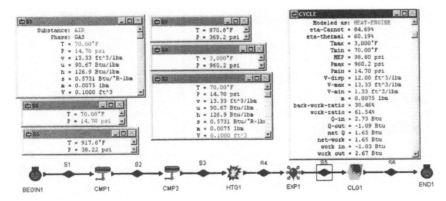

Figure 3.6 Otto engine without turbocharging.

3. Display results.

The results are: $W_{12} = 0$ Btu, $W_{23} = -1.03$ Btu, $Q_{34} = 2.73$ Btu, $W_{45} = 2.67$ Btu, $W_{net} = 1.65$ Btu, $Q_{56} = -1.09$ Btu, MEP $= 98.80$ psia, $\eta = 60.19\%$, and $m = 0.0075$ lbm. (See Fig. 3.6.)

Example 3.4

Determine the heat supplied, work output, MEP, and thermal efficiency of an ideal Otto cycle with a turbocharger that compresses air to 20 psia and with a compression ratio of 10. The highest temperature of the cycle is 3000°F. The volume of the cylinder before compression is 0.1 ft³. What is the mass of air in the cylinder? The atmosphere conditions are 14.7 psia and 70°F.

To solve this problem, we build the cycle as shown in Fig. 3.5. Then, (1) assume isentropic for compression process 1-2, isentropic for compression process 2-3, isochoric for the heating process 3-4, isentropic for expansion process 4-5, and isochoric for the cooling process 5-6; (2) input $p_1 = 14.7$ psia, $T_1 = 70°F$; $p_2 = 20$ psia, $V_2 = 0.1$ ft³ (with turbocharger); compression ratio $= 10$, $T_4 = 3000°F$, $p_6 = 14.7$ psia, and $T_6 = 70°F$; and (3) display results. The results are: $W_{13} = -1.48$ Btu, $Q_{34} = 3.21$ Btu, $W_{45} = 3.52$ Btu, $W_{net} = 2.04$ Btu, $Q_{56} = -1.17$ Btu, MEP $= 96.20$ psia, $\eta = 63.54\%$, and $m = 0.0093$ lbm. (See Fig. 3.7.)

Modern Otto engine designs are affected by environmental constrains as well as by desires to increase gas mileage. Recent design improvements include the use of four valves per cylinder to reduce the restriction to air flow into and out of the cylinder, turbochargers to increase the air and fuel flow to each cylinder, and catalytic converters to

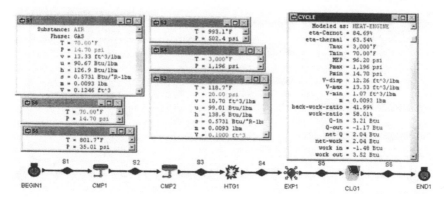

Figure 3.7 Otto engine with turbocharging.

aid the combustion of unburned hydrocarbons that are expelled by the engine, among others.

Review Problems 3.1 Otto Cycle

1. Do Otto heat engines operate on a closed system or an open system? Why?

2. What is the compression ratio of an Otto cycle? How does it affect the thermal efficiency of the cycle?

3. Do you know the compression ratio of your car? Is there any limit to an Otto cycle? Why?

4. Which area represents cycle net work of an Otto cycle plotted on a *T–s* diagram and *p–v* diagram?

5. How do you define MEP (mean effective pressure)?

6. What is engine knock? What causes the engine knock problem?

7. Do you get a better performance using premier gasoline (octane number 93) for your compact car?

8. An engine operates on an Otto cycle with a compression ratio of 8. At the beginning of the isentropic compression process, the volume, pressure, and temperature of the air are 0.01 m³, 110 kPa, and 50°C. At the end of the combustion process, the temperature is 900°C. Find (a) the temperature at the remaining two states of the Otto cycle, (b) the pressure of the gas at the end of the combustion process, (c) the heat added per unit mass to the engine in the combustion chamber, (d) the heat removed per unit mass from the engine to the environment, (e) the compression work per unit mass added, (f) the expansion work per unit mass done, (g) MEP, and (h) thermal cycle efficiency.

9. An ideal Otto Cycle with air as the working fluid has a compression ratio of 9. At the beginning of the compression process, the air is at 290 K and 90 kPa. The peak temperature in the cycle is 1800 K. Determine (a) the pressure and temperature at the end of the expansion process (power stroke), (b) the heat per unit mass added in kJ/kg during the combustion process, (c) net work, (d) thermal efficiency of the cycle, and (e) mean effective pressure in kPa.

10. An ideal Otto engine receives air at 15 psia, 0.01 ft^3, and 65°F. The maximum cycle temperature is 3465°F and the compression ratio of the engine is 7.5. Determine (a) the work added during the compression process, (b) the heat added to the air during the heating process, (c) the work done during the expansion process, (d) the heat removed from the air during the cooling process, and (e) the thermal efficiency of the cycle.

11. An ideal Otto engine receives air at 14.6 psia and 55°F. The maximum cycle temperature is 3460°F and the compression ratio of the engine is 10. Determine (a) the work done per unit mass during the compression process, (b) the heat added per unit mass to the air during the heating process, (c) the work done per unit mass during the expansion process, (d) the heat removed per unit mass from the air during the cooling process, and (e) the thermal efficiency of the cycle.

12. An ideal Otto engine receives air at 100 kPa and 25°C. Work is performed on the air in order to raise the pressure at the end of the compression process to 1378 kPa. 400 kJ/kg of heat is added to the air during the heating process. Determine (a) the work done during the compression process, (b) the compression ratio, (c) the work done during the expansion process, (d) the heat removed from the air during the cooling process, (e) the MEP (mean effective pressure), and (f) the thermal efficiency of the cycle.

13. At the beginning of the compression process of an air-standard Otto cycle, $p = 100$ kPa, $T = 290$ K, and $V = 0.04$ m^3. The maximum temperature in the cycle is 2200 K and the compression ratio is 8. Determine (a) the heat addition, (b) the net work, (c) the thermal efficiency, and (d) the MEP.

ANSWERS: (a) 52.89 kJ; (b) 29.87 kJ; (c) 56.47%; (d) 853.3 kPa.

14. An Otto engine operates with a compression ratio of 8.5. The following information is known:

- Temperature prior to the compression process = 70°F.
- Volume prior to the compression process = 0.05 ft^3.
- Pressure prior to the compression process = 14.7 psia.
- Heat added during the combustion process = 345 Btu/lbm.

    a.   Determine the mass of air in the cylinder.

b. Determine the temperature and pressure at each process endpoint.

c. Find the compression work and expansion work in Btu/lbm.

d. Determine the thermal efficiency.

15. The compression ratio in an Otto cycle is 8. If the air before compression (state 1) is at 80°F and 14.7 psia, and 800 Btu/lbm is added to the cycle and the mass of air contained in the cylinder is 0.02 lbm, find the heat added, heat. removed, work added, work produced, net work produced, MEP, and efficiency of the cycle.

ANSWERS: heat added = 16 Btu; heat removed = − 6.96 Btu; work added = − 2.4 Btu; work produced = 11.43 Btu; net work produced = 9.04 Btu; MEP = 205.4 psia; efficiency of the cycle = 56.47%.

16. The compression ratio in an Otto cycle is 10. If the air before compression (state 1) is at 60°F and 14.7 psia, and 800 Btu/lbm is added to the cycle and the mass of air contained in the cylinder is 0.02 lbm, find the heat added, heat removed, work added, work produced, net work produced, MEP, and efficiency of the cycle.

ANSWERS: heat added = 16 Btu; heat removed = − 6.37 Btu; work added = − 2.69 Btu; work produced = 12.32 Btu; net work produced = 9.63 Btu; MEP = 221.0 psia; efficiency of the cycle = 60.19%.

17. The compression ratio in an Otto cycle is 16. If the air before compression (state 1) is at 60°F and 14.7 psia, and 800 Btu/lbm is added to the cycle and the mass of air contained in the cylinder is 0.02 lbm, find the heat added, heat removed, work added, work produced, net work produced, MEP, and efficiency of the cycle.

ANSWERS: heat added = 16 Btu; heat removed = − 5.28 Btu; work added = − 3.61 Btu; work produced = 14.34 Btu; net work produced = 10.72 Btu; MEP = 236.2 psia; efficiency of the cycle = 67.01%.

18. An Otto engine with a turbocharger operates with a compression ratio of 8.5. The following information is known:

- Temperature prior to the turbo-charging compression process = 70°F.
- Pressure prior to the turbo-charging compression process = 14.7 psia.
- Pressure after the turbo-charging compression process = 20 psia.
- Heat added during the combustion process = 345 Btu/lbm.
- Volume after the compression process = 0.05 ft^3.

a. Determine the mass of air in the cylinder

b. Determine the temperature and pressure at each process endpoint.

    c. Find the compression work and expansion work in Btu/lbm.

    d. Determine the thermal efficiency.

19. An ideal Otto Cycle with a turbocharger using air as the working fluid has a compression ratio of 9. The volume of the cylinder is $0.01\,m^3$. At the beginning of the turbocharging compression process, the air is at 290 K and 90 kPa. The air pressure is 150 kPa after the turbocharging compression process. The peak temperature in the cycle is 1800 K. Determine (a) the pressure and temperature at the end of the expansion process (power stroke), (b) the heat per unit mass added in kJ/kg during the combustion process, (c) net work, (d) thermal efficiency of the cycle, and (e) mean effective pressure in kPa.

3.2 DIESEL CYCLE

The *Diesel cycle* was proposed by Rudolf Diesel in the 1890s. The Diesel cycle as shown in Fig. 3.8 is somewhat similar to the Otto cycle, except that ignition of the fuel–air mixture is caused by spontaneous combustion owing to the high temperature that results from compressing the mixture to a very high pressure. The basic components of the Diesel cycle are the same as those of the Otto cycle, except that the spark plug is replaced by a

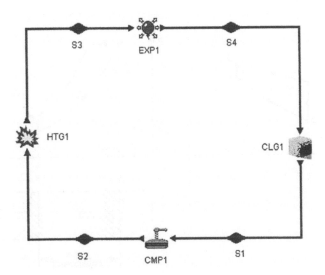

Figure 3.8   Diesel cycle.

fuel injector, and the stroke of the piston is lengthened to provide a larger compression ratio.

The Diesel cycle consists of the following four processes:

1-2 Isentropic compression
2-3 Constant-pressure heat addition
3-4 Isentropic expansion
4-1 Constant-volume heat removal

Since the duration of the heat-addition process is extended, this process is modeled by a constant-pressure process. The p–v and T–s diagrams for the Diesel cycle are illustrated in Fig. 3.9.

Applying the first law and second law of thermodynamics of the closed system to each of the four processes of the cycle yields:

$$W_{12} = \int p\,dV \tag{3.13}$$

$$Q_{12} - W_{12} = m(u_2 - u_1), \qquad Q_{12} = 0 \tag{3.14}$$

$$W_{23} = \int p\,dV = m(p_3 v_3 - p_2 v_2) \tag{3.15}$$

$$Q_{23} = m(u_3 - u_2) + W_{23} = m(h_3 - h_2) \tag{3.16}$$

$$W_{34} = \int p\,dV \tag{3.17}$$

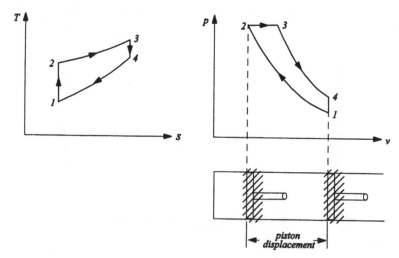

Figure 3.9 Diesel cycle p–v and T–s diagrams.

$$Q_{34} - W_{34} = m(u_4 - u_3), \qquad Q_{34} = 0 \tag{3.18}$$

$$W_{41} = \int p \, dV = 0 \tag{3.19}$$

and

$$Q_{41} - 0 = m(u_1 - u_4) \tag{3.20}$$

The net work (W_{net}), which is also equal to net heat (Q_{net}), is

$$W_{net} = W_{12} + W_{23} + W_{34} = Q_{net} = Q_{23} + Q_{41} \tag{3.21}$$

The thermal efficiency of the cycle is

$$\eta = W_{net}/Q_{23} = Q_{net}/Q_{23} = 1 - Q_{41}/Q_{23} = 1 - (u_4 - u_1)/(h_3 - h_2) \tag{3.22}$$

This expression for thermal efficiency of an ideal Otto cycle can be simplified if air is assumed to be the working fluid with constant specific heats. Equation (3.22) is reduced to

$$\eta = 1 - (T_4 - T_1)/[k(T_3 - T_2)] = 1 - (r)^{1-k}\{[(r_c)^k - 1]/[k(r_c - 1)]\} \tag{3.23}$$

where r is the *compression ratio* and r_c is the *cut-off ratio* for the engine defined by the equation:

$$r = V_1/V_2 \tag{3.24}$$

and

$$r_c = V_3/V_2 \tag{3.25}$$

A comparison of the thermal efficiency of the Diesel cycle with that of the Otto cycle shows that the two thermal cycle efficiencies differ by the quantity in the brackets of Eq. (3.23). This bracket factor is always larger than one, hence the Diesel cycle efficiency is always less than the Otto cycle efficiency operating at the same compression ratio.

Since the fuel is not injected into the cylinder until after the air has been completely compressed in the Diesel cycle, there is no engine knock problem. Therefore, the Diesel engine can be designed to operate at much higher compression ratios and with less refined fuel than those of the Otto cycle. As a result of the higher compression ratio, Diesel engines are slightly more efficient than Otto engines.

Example 3.5

A Diesel engine receives air at $27°C$ and $100\,kPa$. The compression ratio is 18. The amount of heat addition is $500\,kJ/kg$. The mass of air contained in the cylinder is $0.0113\,kg$. Determine (1) the maximum cycle pressure and maximum cycle temperature, (2) the efficiency and work output, and (3) the MEP. Plot the sensitivity diagram of cycle efficiency versus compression ratio.

To solve this problem by CyclePad, we take the following steps:

1. Build:
   a. Take a compression device, a combustion chamber, an expander, and a cooler from the closed-system inventory shop and connect the four devices to form the Diesel cycle.
   b. Switch to analysis mode.
2. Analysis:
   a. Assume a process for each of the four devices (1) compression device as isentropic, (2) combustion chamber as isobaric, (3) expander as isentropic, and (4) cooler as isochoric.
   b. Input the given information: (1) working fluid is air, (2) inlet pressure and temperature of the compression device are $100\,kPa$ and $27°C$, (3) compression ratio of the compression device is 18, (4) heat addition is $500\,kJ/kg$ in the combustion chamber, and (5) mass of air is $0.0113\,kg$.
3. Display results:
   a. Display cycle properties' results. The cycle is a heat engine. The answers are $T_{max} = 1179°C$, $p_{max} = 5720\,kPa$, $\eta = 65.53\%$, MEP $= 403.2\,kPa$, and $W_{net} = 3.72\,kJ$.
   b. Display the sensitivity diagram of cycle efficiency versus compression ratio. (See Figs. 3.10 and 3.11.)

COMMENT: Efficiency increases as compression ratio increases.

Example 3.6

A Diesel engine receives air at $60°F$ and $14.7\,psia$, the compression ratio is 16, the amount of heat addition is $800\,Btu/lbm$, and the mass of air contained in the cylinder is $0.02\,lbm$. Determine the maximum cycle temperature, heat added, heat removed, work added, work produced, net work produced, MEP, and efficiency of the cycle.

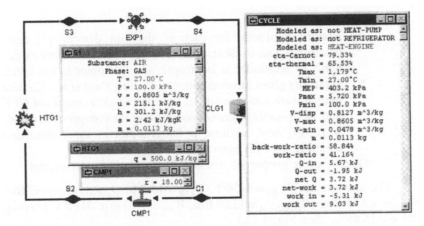

Figure 3.10 Diesel cycle.

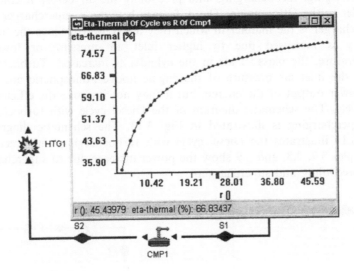

Figure 3.11 Diesel cycle sensitivity analysis.

To solve this problem by CyclePad, we take the following steps:

1. Build:
   a. Take a compression device, a combustion chamber, an expander, and a cooler from the closed-system inventory shop and connect the four devices to form the Diesel cycle.
   b. Switch to analysis mode.

2. Analysis:
   a. Assume a process for each of the four devices: (1) compression device as isentropic, (2) combustion chamber as isobaric, (3) expander as isentropic, and (4) cooler as isochoric.
   b. Input the given information: (1) working fluid is air, (2) inlet temperature and pressure of the compression device are 60°F and 14.7 psia, (3) compression ratio of the compression device is 16, (4) heat addition is 800 Btu/lbm in the combustion chamber, and (5) mass of air is 0.02 lbm.
3. Display cycle properties' results. The cycle is a heat engine.

The answers are $T_{max} = 4454°F$, $Q_{add} = 16$ Btu, $Q_{remove} = -6.97$ Btu, $W_{add} = -3.61$ Btu, $W_{expansion} = 12.65$ Btu, $W_{net} = 9.03$ Btu, MEP = 199 psia, and $\eta = 56.45\%$. (See Fig. 3.12.)

The power output of the Diesel cycle can be increased by supercharging, turbocharging, and precooling the air before it enters the cylinder in the Otto engine. The difference between a supercharger and a turbocharger is the manner in which they are powered. Since the inlet air density is increased due to higher inlet air pressure or lower air temperature, the mass of air in the cylinder is increased. Turbocharging raises the inlet air pressure of the engine above atmospheric and raises the power output of the engine, but it may not improve the efficiency of the cycle. The schematic diagram of the Diesel cycle with turbocharging or supercharging is illustrated in Fig. 3.13. The schematic diagram of Fig. 3.14 illustrates the Diesel cycle with turbocharging and precooling. Examples 3.7, 3.8, and 3.9 show the power increase due to supercharging, and precooling and supercharging.

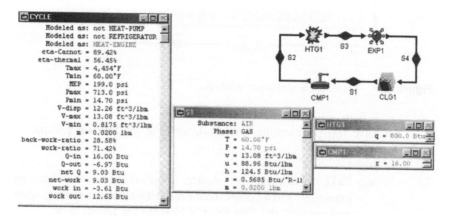

Figure 3.12 Diesel cycle.

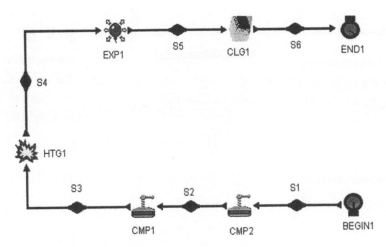

Figure 3.13 Diesel cycle with supercharging.

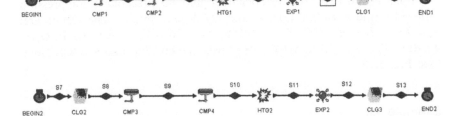

Figure 3.14 Diesel cycle with supercharging and precooling.

Example 3.7

Find the pressure and temperature of each state of an ideal Diesel cycle with a compression ratio of 15 and a cut-off ratio of 2. The cylinder volume before compression is $0.16\,\text{ft}^3$. The atmosphere conditions are $14.7\,\text{psia}$ and $70°F$. Also determine the mass of air in the cylinder, heat supplied, net work produced, MEP, and cycle efficiency.

To solve this problem, we build the cycle as shown in Fig. 3.14. Then, (1) assume isobaric for the precooling process 7-8, isentropic for the compression process 8-9, isentropic for the compression process 9-10, isobaric for the heating process 10-11, isentropic for the expansion process 11-12, and isochoric for the cooling process 12-13; (2) input $p_7 = 14.7\,\text{psia}$,

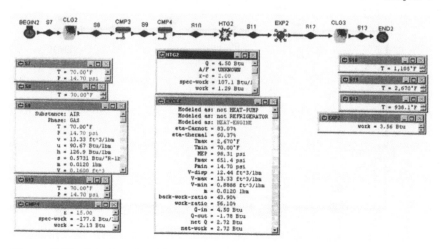

Figure 3.15 Diesel cycle without precooler and without turbocharger.

$T_7 = 70°F$, $p_{13} = 14.7$ psia, $T_{13} = 70°F$, $p_8 = 14.7$ psia, $T_8 = 70°F$, $p_9 = 14.7$ psia, $V_9 = 0.16$ ft^3 (no turbocharger and no precooler); compression ratio = 15, and cut-off ratio = 2; and (3) display results. The results are: $T_8 = 70°F$, $T_9 = 96.87°F$, $T_{10} = 1105°F$, $T_{11} = 2670°F$, $T_{12} = 938.1°F$, $Q_{in} = 4.5$ Btu, $W_{net} = 2.72$ Btu, MEP = 98.31 psia, $\eta = 60.37\%$, and $m = 0.012$ lbm. (See Fig. 3.15.)

Example 3.8

Find the pressure and temperature of each state of an ideal Diesel cycle with a compression ratio of 15 and a cut-off ratio of 2, and a supercharger that compresses fresh air to 20 psia before it enters the cylinder of the engine. The cylinder volume before compression is 0.16 ft^3. The atmosphere conditions are 14.7 psia and 70°F. Also determine the mass of air in the cylinder, heat supplied, net work produced, MEP, and cycle efficiency.

To solve this problem, we build the cycle as shown in Fig. 3.14. Then, (1) assume isobaric for the precooling process 7-8, isentropic for the compression process 8-9, isentropic for the compression process 9-10, isobaric for the heating process 10-11, isentropic for the expansion process 11-12, and isochoric for the cooling process 12-13; (2) input $p_7 = 14.7$ psia, $T_7 = 70°F$, $p_{13} = 14.7$ psia, $T_{13} = 70°F$, $T_8 = 70°F$, $p_9 = 20$ psia, $V_9 = 0.16$ ft^3 (with turbocharger and no precooler), compression ratio = 15, and cut-off ratio = 2; and (3) display results. The results are: $T_8 = 70°F$, $T_9 = 96.87°F$,

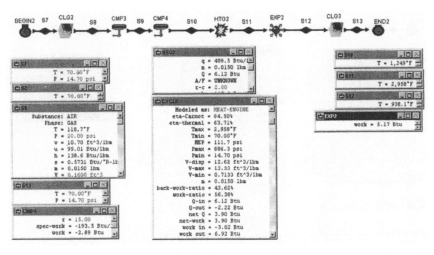

Figure 3.16 Diesel cycle with turbocharger.

$T_{10} = 1249°F$, $T_{11} = 2958°F$, $T_{12} = 938.1°F$, $Q_{in} = 6.12$ Btu, $W_{net} = 3.90$ Btu, MEP $= 106.5$ psia, $\eta = 64.51\%$, and m $= 0.015$ lbm. (See Fig. 3.16.)

Example 3.9

Find the pressure and temperature of each state of an ideal Diesel cycle with a compression ratio of 15 and a cut-off ratio of 2. A precooler that cools the atmospheric air from 70° to 50°F, and a supercharger that compresses fresh air to 20 psia before it enters the cylinder of the engine, are added to the engine. The cylinder volume before compression is 0.16 ft³. The atmosphere conditions are 14.7 psia and 70°F. Also determine the mass of air in the cylinder, heat supplied, net work produced, MEP, and cycle efficiency.

To solve this problem, we build the cycle as shown in Fig. 3.17. Then, (1) assume isobaric for the precooling process 7-8, isentropic for the compression process 8-9, isentropic for the compression process 9-10, isobaric for the heating process 10-11, isentropic for the expansion process 11-12, and isochoric for the cooling process 12-13; (2) input $p_7 = 14.7$ psia, $T_7 = 70°F$, $p_8 = 14.7$ psia, $p_{13} = 14.7$ psia, $T_{13} = 70°F$, $T_8 = 50°F$, $p_9 = 20$ psia, $V_9 = 0.16$ ft³ (with turbocharger and precooler), compression ratio $= 15$, and cut-off ratio $= 2$; and (3) display results. The results are: $T_8 = 50°F$, $p_8 = 14.7$ psia, $T_9 = 96.87°F$, $p_9 = 20$ psia, $T_{10} = 1184°F$, $p_{10} = 886.3$ psia, $T_{11} = 2829°F$, $p_{11} = 886.3$ psia, $T_{12} = 864.8°F$, $p_{12} = 36.76$ psia, $Q_{78} = -0.0745$ Btu, $W_{89} = -0.1247$ Btu, $W_{910} = -2.89$ Btu, $W_{1011} = 1.75$ Btu,

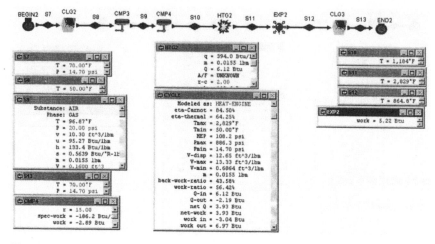

Figure 3.17 Diesel cycle with precooler and turbocharger.

$Q_{1011} = 6.12$ Btu, $W_{1112} = 5.2$ Btu, $W_{net} = 3.93$ Btu, MEP $= 108.2$ psia, $\eta = 64.25\%$, and $m = 0.0155$ lbm.

Review Problems 3.2 Diesel Cycle Analysis and Optimization

1. What is the difference between the compression ratio and cut-off ratio?

2. What is the difference between the Otto and Diesel engine?

3. Does the Diesel engine have sparkling plugs? If yes, for what reason?

4. Does the Diesel engine have engine knock problem? Why?

5. Is the Otto cycle more efficient than a Diesel cycle with the same compression ratio?

6. Why is the Diesel engine usually used for large trucks and the Otto engine usually used for compact cars?

7. Can the Diesel engine afford to have a large compression ratio? Why?

8. Suppose a large amount of power is required. Which engine would you choose between Otto and Diesel? Why?

9. The compression ratio of an air-standard Diesel cycle is 15. At the beginning of the compression stroke, the pressure is 14.7 psia and the temperature is 80°F. The maximum temperature of the cycle is 4040°F. Find the temperature at the end of the compression stroke, the temperature at the beginning of the exhaust process, the heat addition

to the cycle, the net work produced by the cycle, the thermal efficiency, and the MEP of the cycle.

10. An ideal Diesel cycle with a compression ratio of 17 and a cut-off ratio of 2 has an air temperature of 105°F and a pressure of 15 psia at the beginning of the isentropic compression process. Determine (a) the temperature and pressure of the air at the end of the isentropic compression process, (b) the temperature and pressure of the air at the end of the combustion process, and (c) the thermal efficiency of the cycle.

11. An ideal Diesel cycle with a compression ratio of 20 and a cut-off ratio of 2 has a temperature of 105°F and a pressure of 15 psia at the beginning of the compression process. Determine (a) the temperature and pressure of the gas at the end of the compression process, (b) the temperature and pressure of the gas at the end of the combustion process, (c) heat added to the engine in the combustion chamber, (d) heat removed from the engine to the environment, and (e) thermal cycle efficiency.

12. The pressure and temperature at the start of compression in an air Diesel cycle are 101 kPa and 300 K. The compression ratio is 15, and the amount of heat addition is 2000 kJ/kg of air. Determine (a) the maximum cycle pressure and maximum temperature of the cycle, and (b) the cycle thermal efficiency.

13. An ideal Diesel engine receives air at 103.4 kPa and 27°C. Heat added to the air is 1016.6 kJ/kg, and the compression ratio of the engine is 13. Determine (a) the work added during the compression process, (b) the cut-off ratio, (c) the work done during the expansion process, (d) the heat removed from the air during the cooling process, (e) the MEP (mean effective pressure), and (f) the thermal efficiency of the cycle.

14. An ideal Diesel engine receives air at 15 psia and 65°F. Heat added to the air is 160 Btu/lbm, and the compression ratio of the engine is 6. Determine (a) the work added during the compression process, (b) the cut-off ratio, (c) the work done during the expansion process, (d) the heat removed from the air during the cooling process, (e) the MEP, and (f) the thermal efficiency of the cycle.

15. An ideal Diesel engine receives air at 100 kPa and 25°C. The maximum cycle temperature is 1460°C and the compression ratio of the engine is 16. Determine (a) the work done during the compression process, (b) the heat added to the air during the heating process, (c) the work done during the expansion process, (d) the heat removed from the air during the cooling process, and (e) the thermal efficiency of the cycle.

16. A Diesel engine receives air at 60°F and 14.7 psia. The compression ratio is 20, the amount of heat addition is 800 Btu/lbm, and the mass of air contained in the cylinder is 0.02 lbm. Determine the

maximum cycle temperature, heat added, heat removed, work added, work produced, net work produced, MEP, and efficiency of the cycle.

ANSWERS: $T_{max} = 4601°F$; $Q_{add} = 16$ Btu; $Q_{remove} = -6.27$ Btu; $W_{add} = -4.12$ Btu; $W_{expansion} = 13.85$ Btu; $W_{net} = 9.73$ Btu; MEP = 211.7 psia; and $\eta = 60.84\%$.

17. A Diesel engine receives air at 80°F and 14.7 psia. The compression ratio is 20, the amount of heat addition is 800 Btu/lbm, and the mass of air contained in the cylinder is 0.02 lbm. Determine the maximum cycle temperature, heat added, heat removed, work added, work produced, net work produced, MEP, and efficiency of the cycle.

ANSWERS: $T_{max} = 4667°F$, $Q_{add} = 16$ Btu, $Q_{remove} = -6.22$ Btu, $W_{add} = -4.28$ Btu, $W_{expansion} = 14.05$ Btu, $W_{net} = 9.78$ Btu, MEP = 204.7 psia, and $\eta = 61.11\%$.

18. An ideal Diesel engine receives air at 15 psia and 70°F. The air volume is 7 ft^3 before compression. Heat added to the air is 200 Btu/lbm, and the compression ratio of the engine is 11. Determine (a) the work added during the compression process, (b) the maximum temperature of the cycle, (c) the work done during the expansion process, (d) the heat removed from the air during the cooling process, (e) the MEP, and (f) the thermal efficiency of the cycle.

19. A Diesel cycle has a compression ratio of 18. Air-intake conditions (prior to compression) are 72°F and 14.7 psia, and the highest temperature in the cycle is limited to 2500°F to avoid damaging the engine block. Calculate: (a) thermal efficiency, (b) net work, and (c) mean effective pressure; (d) compare engine efficiency with that of a Carnot cycle engine operating between the same temperatures.

20. A Diesel engine is modeled with an ideal Diesel cycle with a compression ratio of 17. The following information is known:
- Temperature prior to the compression process = 70°F.
- Pressure prior to the compression process = 14.7 psia.
- Heat added during the combustion process = 245 Btu/lbm.
    a. Determine the temperature and pressure at each process endpoint.
    b. Solve for the net cycle work (Btu/lbm).
    c. Solve for the thermal efficiency.

21. An ideal Diesel cycle with a compression ratio of 17 and a cut-off ratio of 2 has a temperature of 313 K and a pressure of 100 kPa at the beginning of the isentropic compression process. Use the cold air-standard assumptions and assume that $k = 1.4$. Determine (a) the temperature and pressure of the air at the end of the isentropic compression process and at the end of the combustion process, and (b) the thermal efficiency of the cycle.

22. Find the pressure and temperature of each state of an ideal Diesel cycle with a compression ratio of 15 and a cut-off ratio of 2. A precooler that cools the atmospheric air from 80° to 50°F, and a supercharger that compresses fresh air to 20 psia before it enters the cylinder of the engine, are added to the engine. The cylinder volume before compression is $0.1 \, ft^3$. The atmosphere conditions are 14.7 psia and 80°F. Also determine the mass of air in the cylinder, heat supplied, net work produced, MEP, and cycle efficiency.

ANSWERS: MEP $= 106.5$ psia; $\eta = 64.51\%$.

23. Find the pressure and temperature of each state of an ideal Diesel cycle with a compression ratio of 15 and a cut-off ratio of 2. A precooler that cools the atmospheric air from 80° to 50°F, and a supercharger that compresses fresh air to 25 psia before it enters the cylinder of the engine, are added to the engine. The cylinder volume before compression is $0.1 \, ft^3$. The atmosphere conditions are 14.7 psia and 80°F. Also determine the mass of air in the cylinder, heat supplied, net work produced, MEP, and cycle efficiency.

ANSWERS: MEP $= 116.4$ psia; $\eta = 66.7\%$.

3.3 ATKINSON CYCLE

A cycle called the *Atkinson* cycle is similar to the Otto cycle except that the isochoric exhaust and intake process at the end of the Otto cycle power stroke is replaced by an isobaric process. The schematic diagram of the cycle is shown in Fig. 3.18. The cycle consists of the following four processes:

1-2 Isentropic compression
2-3 Isochoric heat addition
3-4 Isentropic expansion
4-1 Isobaric heat removal

Applying the first and second laws of thermodynamics of the closed system to each of the four processes of the cycle yields:

$$W_{12} = \int p \, dV \tag{3.26}$$

$$Q_{12} - W_{12} = m(u_2 - u_1) \qquad Q_{12} = 0 \tag{3.27}$$

$$W_{23} = \int p \, dV = 0 \tag{3.28}$$

$$Q_{23} - 0 = m(u_3 - u_2) \tag{3.29}$$

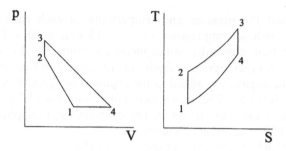

Figure 3.18 Atkinson cycle.

$$W_{34} = \int p \, dV \tag{3.30}$$

$$Q_{34} - W_{34} = m(u_4 - u_3) \qquad Q_{34} = 0 \tag{3.31}$$

$$W_{41} = \int p \, dV = pm(v_1 - v_4) \tag{3.32}$$

and

$$Q_{41} - W_{41} = m(u_1 - u_4) \tag{3.33}$$

The net work (W_{net}), which is also equal to net heat (Q_{net}), is

$$W_{net} = W_{12} + W_{34} + W_{41} = Q_{net} = Q_{23} + Q_{41} \tag{3.34}$$

The thermal efficiency of the cycle is

$$\eta = W_{net}/Q_{23} = Q_{net}/Q_{23} = 1 - Q_{41}/Q_{23} = 1 - (h_4 - h_1)/(u_3 - u_2) \tag{3.35}$$

This expression for thermal efficiency of the cycle can be simplified if air is assumed to be the working fluid with constant specific heat. Equation (3.35) is reduced to

$$\eta = 1 - k(T_4 - T_1)/(T_3 - T_2) \tag{3.36}$$

Example 3.10

Find the pressure and temperature of each state of an ideal Atkinson cycle with a compression ratio of 8. The heat addition to the combustion chamber is 800 Btu/lbm, the atmospheric air is at 14.7 psia and 60°F, and the cylinder contains 0.02 lbm of air. Determine the maximum temperature, maximum pressure, heat supplied, heat removed, work added during the compression processes, work produced during the expansion

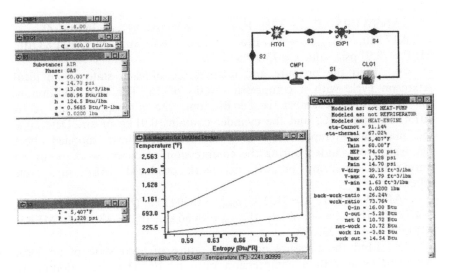

Figure 3.19 Atkinson cycle.

process, net work produced, MEP, and cycle efficiency. Draw the T–s diagram of the cycle.

To solve this problem, we build the cycle as shown in Fig. 3.19. Then, (1) assume isentropic for the compression process 1-2 and the expansion process 3-4, isochoric for the heating process 2-3, and isobaric for the cooling process 4-1; (2) input $p_1 = 14.7$ psia, $T_1 = 60°$F, $m\mathrm{dot} = 0.02$ lbm; $r = 8$ for the compression process 1-2, and $q = 800$ Btu/lbm for the heating process 2-3; and (3) display results. The results are: $T_{\max} = 5407°$F, $p_{\max} = 1328$ psia, $Q_{\mathrm{add}} = 16$ Btu, $Q_{\mathrm{remove}} = -5.28$ Btu, $W_{\mathrm{comp}} = -3.82$ Btu, $W_{\mathrm{expan}} = 14.54$ Btu, $W_{\mathrm{net}} = 10.72$ Btu, MEP $= 74.00$ psia, and $\eta = 67.02\%$.

Review Problems 3.3   Atkinson Cycle

1. What are the four processes of the Atkinson cycle?
2. What is the difference between the Otto cycle and the Atkinson cycle?
3. Find the pressure and temperature of each state of an ideal Atkinson cycle with a compression ratio of 16. The heat addition to the combustion chamber is 800 Btu/lbm, the atmospheric air is at 14.7 psia and 60°F, and the cylinder contains 0.02 lbm of air. Determine the maximum temperature, maximum pressure, heat supplied, heat removed, work added during the compression processes, work produced during the expansion process, net work produced, MEP, and cycle efficiency.

ANSWERS: $T_{max} = 5789°F$, $p_{max} = 2828$ psia, $Q_{add} = 16$ Btu, Q_{remove} $= -4.17$ Btu, $W_{comp} = -4.81$ Btu, $W_{expan} = 16.63$ Btu, $W_{net} = 11.83$ Btu, MEP $= 73.91$ psia, and $\eta = 73.91\%$.

4. Find the pressure and temperature of each state of an ideal Atkinson cycle with a compression ratio of 16. The heat addition in the combustion chamber is 800 Btu/lbm, the atmospheric air is at 101.4 kPa and 18°C, and the cylinder contains 0.01 kg of air. Determine the maximum temperature, maximum pressure, heat supplied, heat removed, work added during the compression processes, work produced during the expansion process, net work produced, MEP, and cycle efficiency.

ANSWERS: $T_{max} = 3121°C$, $p_{max} = 18904$ kPa, $Q_{add} = 18$ kJ, $Q_{remove} = -4.72$ kJ, $W_{comp} = -5.59$ kJ, $W_{expan} = 18.86$ kJ, $W_{net} = 13.28$ kJ, MEP $= 631$ kPa, and $\eta = 73.75\%$.

5. Find the pressure and temperature of each state of an ideal Atkinson cycle with a compression ratio of 10. The heat addition to the combustion chamber is 800 Btu/lbm. The atmospheric air is at 101.4 kPa and 18°C, and the cylinder contains 0.01 kg of air. Determine the maximum temperature, maximum pressure, heat supplied, heat removed, work added during the compression processes, work produced during the expansion process, net work produced, MEP, and cycle efficiency.

ANSWERS: $T_{max} = 2970°C$, $p_{max} = 11289$ kPa, $Q_{add} = 18$ kJ, $Q_{remove} = -5.54$ kJ, $W_{comp} = -4.74$ kJ, $W_{expan} = 17.20$ kJ, $W_{net} = 12.46$ kJ, MEP $= 540.7$ kPa, and $\eta = 69.21\%$.

3.4 DUAL CYCLE

Combustion in the Otto cycle is based on a constant-volume process; in the Diesel cycle, it is based on a constant-pressure process. However, combustion in actual spark-ignition engine requires a finite amount of time if the process is to be complete. For this reason, combustion in the Otto cycle does not actually occur under the constant-volume condition. Similarly, in compression–ignition engines, combustion in the Diesel cycle does not actually occur under the constant-pressure condition, because of the rapid and uncontrolled combustion process.

The operation of the reciprocating internal combustion engines represents a compromise between the Otto and the Diesel cycles, and can be described as a dual combustion cycle. Heat transfer to the system may be considered to occur first at constant volume and then at constant pressure. Such a cycle is called a dual cycle.

The Dual cycle as shown in Fig. 3.20 is composed of the following five processes:

1-2 Isentropic compression
2-3 Constant-volume heat addition
3-4 Constant-pressure heat addition
4-5 Isentropic expansion
5-1 Constant-volume heat removal

Figure 3.21 shows the dual cycle on *p–v* and *T–s* diagrams.

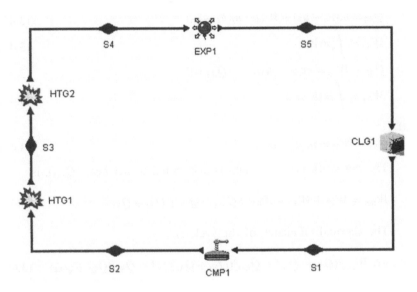

Figure 3.20   Dual cycle.

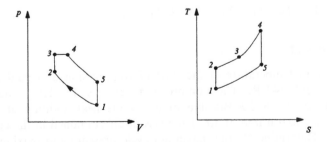

Figure 3.21   Dual cycle on *p–v* and *T–s* diagrams.

Applying the first and second laws of thermodynamics of the closed system to each of the five processes of the cycle yields:

$$W_{12} = \int p\,dV \tag{3.37}$$

$$Q_{12} - W_{12} = m(u_2 - u_1), \qquad Q_{12} = 0 \tag{3.38}$$

$$W_{23} = \int p\,dV = 0 \tag{3.39}$$

$$Q_{23} - 0 = m(u_3 - u_2) \tag{3.40}$$

$$W_{34} = \int p\,dV = m(p_4 v_4 - p_3 v_3) \tag{3.41}$$

$$Q_{34} = m(u_4 - u_3) + W_{34} = m(h_4 - h_3) \tag{3.42}$$

$$W_{45} = \int p\,dV \tag{3.43}$$

$$Q_{45} - W_{45} = m(u_5 - u_4), \qquad Q_{45} = 0 \tag{3.44}$$

$$W_{51} = \int p\,dV = 0 \tag{3.45}$$

and

$$Q_{51} - W_{51} = m(u_1 - u_5) \tag{3.46}$$

The net work (W_{net}), which is also equal to net heat (Q_{net}), is

$$W_{net} = W_{12} + W_{34} + W_{45} = Q_{net} = Q_{23} + Q_{34} + Q_{51} \tag{3.47}$$

The thermal efficiency of the cycle is

$$\eta = W_{net}/(Q_{23} + Q_{34}) = Q_{net}/(Q_{23} + Q_{34}) = 1 - Q_{51}/(Q_{23} + Q_{34}) \tag{3.48}$$

This expression for the thermal efficiency of an ideal Otto cycle can be simplified if air is assumed to be the working fluid with constant specific heat. Equation (3.48) is reduced to:

$$\eta = 1 - (T_5 - T_1)/[(T_3 - T_2) + k(T_4 - T_3)] \tag{3.49}$$

Example 3.11

Pressure and temperature at the start of compression in a dual cycle are 14.7 psia and 540°R. The compression ratio is 15. Heat addition at constant volume is 300 Btu/lbm of air, while heat addition at constant pressure is 500 Btu/lbm of air. The mass of air contained in the cylinder is 0.03 lbm. Determine (1) the maximum cycle pressure and maximum cycle temperature, (2) the efficiency and work output per kilogram of air, and

(3) the MEP. Show the cycle on *T–s* diagram. Plot the sensitivity diagram of cycle efficiency versus compression ratio.

To solve this problem by CyclePad, we take the following steps:

1. Build:
   a. Take a compression device, two combustion chambers, an expander, and a cooler from the closed-system inventory shop and connect the five devices to form the Dual cycle.
   b. Switch to analysis mode.
2. Analysis:
   a. Assume a process for each of the five devices: (1) compression device as isentropic, (2) first combustion chamber as isochoric and second combustion chamber as isobaric, (3) expander as isentropic, and (4) cooler as isochoric.
   b. Input the given information: (1) working fluid is air, (2) inlet pressure and temperature of the compression device are 14.7 psia and 540°R, (3) compression ratio of the compression device is 15, (4) heat addition is 300 Btu/lbm in the isocbaric combustion chamber, (5) heat addition is 500 Btu/lbm in the isobaric combustion chamber, and (6) mass of air contained in the cylinder is 0.03 lbm.
3. Display results:
   a. Display the *T–s* diagram and cycle properties' results. The cycle is a heat engine. The answers are $T_{max} = 5434°R$, $p_{max} = 1367$ psia, $\eta = 63.78\%$, MEP $= 217.3$ psia, and $W_{net} = 15.31$ Btu.
   b. Display the sensitivity diagram of cycle efficiency versus compression ratio. (See Figs. 3.22a–3.22c.)

Review Problems 3.4 Dual Cycle

1. What five processes make up the dual cycle?
2. The combustion process in internal combustion engines as an isobaric or isometric heat-addition process is oversimplistic and not realistic. A real cycle *p–v* diagram of the Otto or Diesel cycle looks like a curve (combination of isobaric and isometric) rather than a linear line. Are the combustion processes in the dual cycle more realistic?
3. Can we consider the Otto or Diesel cycle to be special cases of the dual cycle?
4. Pressure and temperature at the start of compression in a dual cycle are 101 kPa and 15°C. The compression ratio is 8. Heat addition at constant volume is 100 kJ/kg of air, while the maximum temperature of the cycle is limited to 2000°C. The mass of air contained in the cylinder is

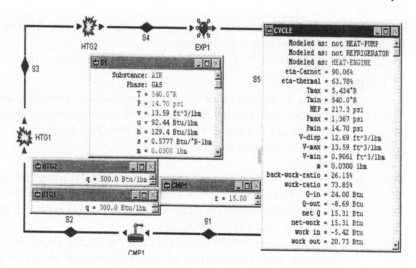

Figure 3.22a Dual cycle.

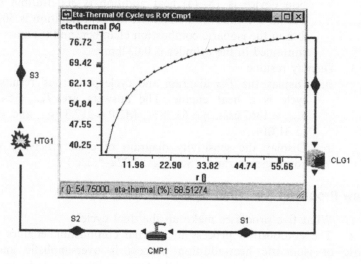

Figure 3.22b Dual cycle sensitivity analysis.

0.01 kg. Determine the maximum cycle pressure, the MEP, heat added, heat removed, compression work added, expansion work produced, net work produced, and efficiency of the cycle.

ANSWERS: $p_{\max} = 2248 \, \text{kPa}$, MEP $= 988.1 \, \text{kPa}$, $Q_{\text{add}} = 15.77 \, \text{kJ}$, $Q_{\text{remove}} = -8.7 \, \text{kJ}$, $W_{\text{comp}} = -2.68 \, \text{kJ}$, $W_{\text{expansion}} = 9.75 \, \text{kJ}$, $W_{\text{net}} = 7.07 \, \text{kJ}$, and $\eta = 44.85\%$.

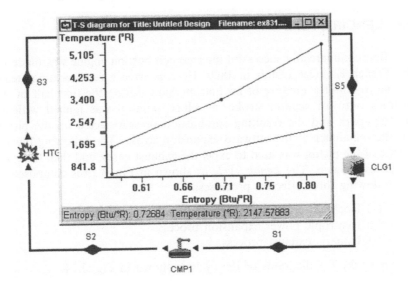

Figure 3.22c Dual cycle *T–s* diagram.

5. Pressure and temperature at the start of compression in a dual cycle are 101 kPa and 15°C. The compression ratio is 12. Heat addition at constant volume is 100 kJ/kg of air, while the maximum temperature of the cycle is limited to 2000°C. The mass of air contained in the cylinder is 0.01 kg. Determine the maximum cycle pressure, the MEP, heat added, heat removed, compression work added, expansion work produced, net work produced, and efficiency of the cycle.

ANSWERS: $p_{max} = 3862$ kPa, MEP $= 1067$ kPa, $Q_{add} = 14.60$ kJ, $Q_{remove} = -6.60$ kJ, $W_{comp} = -3.51$ kJ, $W_{expansion} = 11.51$ kJ, $W_{net} = 8.00$ kJ, and $\eta = 57.48\%$.

6. Pressure and temperature at the start of compression in a dual cycle are 101 kPa and 15°C. The compression ratio is 12. Heat addition at constant volume is 100 kJ/kg of air, while the maximum temperature of the cycle is limited to 2200°C. The mass of air contained in the cylinder is 0.01 kg. Determine the maximum cycle pressure, the MEP, heat added, heat removed, compression work added, expansion work produced, net work produced, and efficiency of the cycle.

ANSWERS: $p_{max} = 3862$ kPa, MEP $= 1189$ kPa, $Q_{add} = 16.60$ kJ, $Q_{remove} = -7.69$ kJ, $W_{comp} = -3.51$ kJ, $W_{comp} = 12.43$ kJ, $W_{net} = 8.92$ kJ, and $\eta = 53.71\%$.

3.5 LENOIR CYCLE

The first commercially successful internal combustion engine was made by the French engineer Lenoir in 1860. He converted a reciprocating steam engine to admit a mixture of air and methane during the first half of the piston's outward suction stroke, at which point it was ignited with an electric spark and the resulting combustion pressure acted on the piston for the remainder of the outward expansion stroke. The following inward stroke of the piston was used to expel the exhaust gases, and then the cycle began over again. The *Lenoir cycle* as shown in Fig. 3.23 is composed of the following three effective processes:

1-2 Isochoric combustion process
2-3 Isentropic power expansion process
3-1 Isobaric exhaust process

The p–v and T–s diagrams of the cycle is shown in Fig. 3.24.

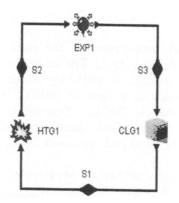

Figure 3.23   Lenoir cycle.

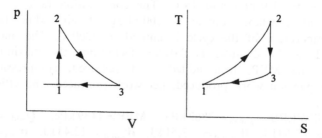

Figure 3.24   Lenoir cycle p–v diagram and T–s diagram.

Applying the first and second laws of thermodynamics of the closed system to each of the three processes of the cycle yields:

$$W_{12} = 0 \tag{3.50}$$

$$Q_{12} - 0 = m(u_2 - u_1) \tag{3.51}$$

$$Q_{23} = 0 \tag{3.52}$$

$$0 - W_{23} = m(u_3 - u_2) \tag{3.53}$$

$$W_{31} = \int p\,\mathrm{d}V = m(p_1 v_1 - p_3 v_3) \tag{3.54}$$

and

$$Q_{31} = m(u_1 - u_3) + W_{31} = m(h_1 - h_3) \tag{3.55}$$

The net work (W_{net}), which is also equal to net heat (Q_{net}), is

$$W_{net} = W_{23} + W_{31} = Q_{net} = Q_{12} + Q_{31} \tag{3.56}$$

The thermal efficiency of the cycle is

$$\eta = W_{net}/Q_{12} = Q_{net}/Q_{12} = 1 + Q_{31}/Q_{12} \tag{3.57}$$

This expression for thermal efficiency of the cycle can be simplified if air is assumed to be the working fluid with constant specific heat. Equation (3.57) is reduced to

$$\eta = 1 - (h_3 - h_1)/(u_2 - u_1) = 1 - kT_2(r_s - 1)/(T_2 - T_1) \tag{3.58}$$

where r_s is the isentropic volume compression ratio, $r_s = v_3/v_1$.

Because the air–fuel mixture was not compressed before ignition, the engine efficiency was very low and fuel consumption was very high. The fuel–air mixture was ignited by an electric spark inside the cylinder.

Example 3.12

The isochoric heating process of a Lenoir engine receives air at 15°C and 101 kPa. The air is heated to 2000°C, and the mass of air contained in the cylinder is 0.01 kg. Determine the pressure at the end of the isochoric heating process, the temperature at the end of the isentropic expansion process, heat added, heat removed, work added, work produced, net work produced, and efficiency of the cycle. Draw the *T–s* diagram of the cycle.

To solve this problem by CyclePad, we take the following steps:

1. Build:
   a. Take a combustion chamber, an expander, and a cooler from the closed-system inventory shop and connect the three devices to form the Lenoir cycle.
   b. Switch to analysis mode.
2. Analysis:
   a. Assume a process for each of the three devices: (1) combustion chamber as isochoric, (2) expander as isentropic, and (3) cooler as isobaric.
   b. Input the given information: (1) working fluid is air, (2) inlet pressure and temperature of the combustion device are 101 kPa and 15°C, (3) temperature at end of the combustion device is 2000°C, and (4) mass of air is 0.01 kg.
3. Display results:
   a. Display cycle properties' results. The cycle is a heat engine. The answers are $T_3 = 986.8°C$, $p_2 = 796.8$ kPa, $Q_{add} = 14.23$ kJ, $Q_{remove} = -9.75$ kJ, $W_{comp} = -2.79$ kJ, $W_{expan} = 7.26$ kJ, $W_{net} = 4.48$ kJ, and $\eta = 31.46\%$.
   b. Display the *T–s* diagram. (See Fig. 3.25.)

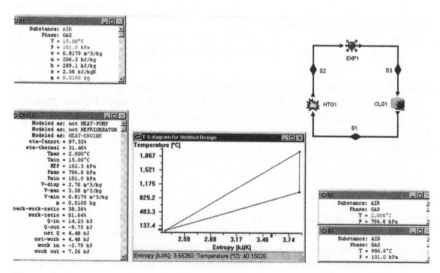

Figure 3.25 Lenoir cycle.

Review Problems 3.5   Lenoir Cycle

1. What are the five processes that make up the Lenoir cycle?

2. The isochoric heating process of a Lenoir engine receives air at 15°C and 101 kPa. The air is heated to 2200°C, and the mass of air contained in the cylinder is 0.01 kg. Determine the pressure at the end of the isochoric heating process, the temperature at the end of the isentropic expansion process, heat added, heat removed, work added, work produced, net work produced, and efficiency of the cycle.

ANSWERS: $T_3 = 1065°C$, $p_2 = 866.9$ kPa, $Q_{add} = 15.66$ kJ, $Q_{remove} = -10.54$ kJ, $W_{comp} = -3.01$ kJ, $W_{expan} = 8.13$ kJ, $W_{net} = 5.12$ kJ, and $\eta = 32.72\%$.

3. The isochoric heating process of a Lenoir engine receives air at 60°F and 14.7 psia. The air is heated to 4000°F, and the mass of air contained in the cylinder is 0.02 lbm. Determine the pressure at the end of the isochoric heating process, the temperature at the end of the isentropic expansion process, heat added, heat removed, work added, work produced, net work produced, and efficiency of the cycle.

ANSWERS: $T_3 = 1953°F$, $p_2 = 126.2$ psia, $Q_{add} = 13.49$ Btu, $Q_{remove} = -9.08$ Btu, $W_{comp} = -2.59$ Btu, $W_{expan} = 7.01$ Btu, $W_{net} = 4.41$ Btu, and $\eta = 32.72\%$.

4. The isochoric heating process of a Lenoir engine receives air at 80°F and 14.7 psia. The air is heated to 4500°F, and the mass of air contained in the cylinder is 0.02 lbm. Determine the pressure at the end of the isochoric heating process, the temperature at the end of the isentropic expansion process, heat added, heat removed, work added, work produced, net work produced, and efficiency of the cycle.

ANSWERS: $T_3 = 2172°F$, $p_2 = 135.1$ psia, $Q_{add} = 15.13$ Btu, $Q_{remove} = -10.03$ Btu, $W_{comp} = -2.86$ Btu, $W_{expan} = 7.97$ Btu, $W_{net} = 5.11$ Btu, and $\eta = 33.74\%$.

3.6   STIRLING CYCLE

The *Stirling cycle* is composed of the following four processes:

1-2 Isothermal compression
2-3 Constant-volume heat addition
3-4 Isothermal expansion
4-1 Constant-volume heat removal

The Stirling cycle engine is an external combustion engine. Figure 3.26 shows the Stirling cycle on p–v and T–s diagrams.

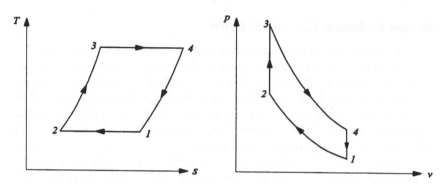

Figure 3.26 Stirling cycle on *p–v* and *T–s* diagrams.

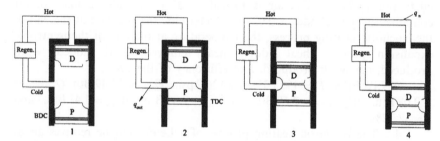

Figure 3.27 Stirling cycle operation.

During the isothermal compression process 1-2, heat is rejected to maintain a constant temperature T_L. During the isothermal expansion process 3-4, heat is added to maintain a constant temperature T_H. There are also heat interactions along the constant-volume heat addition process 2-3 and the constant-volume heat removal process 4-1. The quantities of heat in these two constant-volume processes are equal but opposite in direction.

The operation of the Stirling cycle engine is shown in Fig. 3.27. There are two pistons in the cylinder. One is a power piston (P) and the other is the displace piston (D). The purpose of the displace piston is to move the working fluid around from one space to another space through the regenerator. At state 1, the power piston is at BDC (bottom dead center), with the displacer at its TDC (top dead center). The power piston moves from its BDC to TDC to compress the working fluid during the compression process 1-2. From 1-2, the working fluid in the cylinder is in contact with the low-temperature reservoir, so the temperature remains constant ($T_1 = T_2$) and heat is removed. During the heating process 2-3, the displacer moves downward, pushing the working fluid through the regenerator where it picks up heat to reach T_3. During the expansion process 3-4, the working fluid in the cylinder is in contact with the high-temperature reservoir, so the temperature remains constant $(T_3 = T_4)$ and heat is added. During the cooling process 4-1, the displacer moves upward, pushing the working fluid through the regenerator where it removes heat to reach T_1.

Applying the first and second laws of thermodynamics of the closed system to each of the four processes of the cycle yields:

$$W_{12} = \int p\,dV, \quad Q_{12} = \int T\,dS = T_1(S_2 - S_1) \tag{3.59}$$

$$Q_{12} - W_{12} = m(u_2 - u_1) = 0 \tag{3.60}$$

$$W_{23} = \int p\,dV = 0 \tag{3.61}$$

$$Q_{23} - 0 = m(u_3 - u_2) \tag{3.62}$$

$$W_{34} = \int p\,dV, \quad Q_{34} = \int T\,dS = T_3(S_4 - S_3) \tag{3.63}$$

$$Q_{34} - W_{34} = m(u_4 - u_3) = 0 \tag{3.64}$$

$$W_{41} = \int p\,dV = 0 \tag{3.65}$$

and
$$Q_{41} - 0 = m(u_1 - u_4) \tag{3.66}$$

The net work (W_{net}), which is also equal to net heat (Q_{net}), is

$$W_{net} = W_{12} + W_{34} = Q_{net} = Q_{12} + Q_{23} + Q_{34} + Q_{41} \tag{3.67}$$

The thermal efficiency of the cycle is

$$\eta = W_{net}/(Q_{12} + Q_{41}) \tag{3.68}$$

Example 3.13

A Stirling cycle operates with 0.1 kg of hydrogen as a working fluid between 1000°C and 30°C. The highest and the lowest pressures during the cycle are 3 MPa and 500 kPa. Determine the heat and work added in each of the four processes, net work, and cycle efficiency.

To solve this problem by CyclePad, we take the following steps:

1. Build:
   a. Take a compression device, a combustion chamber, an expander, and a cooler from the closed-system inventory shop and connect the four devices to form the Stirling cycle.
   b. Switch to analysis mode.

2. Analysis:
    a. Assume a process for each of the four devices: (1) compression device as isothermal, (2) combustion chamber as isochoric, (3) expander as isothermal, and (4) cooler as isochoric.
    b. Input the given information: (1) working fluid is helium, (2) inlet pressure and temperature of the compression device are 500 kPa and 30°C and $m = 0.1$ kg, (3) inlet pressure and temperature of the expander are 3 MPa and 1000°C.
3. Display the cycle properties' results. The cycle is a heat engine.

The answers are: $Q_{12} = W_{12} = -22.46$ kJ, $Q_{23} = 300.7$ kJ, $Q_{34} = W_{34} = 94.33$ kJ, $Q_{41} = -300.7$ kJ, $W_{net} = 71.87$ kJ, $Q_{in} = 395.0$ kJ, and $\eta = 18.19\%$. (See Fig. 3.28.)

The Stirling cycle is an attempt to achieve Carnot efficiency by the use of an ideal regenerator.

A device called a regenerator can be used to absorb heat during process 4-1 (Q_{41}) and ideally delivering the same quantity of heat during process 2-3 (Q_{23}). These two quantities of heat are represented by the areas underneath of the process 4-1 and process 2-3 of the T–s diagram in Fig. 3.26. Using the ideal regenerator, Q_{41} is not counted as a part of the heat input. The efficiency of the Stirling cycle can be reduced from Eq. (3.68) to

$$\eta = W_{net}/Q_{12} = 1 - T_3/T_1 \tag{3.69}$$

In this respect, the Stirling cycle has the same efficiency as the Carnot cycle.

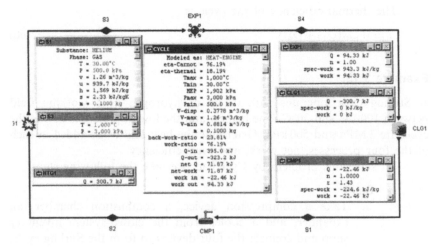

Figure 3.28 Stirling cycle.

Figure 3.29 Regenerative Stirling cycle.

The regenerative Stirling cycle is illustrated in Fig. 3.29. In this figure, the combination of heater #1 and cooler #1 is equivalent to the regenerator. Heat removed from the cooler #1 is added to the heater #1. Since this energy transfer occurs within the cycle internally, the amount of heat added to heater #1 from cooler #1 is not a part of the heat added to the cycle from its surrounding heat reservoirs. Therefore,

$$Q_{in} = Q_{12} \tag{3.70}$$

Example 3.14 illustrates the analysis of the regenerative Stirling cycle.

Practical attempts to follow the Stirling cycle present difficulties primarily due to the difficulty of achieving isothermal compression and isothermal expansion in a machine operating at a reasonable speed.

Example 3.14

A regenerative Stirling cycle operates with 0.1 kg of hydrogen as a working fluid between 1000°C and 30°C. The highest and the lowest pressures during the cycle are 3 MPa and 500 kPa. The temperature at the exit of the regenerator (heater #1) and inlet to the heater #2 is 990°C and the temperature at the exit of the regenerator (cooler #1) and inlet to the cooler #2 is 40°C. Determine the heat and work added in each of the four processes, net work, and cycle efficiency.

To solve this problem by CyclePad, we take the following steps:

1. Build:
   a. Take a compression device, a heater (cold-side regenerator), combustion chamber, an expander and two coolers (cooler #1 is the hot-side regenerator) from the closed-system inventory shop and connect the six devices to form the regenerative Stirling cycle as shown in Fig. 3.27.
   b. Switch to analysis mode.
2. Analysis:
   a. Assume a process for each of the six devices: (1) compression device as isothermal, (2) both heaters as isochoric, (3) expander as isothermal, and (4) both coolers as isochoric.
   b. Input the given information: (1) working fluid is helium, (2) inlet pressure and temperature of the compression device are 500 kPa and 30°C and $m = 0.1$ kg, (3) inlet pressure and

temperature of the expander are 3000 kPa and 1000°C, (4) temperature at the exit of the regenerator (heater #1) = 990°C, and (5) temperature at the exit of the regenerator (cooler #1) = 40°C.

3. Display the cycle properties' results. The cycle is a heat engine.

The answers are: $Q_{12} = W_{12} = -22.46$ kJ, $Q_{23} = Q_{htr\#1} = -Q_{clr\#1} = Q_{regenerator} = 297.6$ kJ, $Q_{34} = 3.1$ kJ, $Q_{45} = W_{45} = 94.33$ kJ, $Q_{56} = -3.1$ kJ, $W_{net} = 71.87$ kJ, $Q_{in} = 94.33 + 3.1 = 97.43$ kJ, and $\eta = 71.87/97.43 = 73.77\%$. (See Fig. 3.30.)

COMMENT: The regenerator used in this example is not ideal. Yet, the regenerator raises the cycle efficiency almost to the Carnot efficiency.

Review Problems 3.6   Stirling Cycle

1. What are the four processes of the basic Stirling cycle?
2. What is a regenerator?
3. What is a regenerative Stirling cycle?
4. What would be the cycle efficiency of the Stirling cycle with an ideal regenerator?
5. A cycle is executed in a closed system with 4 g of air and is composed of the following three processes:

> 1-2 Constant-volume heating from 100 kPa and 20°C to 400 kPa
> 2-3 Isentropic expansion to 100 kPa
> 3-1 Constant pressure cooling to 100 kPa and 20°C

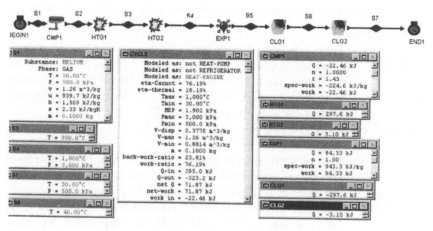

Figure 3.30 Regenerative Stirling cycle.

Determine the net work per cycle, the thermal cycle efficiency, and MEP. Show the cycle on a *T–s* diagram.

6. A proposed air standard piston–cylinder arrangement cycle consists of an isentropic compression process, a constant-volume heat addition process, an isentropic expansion process, and a constant-pressure heat-rejection process. The compression ratio (v_1/v_2) during the isentropic compression process is 8.5. At the beginning of the compression process, $P = 100$ kPa and $T = 300$ K. The constant-volume specific heat addition is 1400 kJ/kg. Assume constant specific heats at 25°C.

    a. Draw the cycle (with labels) on a *T–s* diagram.

    b. Find the specific heat rejection during the constant-pressure process.

    c. Find the specific net work for the cycle.

    d. Find the thermal efficiency for the cycle.

    e. Find the mean effective pressure for the cycle.

7. A Stirling cycle operates with 1 lbm of helium as a working fluid between 1800°R and 540°R. The highest and lowest pressures during the cycle are 450 and 75 psia. Determine the heat added, net work, and cycle efficiency.

    ANSWERS: $W_{net} = 367.4$ Btu, $Q_{in} = 1458$ Btu, and $\eta = 25.20\%$.

8. A Stirling cycle operates with 0.1 kg of air as a working fluid between 1000°C and 30°C. The highest and lowest pressures during the cycle are 3 MPa and 500 kPa. Determine the heat and work added in each of the four processes, net work, and cycle efficiency.

    ANSWERS: $Q_{12} = W_{12} = -3.10$ kJ, $Q_{23} = 69.52$ kJ, $Q_{34} = W_{34} = 13.02$ kJ, $Q_{41} = -69.52$ kJ, $W_{net} = 9.92$ kJ, $Q_{in} = 82.54$ kJ, and $\eta = 12.02\%$.

9. A regenerative Stirling cycle operates with 0.1 kg of air as a working fluid between 1000°C and 30°C. The highest and lowest pressures during the cycle are 3 MPa and 500 kPa. The temperature at the exit of the regenerator (heater #1) and inlet to the heater #2 is 990°C and the temperature at the exit of the regenerator (cooler #1) and inlet to the cooler #2 is 40°C. Determine the heat and work added in each of the four processes, net work, and cycle efficiency.

3.7 MILLER CYCLE

An alternative to lowering the compression ratio and simultaneously the expansion ratio of an Otto or Diesel cycle is to lower compression ratio only while the expansion ratio is kept as the original. *Miller* (Miller, R.H., Supercharging and internal cooling cycle for high output. *Transactions of the American Society of Mechanical Engineers*, vol. 69, pp. 453–457, 1947) proposed a cycle having the following characteristics:

1. Effective compression stroke is shorter than expansion stroke
2. Increased charging pressure
3. Variable valve timing

Miller proposed the use of early intake valve closing to provide internal cooling before compression so as to reduce compression work. Miller further proposed increasing the boost pressure to compensate for the reduced inlet duration. By proper selection of boost pressure and variation of intake valve closing time, Miller showed that turbocharged engines could maintain sea-level power while operating over varying altitudes.

A modified Otto cycle is known as the *Miller–Otto cycle* whose *p–v* and *T–s* diagrams are shown in Fig. 3.31. A modified Diesel cycle is known as the *Miller–Diesel cycle* whose *p–v* and *T–s* diagrams are shown in Fig. 3.32.

A four-stroke Miller–Otto cycle without supercharger and intercooler is shown in Fig. 3.33. The intake valve is closed late at state 3.

A four-stroke Miller–Otto cycle with supercharger is shown in Fig. 3.34. The intake valve is closed late at state 3.

Similarly, an extended expansion stroke is desirable in four-stroke spark ignition and Diesel engines from the viewpoint of providing an increase in thermal cycle efficiency and, for prescribed air and fuel flow rates, an increase in engine output. Spark-ignition engines modified to achieve extended expansion within the engine cylinder are termed *Otto–Atkinson cycles* (Ma, T.H., Recent advances in variable valve timing. *Automotive Engine Alternatives*, Ed. by R.L. Evans, Plenum Press,

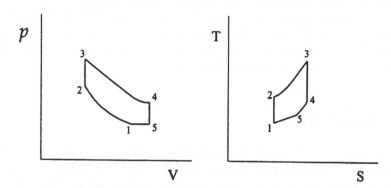

Figure 3.31   Miller–Otto cycle.

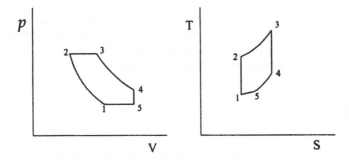

Figure 3.32   Miller–Diesel cycle.

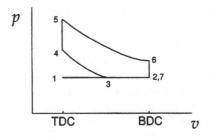

Figure 3.33   Miller–Otto cycle without supercharger and intercooler.

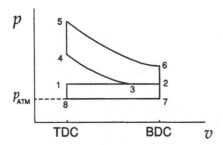

Figure 3.34   Miller–Otto cycle with supercharger.

pp. 235–252, 1986). By analogy, Diesel engines modified to achieve extended expansion within the engine cylinder are termed *Diesel–Atkinson cycles* (Kentfield, J.A.C., Diesel engines with extended strokes. *SAE Transaction Journal of Engines*, vol. 98, pp. 1816–1825, 1989).

Variable valve timing is being developed to improve the performance and reduce the pollution emissions from internal combustion heat engines for automobiles and trucks. A unique benefit for these engines is that changing the timing of the intake valves can be used to control the engine's compression ratio. These engines can be designed to have a conventionally high compression ratio for satisfactory cold-starting characteristics and a reduced compression ratio for better cycle efficiency, exhaust gas emissions, and noise characteristics.

Example 3.15

Determine the temperature at the end of the compression process, compression work, expansion work, and thermal efficiency of an ideal Otto cycle. The volumes of the cylinder before and after compression are 3 liters and 0.3 liter. Heat added to the air in the combustion chamber is 800 kJ/kg. What is the mass of air in the cylinder? The atmosphere conditions are 101.3 kPa and 20°C.

To solve this problem, we build the cycle as shown in Fig. 3.4a. Then, (1) assume isentropic for the compression process 1-2, isochoric for the heating process 2-3, isentropic for the expansion process 3-4, and isochoric for the cooling process 4-5; (2) input $p_1 = 101.3$ kPa, $T_1 = 20°C$, $V_1 = 3$ liters, $V_2 = 0.3$ liter, heat added to the combustion chamber is 800 kJ/kg, $p_5 = 101.3$ kPa, and $T_5 = 20°C$; and (3) display results. The results are: $T_2 = 463.2°C$, $W_{12} = -1.15$ kJ, $Q_{23} = 2.89$ kJ, $W_{34} = 2.89$ kJ, $W_{net} = 1.74$ kJ, $\eta = 60.19\%$, and $m = 3.62$ g as shown in Fig. 3.35.

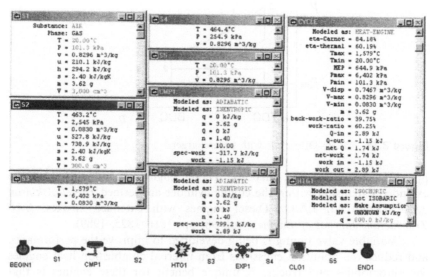

Figure 3.35 Otto cycle.

Example 3.16

Determine the temperature at the end of the compression process, compression work, expansion work, and thermal efficiency of an Otto–Miller cycle. The volumes of the cylinder before and after compression are 3 liters and 0.3 liter. Heat added to the air in the combustion chamber is 800 kJ/kg. A supercharger and an intercooler are used. The supercharger pressure is 180 kPa and the temperature at the end of the intercooler is 20°C. The intake valve closes at 2.8 liters. The end temperature of the cooling process of the cycle is 20°C. What is the mass of air in the cylinder? The atmosphere conditions are 101.3 kPa and 20°C.

To solve this problem, we build the cycle as shown in Fig. 3.36. Then, (1) assume isentropic for both compression processes, isochoric for the heating process, isentropic for the expansion process, and isochoric for both cooling processes; (2) input $p_1 = 101.3$ kPa, $T_1 = 20$°C, $p_2 = 180$ kPa, $T_3 = 20$°C, $V_3 = 2.8$ liters, $V_4 = 0.3$ liter, heat added in the combustion chamber is 800 kJ/kg, $V_6 = 3$ liters, $T_7 = 20$°C; and (3) display results. The results are: $T_4 = 443.2$°C, $W_{comp} = -1.73$ kJ, $Q_{add} = 4.07$ kJ, $W_{exp} = 4.02$ kJ, $W_{net} = 2.29$ kJ, $\eta = 2.29/4.07 = 56.27\%$, and $m = 5.09$ g as shown in Fig. 3.36. Notice that if the supercharger is operated by the exhaust gas, then $\eta = (4.02 - 1.54)/4.07 = 60.93\%$.

Review Problems 3.7   Miller Cycle

1. What is the idea of the Miller cycle?
2. What are the benefits of the Miller cycle?

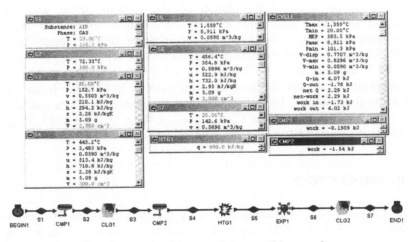

Figure 3.36 Miller–Otto cycle with supercharger and intercooler.

3. Determine the temperature at the end of the compression process, compression work, expansion work, and thermal efficiency of an Otto–Miller cycle. The volumes of the cylinder before and after compression are 3 liters and 0.3 liter. Heat added to the air in the combustion chamber is 800 kJ/kg. A supercharger and an intercooler are used. The supercharger pressure is 180 kPa and the temperature at the end of the intercooler is 20°C. The intake valve closes at 2.5 liters. The end temperature of the cooling process of the cycle is 20°C. What is the mass of air in the cylinder? The atmosphere conditions are 101.3 kPa and 20°C.

ANSWERS: $T_4 = 411.4$°C, $W_{comp} = -1.44$ kJ, $Q_{add} = 3.63$ kJ, $W_{exp} = 3.53$ kJ, $W_{net} = 2.08$ kJ, $\eta = 57.3\%$, and $m = 4.54$ g.

4. Determine the temperature at the end of the compression process, compression work, expansion work, and thermal efficiency of an Otto–Miller cycle. The volumes of the cylinder before and after compression are 3 liters and 0.3 liter. Heat added to the air in the combustion chamber is 800 kJ/kg. A supercharger and an intercooler used. The supercharger pressure is 200 kPa and the temperature at the end of the intercooler is 20°C. The intake valve closes at 2.0 liters. The end temperature of the cooling process of the cycle is 20°C. What is the mass of air in the cylinder? The atmosphere conditions are 101.3 kPa and 20°C.

ANSWERS: $T_4 = 353$°C, $W_{comp} = -1.11$ kJ, $Q_{add} = 3.14$ kJ, $W_{exp} = 2.95$ kJ, $W_{net} = 1.83$ kJ, $\eta = 58.28\%$, and $m = 3.92$ g.

5. Determine the temperature at the end of the compression process, compression work, expansion work, and thermal efficiency of an Otto–Miller cycle. The volumes of the cylinder before and after compression are 3 liters and 0.3 liter. Heat added to the air in the combustion chamber is 800 kJ/kg. A supercharger and an intercooler are used. The supercharger pressure is 180 kPa and the temperature at the end of the intercooler is 20°C. The intake valve closes at 2 liters. The end temperature of the cooling process of the cycle is 20°C. What is the mass of air in the cylinder? The atmosphere conditions are 101.3 kPa and 20°C.

ANSWERS: $T_4 = 411.4$°C, $W_{comp} = -1.44$ kJ, $Q_{add} = 3.63$ kJ, $W_{exp} = 3.53$ kJ, $W_{net} = 2.08$ kJ, $\eta = 57.3\%$, and $m = 4.54$ g.

3.8 WICKS CYCLE

The Carnot cycle is the ideal cycle only for the conditions of constant-temperature hot and cold surrounding thermal reservoirs. However, such conditions do not exist for fuel-burning engines. For these engines, the combustion products are artificially created as a finite-size hot reservoir that releases heat over the entire temperature range from its maximum to ambient temperature. The natural environment in terms of air or water bodies is the cold reservoir and can be considered as an infinite reservoir relative to the engine. Thus, an ideal fuel-burning engine should operate reversibly between a finite-size hot reservoir and an infinite-size cold reservoir. Wicks (Wicks, F., The thermodynamic theory and design of an ideal fuel burning engine. *Proceedings of the Intersociety Engineering Conference of Energy Conversion*, vol. 2, pp. 474–481, 1991) proposed a three-process ideal fuel-burning engine consisting of isothermal compression, isochoric heat addition, and adiabatic expansion processes. The schematic *Wicks cycle* is shown in Fig. 3.37. The *p–v* and *T–s* diagrams of the cycle are shown in Fig. 3.38, and an example of the cycle is given in Example 3.17.

Example 3.17

Air is compressed from 14.7 psia and 500°R isothermally to 821.8 psia, heated isochorically to 2500°R, and then expanded isentropically to 14.7 psia in a Wicks cycle. Determine the heat added, heat removed, work added, work produced, net work, and cycle efficiency.

To solve this problem by CyclePad, we take the following steps:

1. Build:
    a. Take a begin, an end, a compression device, a heater, and an expander from the closed-system inventory shop

BEGIN1   S1   CMP1   S2   HTG1   S3   EXP1   S4   END1

Figure 3.37 Wicks cycle.

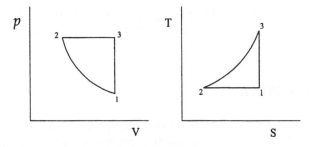

Figure 3.38 *p–v* and *T–s* diagrams of the Wicks cycle.

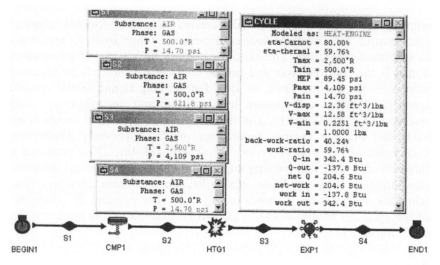

Figure 3.39 Wicks cycle.

and connect them to form the Wicks cycle as shown in Fig. 3.37.

    b. Switch to analysis mode.

2. Analysis:

    a. Assume a process for each of the three devices: (1) compression device as isothermal, (2) heater as isochoric, and (3) expander as isentropic.

    b. Input the given information: (1) working fluid is air, (2) the starting pressure and temperature of the compression device are 14.7 psia and 500°R, and $m = 1$ lbm, (3) the end temperature of the heater expander is 2500°R, and (4) the end pressure of the expander is 14.7 psia.

3. Display the cycle properties' results. The cycle is a heat engine.

The answers are: $Q_{in} = 342.4$ Btu, $Q_{out} = -137.8$ Btu, $W_{in} = -137.8$ Btu, $W_{out} = 342.4$ Btu, $W_{net} = 204.6$ Btu, MEP = 89.45 psia, and $\eta = 59.76\%$. (See Fig. 3.39.)

3.9 RALLIS CYCLE

The *Rallis cycle* is defined by two isothermal processes at temperatures T_H and T_L separated by two regenerative processes that are part constant volume and part constant pressure in any given combination. The Stirling cycle is a special case of the Rallis cycle. Many other Rallis cycles can be defined, which have no identifying names.

A conceptual arrangement of a Rallis heat engine is shown in Fig. 3.40. The p–v and T–s diagrams for the cycle are shown in Fig. 3.41. T_H is the heat source temperature and T_L is the heat sink temperature. The cycle is composed of the following six processes:

1-2 Isobaric cooling
2-3 Isothermal compression at T_L
3-4 Constant-volume heat addition
4-5 Isobaric heating
5-6 Isothermal expansion at T_H
6-1 Constant-volume heat removal

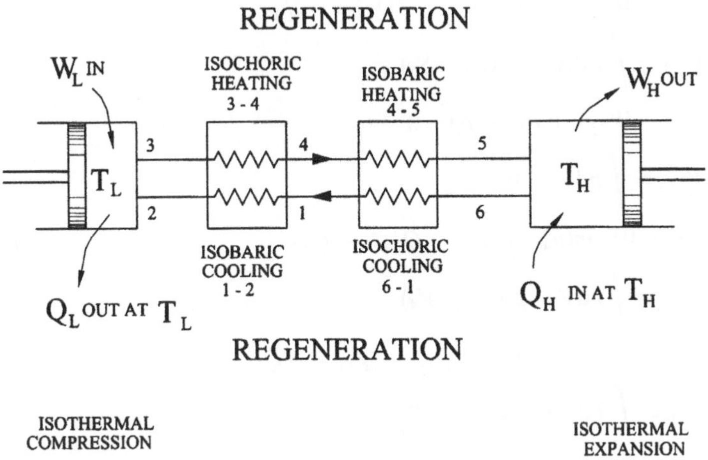

REGENERATION

Figure 3.40 Rallis cycle.

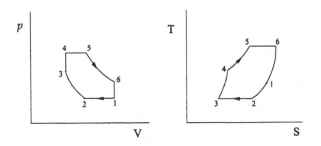

Figure 3.41 p–v and T–s diagram of Rallis cycle.

During the isothermal compression process 2-3, heat is rejected to maintain a constant temperature T_L. During the isothermal expansion process 5-6, heat is added to maintain a constant temperature T_H. There are heat interactions along the constant-volume heat-addition process 3-4 and the constant-volume heat-removal process 6-1; the quantities of heat in these two constant-volume processes are equal, but opposite in direction. There are also heat interactions along the constant-pressure heat addition process 4-5 and the constant-pressure heat-removal process 1-2. The quantities of heat in these two constant-pressure processes are equal, but opposite in direction.

Applying the first and second laws of thermodynamics of the closed system to each of the six processes of the cycle yields:

$$W_{12} = \int p\,dV = p_1(V_2 - V_1) \tag{3.71}$$

$$Q_{12} - W_{12} = m(u_2 - u_1) \qquad Q_{12} = -Q_{34} \tag{3.72}$$

$$Q_{23} = \int T\,dS = T_L(S_3 - S_2) \tag{3.73}$$

$$Q_{23} - W_{23} = m(u_3 - u_2) = 0 \tag{3.74}$$

$$W_{34} = \int p\,dV = 0 \tag{3.75}$$

$$Q_{34} - 0 = m(u_4 - u_3), \qquad Q_{34} = -Q_{12} \tag{3.76}$$

$$W_{45} = \int p\,dV = p_4(V_5 - V_4) \tag{3.77}$$

$$Q_{45} - W_{45} = m(u_5 - u_4), \qquad Q_{45} = -Q_{61}$$

$$Q_{56} = \int T\,dS = T_H(S_6 - S_5) \tag{3.78}$$

$$Q_{56} - W_{56} = m(u_6 - u_5) = 0 \tag{3.79}$$

$$W_{61} = \int p\,dV = 0 \tag{3.80}$$

and

$$Q_{61} - 0 = m(u_1 - u_6), \qquad Q_{61} = -Q_{45} \tag{3.81}$$

The net work (W_{net}), which is also equal to the net heat (Q_{net}), is

$$\begin{aligned}
W_{net} &= W_{12} + W_{23} + W_{45} + W_{56} = Q_{net} \\
&= Q_{12} + Q_{23} + Q_{34} + Q_{45} + Q_{56} + Q_{61} \\
&= Q_{23} + Q_{56} \tag{3.82}
\end{aligned}$$

The thermal efficiency of the cycle is

$$\eta = W_{net}/Q_{56} \tag{3.83}$$

Example 3.18

A Rallis heat engine is shown in Fig. 3.42a. The mass of helium contained in the cylinder is 0.1 lbm. The six processes are:

1-2 Isobaric cooling
2-3 Isothermal compression at T_L
3-4 Constant-volume heat addition
4-5 Isobaric heating
5-6 Isothermal expansion at T_H
6-1 Constant-volume heat removal

The following information is given: $p_2 = 15$ psia, $T_2 = 60°F$, $q_{34} = 60$ Btu/lbm, $q_{12} = -60$ Btu/lbm, $p_5 = 100$ psia, and $T_5 = 800°F$.

Determine the pressure and temperature of each state of the cycle, work and heat of each process, work input, work output, net work output, heat added, heat removed, MEP, and cycle efficiency. Draw the T–s diagram of the cycle.

To evaluate this example by CyclePad, we take the following steps:

1. Build:
   a. Take a compression device, two heaters, an expander, and two coolers from the closed-system inventory shop and

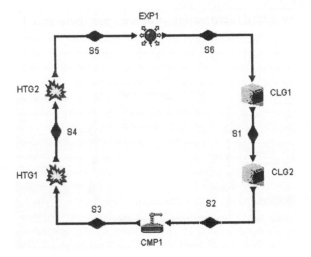

Figure 3.42a Rallis heat engine.

connect the six devices to form the cycle as shown in Fig. 3.42a.
  b. Switch to analysis mode.
  2. Analysis:
  a. Assume a process for each of the six devices: (1) compression device as isothermal, (2) one heater as isochoric and the other as isobaric, (3) expander as isothermal, and (4) one cooler as isochoric and the other as isobaric.
  b. Input the given information: working fluid is air, $m = 0.1$ lbm, $p_2 = 15$ psia, $T_2 = 60°$F, $q_{34} = 60$ Btu/lbm, $q_{12} = -60$ Btu/lbm, $p_5 = 100$ psia, and $T_5 = 800°$F as shown in Fig. 3.42b.
  3. Display the cycle properties' results. The cycle is a heat engine.

The results are: $p_1 = 15$ psia, $T_1 = 108.5°$F, $p_2 = 15$ psia, $T_2 = 60°$F, $p_3 = 45.11$ psia, $T_3 = 60°$F, $p_4 = 45.11$ psia, $T_4 = 108.5°$F, $p_5 = 100$ psia, $T_5 = 800°$F, $p_6 = 33.25$ psia, $T_6 = 800°$F; $q_{12} = -60$ Btu/lbm, $w_{12} = -24.07$ Btu/lbm, $q_{23} = -283.8$ Btu/lbm, $w_{23} = -283.8$ Btu/lbm, $q_{34} = 60$ Btu/lbm, $w_{34} = 2.41$ Btu/lbm, $q_{45} = 512$ Btu/lbm, $w_{45} = 0$ Btu/lbm, $q_{56} = 688$ Btu/lbm, $w_{56} = 688$ Btu/lbm, $q_{61} = -512$ Btu/lbm, $w_{61} = 0$ Btu/lbm; $Q_{in} = 126$ Btu, $Q_{out} = -85.58$ Btu, $Q_{net} = 40.42$ Btu, $W_{in} = -30.79$ Btu, $W_{out} = 71.21$ Btu, $W_{net} = 40.42$ Btu, MEP = 30.91 psia, and $\eta = 32.08\%$ as shown in Fig. 3.42c. The T–s diagram of the cycle is shown in Fig. 3.43.

Review Problems 3.9 Rallis Cycle

1. What is the Rallis cycle?
2. How many regenerating processes are there in a Rallis cycle?

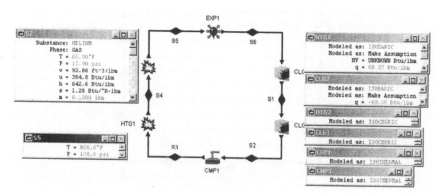

Figure 3.42b Rallis heat engine input.

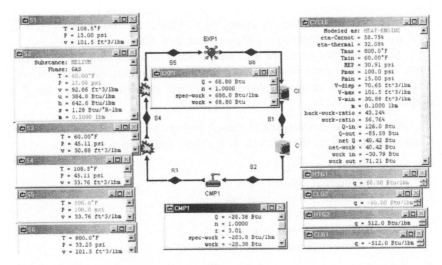

Figure 3.42c Rallis cycle output results.

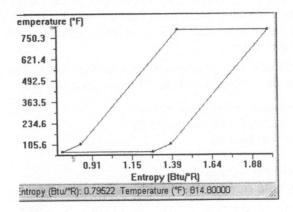

Figure 3.43 Rallis cycle *T–s* diagram.

3. A Rallis heat engine is shown in Fig. 3.42a. The mass of air contained in the cylinder is 0.1 lbm. The six processes are:

1-2 Isobaric cooling
2-3 Isothermal compression at T_L
3-4 Constant-volume heat addition
4-5 Isobaric heating
5-6 Isothermal expansion at T_H
6-1 Constant-volume heat removal

The following information is given: $p_2 = 15$ psia, $T_2 = 60°F$, $q_{34} = 60$ Btu/lbm, $q_{12} = -60$ Btu/lbm, $p_5 = 100$ psia, and $T_5 = 800°F$.

Determine the pressure and temperature of each state of the cycle, work and heat of each process, work input, work output, net work output, heat added, heat removed, MEP, and cycle efficiency. Draw the T–s diagram of the cycle.

4. A Rallis heat engine is shown in Fig. 3.42a. The mass of carbon dioxide contained in the cylinder is 0.1 lbm. The six processes are:

> 1-2 Isobaric cooling
> 2-3 Isothermal compression at T_L
> 3-4 Constant-volume heat addition
> 4-5 Isobaric heating
> 5-6 Isothermal expansion at T_H
> 6-1 Constant-volume heat removal

The following information is given: $p_2 = 15$ psia, $T_2 = 60°F$, $q_{34} = 60$ Btu/lbm, $q_{12} = -60$ Btu/lbm, $p_5 = 100$ psia, and $T_5 = 800°F$.

Determine the pressure and temperature of each state of the cycle, work and heat of each process, work input, work output, net work output, heat added, heat removed, MEP, and cycle efficiency. Draw the T–s diagram of the cycle.

5. A Rallis heat engine is shown in Fig. 3.42a. The mass of air contained in the cylinder is 0.1 lbm. The six processes are:

> 1-2 Constant-volume heat removal
> 2-3 Isothermal compression at T_L
> 3-4 Isobaric heating
> 4-5 Constant-volume heat addition
> 5-6 Isothermal expansion at T_H
> 6-1 Isobaric cooling

The following information is given: $p_2 = 15$ psia, $T_2 = 60°F$, $q_{34} = 60$ Btu/lbm, $q_{12} = -60$ Btu/lbm, $p_5 = 100$ psia, and $T_5 = 800°F$.

Determine the pressure and temperature of each state of the cycle, work and heat of each process, work input, work output, net work output, heat added, heat removed, MEP, and cycle efficiency. Draw the T–s diagram of the cycle.

3.10 DESIGN EXAMPLES

Although the Carnot cycle is useful in determining the ideal behavior of an ideal heat engine, it is not a practical cycle to use in the design of heat engines. There are different reasons for developing cycles other than the Carnot cycle. These reasons includes the characteristics of the energy source available, working fluid chosen for the cycle, material limitations in the hardware, and other practical consideration.

CyclePad is a powerful tool for cycle design and analysis. Due to its capabilities, the software allows users to view the cycle effects of varying design input parameters at once. The following examples illustrate the design of several closed-system gas power cycles.

Example 3.19

A six-process internal combustion engine as shown in Fig. 3.44a is proposed by a junior engineer. Air mass contained in the cylinder is 0.01 kg. The six processes are:

1-2 Isentropic compression
2-3 Isochoric heating
3-4 Isobaric heating
4-5 Isentropic expansion
5-6 Isochoric cooling
6-1 Isobaric cooling

The following information is given: $p_1 = 100\,\text{kPa}$, $T_1 = 20°\text{C}$, $V_1 = 10V_2$, $q_{23} = 600\,\text{kJ/kg}$, $q_{34} = 400\,\text{kJ/kg}$, and $T_5 = 400°\text{C}$.

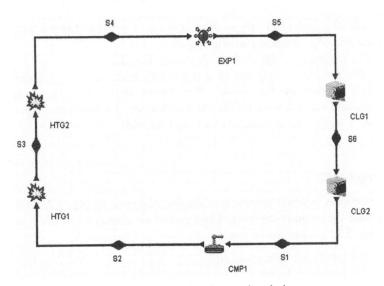

Figure 3.44a Six-process internal combustion engine design.

Determine the pressure and temperature of each state of the cycle, work and heat of each process, work input, work output, net work output, heat added, heat removed, MEP, and cycle efficiency.

To evaluate this design by CyclePad, we take the following steps:

1. Build:
   a. Take a compression device, two heaters, an expander, and two coolers from the closed-system inventory shop and connect the six devices to form the cycle as shown in Fig. 3.44a.
   b. Switch to analysis mode.

2. Analysis:
   a. Assume a process for each of the six devices: (1) compression device as isentropic, (2) one heater as isochoric and the other as isobaric, (3) expander as isentropic, and (4) one cooler as isochoric and the other as isobaric.
   b. Input the given information: working fluid is air, $m = 0.01$ kg, $p_1 = 100$ kPa, $T_1 = 20°C$, $V_1 = 10V_2$, $q_{23} = 600$ kJ/kg, $q_{34} = 400$ kJ/kg, and $T_5 = 400°C$.

3. Display the cycle properties' results. The cycle is a heat engine.

The results are: $p_1 = 100$ kPa, $T_1 = 20°C$, $p_2 = 2512$ kPa, $T_2 = 463.2°C$, $p_3 = 5368$ kPa, $T_3 = 1300°C$, $p_4 = 5368$ kPa, $T_4 = 1699°C$, $p_5 = 124.7$ kPa, $T_5 = 400°C$, $p_6 = 100$ kPa, $T_6 = 266.6°C$, $Q_{12} = 0$, $W_{12} = -3.18$ kJ, $Q_{23} = 6$ kJ, $W_{23} = 0$, $Q_{34} = 4$ kJ, $W_{34} = 1.14$ kJ, $Q_{45} = 0$, $W_{45} = 9.31$ kJ, $Q_{56} = -0.9558$ kJ, $W_{56} = 0$, $Q_{61} = -2.47$ kJ, $W_{61} = -0.7071$ kJ, $W_{add} = -3.88$ kJ, $W_{out} = 10.45$ kJ, $W_{net} = 6.57$ kJ, $Q_{in} = 10$ kJ, $Q_{out} = -3.43$ kJ, MEP $= 448.9$ kPa, and $\eta = 65.69\%$ as shown in Fig. 3.44b.

The *T–s* diagram of the cycle is shown in Fig. 3.44c.

The sensitivity diagram of η (cycle efficiency) versus r (compression ratio) is plotted in Fig. 3.44d. The figure shows that the larger the compression ratio, the better the cycle efficiency. To improve the proposed engine, a larger compression ratio could be used.

Example 3.20

A six-process internal combustion engine as shown in Fig. 3.45a has been proposed by a junior engineer. The mass of air contained in the cylinder is 0.01 kg. The six processes are:

1-2 Isentropic compression
2-3 Isochoric heating
3-4 Isobaric heating

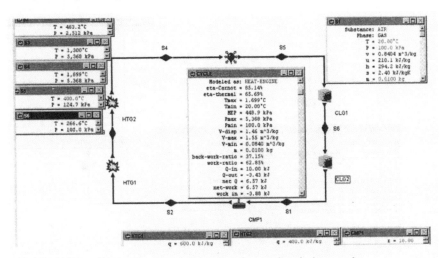

Figure 3.44b Six-process internal combustion engine design results.

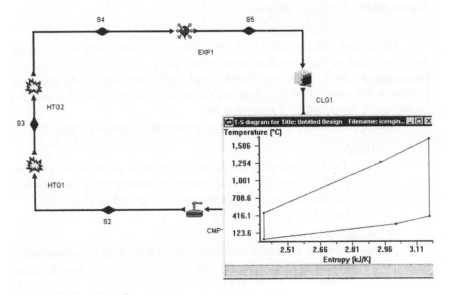

Figure 3.44c *T–s* diagram.

4-5 Isentropic expansion
5-6 Isobaric cooling
6-1 Isochoric cooling

The following information is given: $p_1 = 100\,\text{kPa}$, $T_1 = 20°\text{C}$, $V_1 = 10V_2$, $q_{23} = 600\,\text{kJ/kg}$, $q_{34} = 400\,\text{kJ/kg}$, and $T_5 = 400°\text{C}$.

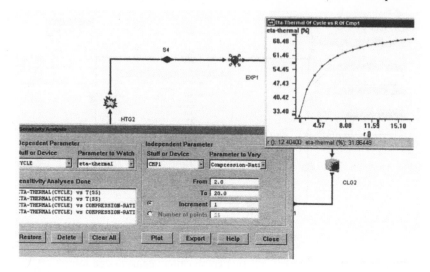

Figure 3.44d Sensitivity diagram.

Determine the pressure and temperature of each state of the cycle, work and heat of each process, work input, work output, net work output, heat added, heat removed, MEP, and cycle efficiency. Notice that processes 5-6 and 6-1 of the cycle are different from processes 5-6 and 6-1 of the cycle proposed in Example 3.19.

To evaluate this design by CyclePad, we take the following steps:

1. Build:
   a. Take a compression device, two heaters, an expander and two coolers from the closed-system inventory shop and connect the six devices to form the cycle as shown in Fig. 3.45a.
   b. Switch to analysis mode.
2. Analysis:
   a. Assume a process for each of the six devices: (1) compression device as isentropic, (2) one heater as isochoric and the other as isobaric, (3) expander as isentropic, and (4) one cooler as isochoric and the other as isobaric.
   b. Input the given information: working fluid is air, $m = 0.01$ kg, $p_1 = 100$ kPa, $T_1 = 20°C$, $V_1 = 10V_2$, $q_{23} = 600$ kJ/kg, $q_{34} = 400$ kJ/kg, and $T_5 = 400°C$.
3. Display the cycle properties' results. The cycle is a heat engine.

The results are: $p_1 = 100$ kPa, $T_1 = 20°C$, $p_2 = 2512$ kPa, $T_2 = 463.2°C$, $p_3 = 5368$ kPa, $T_3 = 1300°C$, $p_4 = 5368$ kPa, $T_4 = 1699°C$, $p_5 = 124.7$ kPa,

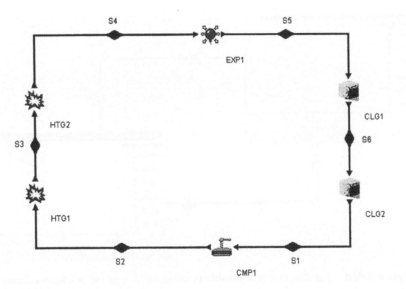

Figure 3.45a Six-process internal combustion engine.

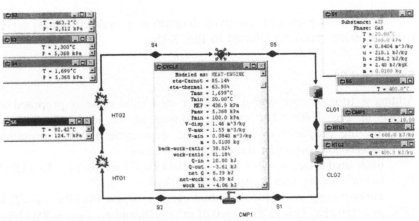

Figure 3.45b Six-process internal combustion engine design results.

$T_5 = 400°C$, $p_6 = 124.7$ kPa, $T_6 = 92.42°C$, $Q_{12} = 0$, $W_{12} = -3.18$ kJ, $Q_{23} = 6$ kJ, $W_{23} = 0$, $Q_{34} = 4$ kJ, $W_{34} = 1.14$ kJ, $Q_{45} = 0$, $W_{45} = 9.31$ kJ, $Q_{56} = -3.09$ kJ, $W_{56} = -0.8818$, $Q_{61} = -0.5191$ kJ, $W_{61} = 0$ kJ, $W_{add} = -4.06$ kJ, $W_{out} = 10.45$ kJ, $W_{net} = 6.39$ kJ, $Q_{in} = 10$ kJ, $Q_{out} = -3.61$ kJ, MEP $= 436.9$ kPa, and $\eta = 63.95\%$. (See Fig. 3.45b.)

It is observed that both the cycle efficiency and MEP of this cycle are less than those of the proposed cycle given by Example 3.19.

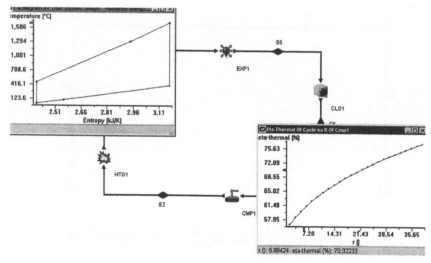

Figure 3.45c *T–s* diagram and sensitivity diagram of η (cycle efficiency) versus *r* (compression ratio).

The *T–s* diagram and sensitivity diagram of η (cycle efficiency) versus *r* (compression ratio) are plotted in Fig. 3.45c.

Example 3.21

Adding a turbocharger and a precooler to a dual cycle is proposed as shown in Fig. 3.46a. The cylinder volume of the engine is $0.01\,\mathrm{m}^3$. Evaluate the proposed cycle. The basic dual cycle and the proposed turbocharger and precooler dual cycle information is:

Basic dual cycle: $p_1 = p_2 = p_3 = p_8 = 101\,\mathrm{kPa}$, $T_1 = T_2 = T_3 = T_8 = 15°\mathrm{C}$, $V_3 = 0.01\,\mathrm{m}^3$, $r = 10$, and $q_{45} = q_{56} = 300\,\mathrm{kJ/kg}$.

Turbocharger and precooler dual cycle: $p_1 = p_8 = 101\,\mathrm{kPa}$, $T_1 = T_8 = 15°\mathrm{C}$, $p_2 = 150\,\mathrm{kPa}$, $T_3 = 15°\mathrm{C}$, $V_3 = 0.01\,\mathrm{m}^3$, $r = 10$, and $q_{45} = q_{56} = 300\,\mathrm{kJ/kg}$. To evaluate this proposed cycle by CyclePad, we take the following steps:

1. Build:
   a. Take two compression devices, two heaters, an expander, and two coolers from the closed-system inventory shop and connect the devices to form the cycle.
   b. Switch to analysis mode.
2. Analysis:
   a. Assume a process for each of the seven devices: (1) compression devices as isentropic, (2) one heater as isochoric

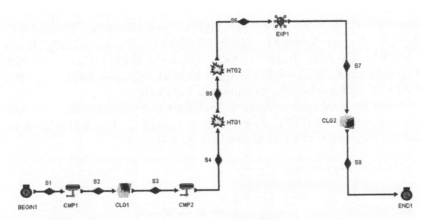

Figure 3.46a Turbocharger and precooler dual cycle.

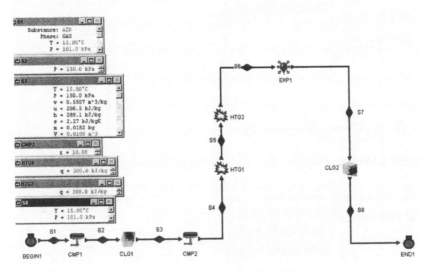

Figure 3.46b Turbocharger and precooler dual cycle input.

and the other as isobaric, (3) expander as isentropic, and (4) one cooler as isochoric and the other as isobaric.

b. Input the given information: working fluid is air, $p_1 = p_8 = 101$ kPa, $T_1 = T_8 = 15°C$, $p_2 = 150$ kPa, $T_3 = 15°C$, $V_3 = 0.01$ m^3, $r = 10$, and $q_{45} = q_{56} = 300$ kJ/kg. (See Fig. 3.46b.)

3. Display the cycle properties' results. The cycle is a heat engine.

The results are: $p_1 = 100$ kPa, $T_1 = 15°C$, $p_2 = 150$ kPa, $T_2 = 49.47°C$, $p_3 = 150$ kPa, $T_3 = 15°C$, $p_4 = 3768$ kPa, $T_4 = 450.7°C$, $p_5 = 5947$ kPa,

$T_5 = 869.2°C$, $p_6 = 5947$ kPa, $T_6 = 1168°C$, $p_7 = 188.4$ kPa, $T_7 = 264.4°C$, $Q_{12} = 0$, $W_{12} = -0.4486$ kJ, $Q_{23} = -0.6281$ kJ, $W_{23} = -0.1795$, $W_{34} = -5.67$ kJ, $Q_{45} = 0$, $W_{45} = 11.76$ kJ, $Q_{56} = -3.25$ kJ, $W_{add} = -6.30$ kJ, $W_{out} = 13.32$ kJ, $W_{net} = 7.02$ kJ, $Q_{in} = 10.89$ kJ, $Q_{out} = -3.87$ kJ, MEP = 506.8 kPa, and $\eta = 64.44\%$ as shown in Fig. 3.46c.

For the dual cycle without turbocharger and precooler, the input are: $p_1 = p_2 = p_3 = p_8 = 101$ kPa, $T_1 = T_2 = T_3 = T_8 = 15°C$, $V_3 = 0.01$ m^3, $r = 10$, and $q_{45} = q_{56} = 300$ kJ/kg as shown in Fig. 3.46d.

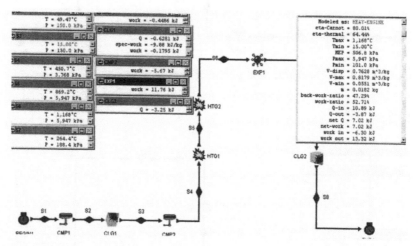

Figure 3.46c Turbocharger and precooler dual cycle result.

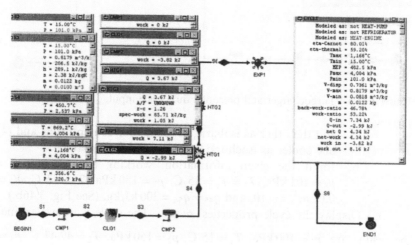

Figure 3.46d Dual cycle without result turbocharger and precooler.

The output results are: $p_1 = p_2 = p_3 = 101$ kPa, $T_1 = T_2 = T_3 = 15°C$, $p_4 = 2537$ kPa, $T_4 = 450.7°C$, $p_5 = 4004$ kPa, $T_5 = 869.2°C$, $p_6 = 4004$ kPa, $T_6 = 1168°C, p_7 = 220.7$ kPa, $T_7 = 355.6°C, Q_{12} = 0, W_{12} = -0$ kJ, $Q_{23} = -0$ kJ, $W_{23} = -0$ kJ, $W_{34} = -3.82$ kJ, $Q_{45} = 0$, $W_{45} = 7.11$ kJ, $Q_{56} = -2.99$ kJ, $W_{add} = -3.82$ kJ, $W_{out} = 8.16$ kJ, $W_{net} = 4.34$ kJ, $Q_{in} = 7.34$ kJ, $Q_{out} = -2.94$ kJ, MEP $= 482.5$ kPa, and $\eta = 59.20\%$ as shown in Fig. 3.46d.

It is observed that both the cycle efficiency and MEP of the proposed cycle are better than those of the dual cycle without turbocharger and precooler.

Review Problems 3.10   Design

1. A six-process internal combustion engine as shown in Fig. 3.45a has been proposed by a junior engineer. The mass of air contained in the cylinder is 0.01 kg. The six processes are:

    1-2 Isentropic compression
    2-3 Isochoric heating
    3-4 Isobaric heating
    4-5 Isentropic expansion
    5-6 Isobaric cooling
    6-1 Isochoric cooling

The following information is given as shown in Fig. 3.45b: $p_1 = 100$ kPa, $T_1 = 15°C$, $V_1 = 8V_2$, $q_{23} = 300$ kJ/kg, $q_{34} = 400$ kJ/kg, and $T_5 = 400°C$.

Determine the pressure and temperature of each state of the cycle, work and heat of each process, work input, work output, net work output, heat added, heat removed, MEP, and cycle efficiency.

2. Adding a turbocharger and a precooler to a dual cycle is proposed as shown in Fig. 3.46a. The cylinder volume of the engine is 0.01 m^3. Evaluate the proposed cycle. The basic Dual cycle and the proposed turbocharger and precooler dual cycle information is:

    Basic dual cycle: $p_1 = p_2 = p_3 = p_8 = 101$ kPa, $T_1 = T_2 = T_3 = T_8 = 15°C$, $V_3 = 0.01$ m^3, $r = 8$, and $q_{45} = q_{56} = 300$ kJ/kg.
    Turbocharger and precooler dual cycle: $p_1 = p_8 = 101$ kPa, $T_1 = T_8 = 15°C$, $p_2 = 120$ kPa, $T_3 = 15°C$, $V_3 = 0.01$ m^3, $r = 10$, and $q_{45} = q_{56} = 300$ kJ/kg.

3.11   SUMMARY

Heat engines that use gases as the working fluid in a closed-system model were discussed in this chapter. The Otto, Diesel, Miller, and dual cycles are internal combustion engines. The Stirling cycle is an external combustion engine.

The Otto cycle is a spark-ignition reciprocating engine consisting of an isentropic compression process, a constant-volume combustion process, an isentropic expansion process, and a constant-volume cooling process. The thermal efficiency of the Otto cycle depends on its compression ratio. The compression ratio is defined as $r = V_{max}/V_{min}$. The Otto cycle efficiency is limited by the compression ratio because of the engine knock problem.

The Diesel cycle is a compression–ignition reciprocating engine consisting of an isentropic compression process, a constant-pressure combustion process, an isentropic expansion process, and a constant-volume cooling process. The thermal efficiency of the Otto cycle depends on its compression ratio and cut-off ratio. The compression ratio is defined as $r = V_{max}/V_{min}$. The cut-off ratio is defined as $r_{cutoff} = V_{combustion\ off}/V_{min}$.

The dual cycle involves two heat-addition processes, one at constant volume and one at constant pressure. It behaves more like an actual cycle than either the Otto or Diesel cycles.

The Lenoir cycle was the first commercially successful internal combustion engine.

The Stirling cycle and Wicks cycle are attempts to achieve the efficiency of the Carnot cycle.

The Miller cycle uses variable valve timing for compression-ratio control to improve the performance of internal combustion engines.

4

Gas Open-System Cycles

4.1 BRAYTON OR JOULE CYCLE

The *ideal Brayton gas turbine cycle* (sometimes called Joule cycle) is named after an American engineer, George Brayton, who proposed the cycle in the 1870s. The gas turbine cycle consists of four processes: an isentropic compression process 1-2, a constant-pressure combustion process 2-3, an isentropic expansion process 3-4, and a constant-pressure cooling process 4-1. The p–v and T–s diagrams for an ideal Brayton cycle are illustrated in Fig. 4.1.

The gas turbine cycle may be either closed or open. The more common cycle is the open one, in which atmospheric air is continuously drawn into the compressor, heat is added to the air by the combustion of fuel in the combustion chamber, and the working fluid expands through the turbine and exhausts to the atmosphere. A schematic diagram of an *open Brayton cycle*, which is assumed to operate steadily as an open system, is shown in Fig. 4.2.

In the closed cycle, the heat is added to the fluid in a heat exchanger from an external heat source, such as a nuclear reactor, and the fluid is cooled in another heat exchanger after it leaves the turbine and before it enters the compressor. A schematic diagram of a *closed Brayton cycle* is shown in Fig. 4.3.

Applying the first and second laws of thermodynamics for an open system to each of the four processes of the Brayton cycle yields:

$$Q_{12} = 0 \tag{4.1}$$

$$W_{12} = m(h_1 - h_2) \tag{4.2}$$

$$W_{23} = 0 \tag{4.3}$$

$$Q_{23} - 0 = m(h_3 - h_2) \tag{4.4}$$

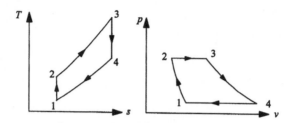

Figure 4.1 Brayton cycle *p–v* and *T–s* diagrams.

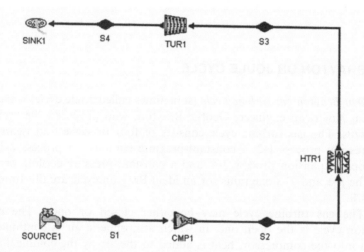

Figure 4.2 Open Brayton cycle.

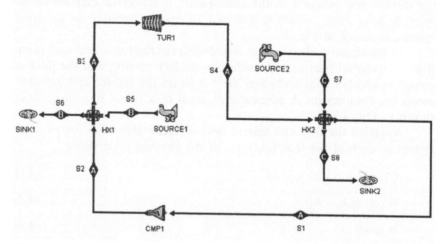

Figure 4.3 Closed Brayton cycle.

$$Q_{34} = 0 \tag{4.5}$$

$$W_{34} = m(h_3 - h_4) \tag{4.6}$$

$$W_{41} = 0 \tag{4.7}$$

and

$$Q_{41} - 0 = m(h_1 - h_4) \tag{4.8}$$

The net work (W_{net}), which is also equal to net heat (Q_{net}), is

$$W_{net} = W_{12} + W_{34} = Q_{net} = Q_{23} + Q_{41} \tag{4.9}$$

The thermal efficiency of the cycle is

$$\eta = W_{net}/Q_{23} = Q_{net}/Q_{23} = 1 - Q_{41}/Q_{23} = 1 - (h_4 - h_1)/(h_3 - h_2) \tag{4.10}$$

This expression for thermal efficiency of an ideal Brayton cycle can be simplified if air is assumed to be the working fluid with constant specific heats. Equation (4.3) is reduced to

$$\eta = 1 - (T_4 - T_1)/(T_3 - T_2) = 1 - (r_p)^{(k-1)/k} \tag{4.11}$$

where r_p is the *pressure compression ratio* for the compressor defined by the equation:

$$r_p = p_2/p_1 \tag{4.12}$$

The highest temperature in the cycle occurs at the end of the combustion process (state 3), and it is limited by the maximum temperature that the turbine blade can withstand. The maximum temperature does have an effect on the optimal performance of the gas turbine cycle.

In the gas turbine cycle, the ratio of the compressor work to the turbine work is called *back-work ratio*. The back-work ratio is very high, usually more than 40%.

Example 4.1

An engine operates on the open Brayton cycle and has a compression ratio of 8. Air, at a mass flow rate of 0.1 kg/sec, enters the engine at 27°C and 100 kPa. The amount of heat addition is 1 MJ/kg. Determine the efficiency, compressor power input, turbine power output, back-work ratio, and net power of the cycle. Show the cycle on a *T–s* diagram. Plot the sensitivity diagram of cycle efficiency versus compression ratio.

To solve this problem by CyclePad, we take the following steps:

1. Build:
   a. Take a compressor, a combustion chamber (heater), a turbine, and a cooler from the open-system inventory shop and connect the four devices to form the open Brayton cycle.
   b. Switch to analysis mode.
2. Analysis:
   a. Assume a process for each of the four devices: (1) compressor device as isentropic, (2) combustion chamber and cooler as isobaric, and (3) turbine as isentropic.
   b. Input the given information: (1) working fluid is air, (2) inlet pressure and temperature of the compression device are 100 kPa and 27°C, (3) compression ratio of the compressor is 8, (4) heat addition is 1 MJ/kg in the combustion chamber, (5) mass flow rate of air is 0.1 kg/sec, and (6) exit pressure of the turbine is 100 kPa.
3. Display:
   a. Display the *T–s* diagram and cycle properties, results. The cycle is a heat engine. The answers are $\eta = 44.80\%$, compressor power $= -24.44$ kW, turbine power $= 69.23$ kW, back-work ratio $= 35.30\%$, and net power $= 69.23$ kW.
   b. Display the sensitivity diagram of cycle efficiency versus compression ratio. (see Figs. 4.4 and 4.5.)

COMMENT: The cycle efficiency increases as compressor ratio increases.

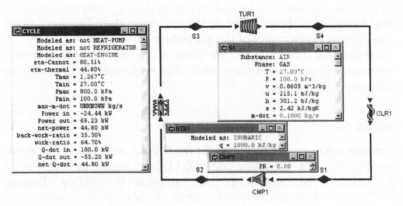

Figure 4.4 Open Brayton cycle.

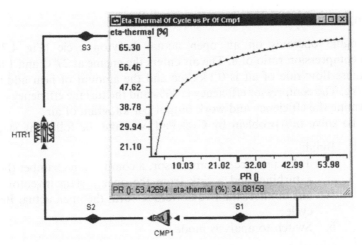

Figure 4.5 Open Brayton cycle sensitivity analysis.

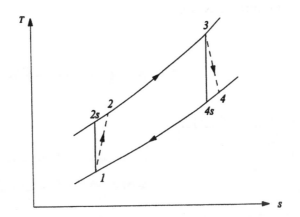

Figure 4.6 Actual Brayton cycle T–s diagram.

For *actual Brayton cycles*, many irreversibilities in various components are present. The T–s diagram of an actual Brayton cycle is shown in Fig. 4.6. The major irreversibilities occur within the turbine and compressor. To account for these irreversibility effects, turbine efficiency and compressor efficiency must be used in computing the actual work produced or consumed. The effect of irreversibilities on the thermal efficiency of a Brayton cycle is illustrated in the following example.

Example 4.2

An engine operates on an open actual Brayton cycle (Fig 4.7) and has a compression ratio of 8. The air enters the engine at 27°C and 100 kPa. The mass flow rate of air is 0.1 kg/sec and the amount of heat addition is 1 MJ/kg. The compressor efficiency is 86% and the turbine efficiency is 89%. Determine the efficiency and work output per kilogram of air.

To solve this problem by CyclePad, we take the following steps:

1. Build:
   a. Take a source, a compressor, a combustion chamber (heater), a turbine, and a sink from the open-system inventory shop and connect the five devices to form the open actual Brayton cycle.
   b. Switch to analysis mode.
2. Analysis:
   a. Assume a process for three of the devices: (1) compressor device as adiabatic, (2) combustion chamber as isobaric, and (3) turbine as adiabatic.

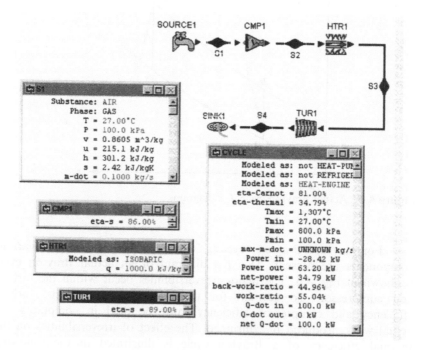

Figure 4.7 Open actual Brayton cycle.

b. Input the given information: (1) working fluid is air, (2) inlet pressure and temperature of the compression device are 100 kPa and 27°C, (3) compression ratio of the compressor is 8, (4) heat addition is 1000 kJ/kg in the combustion chamber, (5) compressor efficiency is 86% and the turbine efficiency is 89%, (6) mass flow rate of air is 0.1 kg/sec, and (7) exit pressure of the turbine is 100 kPa.

3. Display the cycle properties' results. The cycle is a heat engine.

The answers are $\eta = 34.79\%$ and net power $= 34.79$ kW.

Example 4.3

An engine operates on the closed Brayton cycle (Fig 4.8) and has a compression ratio of 8. Helium enters the engine at 47°C and 200 kPa. The mass flow rate of helium is 1.2 kg/sec and the amount of heat addition is 1 MJ/kg. Determine the highest temperature of the cycle, the turbine power produced, the compressor power required, the back-work ratio, the rate of heat added, and the cycle efficiency.

To solve this problem by CyclePad, we take the following steps:

1. Build:
    a. Take a compressor, a combustion chamber (heater), a turbine, and a cooler from the open-system inventory shop and connect the four devices to form the closed Brayton cycle.
    b. Switch to analysis mode.

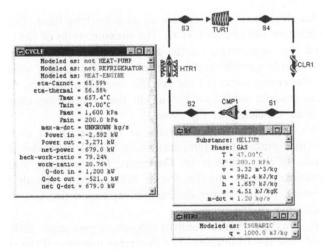

Figure 4.8   Closed Brayton cycle.

2. Analysis:
   a. Assume a process for each of the four devices: (1) compressor device as adiabatic and isentropic, (2) combustion chamber as isobaric, (3) turbine as adiabatic and isentropic, and (4) cooler as isobaric.
   b. Input the given information: (1) working fluid is helium, (2) inlet pressure and temperature of the compression device are 200 kPa and 47°C, (3) compressor exit pressure is 1600 kPa, (4) mass flow rate of helium is 1.2 kg/sec, and (5) heat addition is 1 MJ/kg in the combustion chamber.
3. Display the cycle properties' results. The cycle is a heat engine.

The answers are $T_{max} = 657.4°C$, turbine power $= 3271$ kW, compressor power $= -2592$ kW, back-work ratio $= 79.24\%$, Qdot$_{in} = 1200$ kW, and $\eta = 56.58\%$.

Review Problems 4.1   Brayton Cycle

1. The maximum and minimum temperatures and pressures of a 40 MW turbine shaft output power ideal air Brayton power plant are 1200 K (T_3), 0.38 MPa (P_3), 290 K (T_1), and 0.095 MPa (P_1), respectively. Determine the temperature at the exit of the compressor (T_2), the temperature at the exit of the turbine (T_4), the compressor work, the turbine work, the heat added, the mass rate of flow of air, the back-work ratio (the ratio of compressor work to the turbine work), and the thermal efficiency of the cycle.

2. An ideal Brayton cycle uses air as a working fluid. The air enters the compressor at 100 kPa and 37°C. The pressure ratio of the compressor is 12:1, and the temperature of the air as it leaves the turbine is 497°C. Assuming variable specific heats, determine (a) the specific work required to operate the compressor, (b) the specific work produced by the turbine, (c) the heat transfer added to the air in the combustion chamber, and (d) the thermal efficiency of the cycle.

3. An ideal Brayton engine receives 1 lbm/sec of air at 15 psia and 70°F. The maximum cycle temperature is 1300°F and the compressor pressure ratio of the engine is 10. Determine (a) the power added during the compression process, (b) the rate of heat added to the air during the heating process, (c) the power during the expansion process, and (d) the thermal efficiency of the cycle.

4. A Brayton engine receives air at 15 psia and 70°F. The air mass rate of flow is 4.08 lbm/sec. The discharge pressure of the compressor is 78 psia. The maximum cycle temperature is 1740°F and the air turbine

discharge temperature is 1161°F. Determine (a) the power added during the compression process, (b) the rate of heat added to the air during the heating process, (c) the power produced during the expansion process, (d) the turbine efficiency, and (e) the thermal efficiency of the cycle.

5. A Brayton engine receives air at 103 kPa and 27°C. The maximum cycle temperature is 1050°C and the compressor discharge pressure is 1120 kPa. The compressor efficiency is 85% and the turbine efficiency is 82%. Determine (a) the work added during the compression process, (b) the heat added to the air during the heating process, (c) the work done during the expansion process, and (d) the thermal efficiency of the cycle.

6. An ideal Brayton engine receives air at 15 psia and 80°F. The maximum cycle temperature is 1800°F and the compressor discharge pressure is 225 psia. The mass rate flow of air is 135 lbm/sec. Determine (a) the power added during the compression process, (b) the rate of heat added to the air during the heating process, (c) the power produced during the expansion process, (d) the net power produced by the engine, (e) the back-work ratio, and (f) the thermal efficiency of the cycle.

7. An ideal Brayton engine receives air at 14.7 psia and 60°F. The maximum cycle temperature is 1750°F and the compressor discharge pressure is 147 psia. The mass rate flow of air is 15 lbm/sec. Determine (a) the power added during the compression process, (b) the heat added to the air during the heating process, (c) the power produced during the expansion process, (d) the net power produced by the engine, (e) the back-work ratio, and (f) the thermal efficiency of the cycle.

8. A Brayton engine receives air at 14 psia and 70°F. The air mass rate of flow is 16 lbm/sec. The maximum cycle temperature is 1500°F. The compressor efficiency is 79% and the turbine efficiency is 90%. Determine (a) the work done during the compression process, (b) the heat added to the air during the heating process, (c) the work done during the expansion process, (d) the turbine efficiency, and (e) the thermal efficiency of the cycle.

9. A Brayton engine receives air at 100 kPa and 25°C. The maximum cycle temperature is 1082°C and the compressor discharge pressure is 1300 kPa. The compressor efficiency is 87%. The air temperature at the turbine exit is 536°C. Determine (a) the work done during the compression process, (b) the heat added to the air during the heating process, (c) the work done during the expansion process, (d) the turbine efficiency, and (e) the thermal efficiency of the cycle.

10. An ideal air Brayton cycle has air entering the compressor at a temperature of 310 K and a pressure of 100 kPa. The pressure ratio across the compressor is 12, and the temperature of the air leaving the

turbine is 780 K. Assuming variable specific heats, determine (a) the compressor work required, (b) the turbine work produced, (c) the heat added during the combustion, and (d) the thermal efficiency of the cycle.

11. Air enters the compressor of an ideal Brayton cycle at 100 kPa and 300 K with a volumetric flow rate of $5 \, m^3/sec$. The compressor pressure ratio is 10. The turbine inlet temperature is 1400 K. Determine (a) the thermal efficiency of the cycle, (b) the back-work ratio, and (c) the net power developed.

12. Air enters the compressor of a simple gas turbine at 14.7 psia and 520 R, and has a volumetric flow rate of $1000 \, ft^3/sec$. (The compressor discharge pressure is 260 psia,) and the turbine inlet temperature is 2000 F. The turbine efficiency is 87% and the compressor efficiency is 83%. Determine (a) the thermal efficiency of the cycle, (b) the back-work ratio, and (c) the net power developed.

4.2 SPLIT-SHAFT GAS TURBINE CYCLE

Gas turbines can be arranged either in single-shaft or split-shaft types. The single-shaft arrangement requires the turbine to provide power to drive both the compressor and the load. This means that the compressor is influenced by the load. The compressor efficiency is a function of the speed. When the load is increased, the compressor speed is slowed down, which is not desirable. It is very desirable to make the gas turbine a reliable shaft-driven propulsion system; therefore, the compressor speed must be held constant. Hence, *split-shaft gas turbines* are developed. In this arrangement, there are two turbines each with its own independent shaft. The sole function of the first turbine is to drive the compressor at a steady speed without being influenced by the load. The net power of the gas turbine is produced by the second turbine as shown by Fig. 4.9.

Example 4.4

An engine operates on the split-shaft actual open Brayton cycle (Fig. 4.10) and has a compression ratio of 8. The air enters the engine at 27°C and 100 kPa. The mass flow rate of air is 0.1 kg/sec, and the amount of heat addition is 1000 kJ/kg. The compressor efficiency is 86% and the efficiency is 89% for both turbines. Determine the highest temperature of the cycle, the turbine power produced, the compressor power required, the back-work ratio, the rate of heat added, the pressure and temperature between the two turbines, and the cycle efficiency.

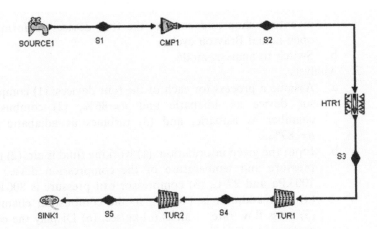

Figure 4.9 Split-shaft gas turbine.

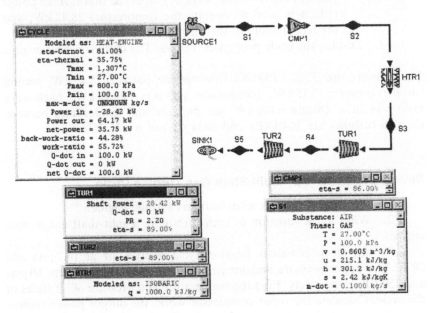

Figure 4.10 Split-shaft open Brayton cycle.

To solve this problem by CyclePad, we take the following steps:

1. Build:
   a. Take a source, a compressor, a combustion chamber (heater), two turbines, and a sink from the open-system

inventory shop and connect the four devices to form the open actual Brayton cycle.

    b. Switch to analysis mode.

2. Analysis:

    a. Assume a process for each of the four devices: (1) compressor device as adiabatic and $\eta = 86\%$, (2) combustion chamber as isobaric, and (3) turbines as adiabatic and $\eta = 89\%$.

    b. Input the given information: (1) working fluid is air, (2) inlet pressure and temperature of the compression device are 100 kPa and 27°C, (3) compressor exit pressure is 800 kPa, (4) heat addition is 1 MJ/kg in the combustion chamber, (5) mass flow rate of air is 0.1 kg/sec, (6) Display the compressor and find the power required to run the compressor (the finding is -28.42 kW), (7) input the first turbine power (which is used to operate the compressor) 28.42 kW, and (8) exit pressure of the turbine is 100 kPa.

3. Display the cycle properties' results. The cycle is a heat engine.

The answers are $T_{max} = 1307°C$, first turbine power $= 28.42$ kW, second turbine power $= 35.75$ kW, compressor power $= -28.42$ kW, back-work ratio $= 44.28\%$, Qdotin $= 100$ kW, the pressure and temperature between the two turbines are 364.1 kPa and 1024°C, and $\eta = 35.75\%$.

Review Problems 4.2 Split-Shaft Gas Turbine Cycle

    1. Why do we need a split-shaft gas turbine engine?

    2. What is the function of each turbine in a split-shaft gas turbine engine?

    3. An ideal split-shaft Brayton cycle receives air at 14.7 psia and 70°F. The upper pressure and temperature limits of the cycle are 60 psia and 1500°F, respectively. Find the temperature and pressure of all states of the cycle. Calculate the input compressor work, the output power turbine work, heat supplied in the combustion chamber, and the thermal efficiency of the cycle, based on variable specific heats.

    4. An actual split-shaft Brayton cycle receives air at 14.7 psia and 70°F. The upper pressure and temperature limit of the cycle are 60 psia and 1500°F, respectively. The turbine efficiency is 85% for both turbines. The compressor efficiency is 80%. Find the temperature and pressure of all states of the cycle. The mass flow rate of air is 1 lbm/sec. Calculate the input compressor power, the output power turbine power, rate of heat

supplied in the combustion chamber, and the thermal efficiency of the cycle, based on variable specific heats.

ANSWERS: $p_2 = 60$ psia, $T_2 = 397.5°$F, $p_3 = 60$ psia, $T_3 = 1500°$F, $p_5 = 14.7$ psia, $T_5 = 940.6°$F, $W\text{dot}_c = -111.0$ hp, $W\text{dot}_{t1} = 111.0$ hp, $W\text{dot}_{t2} = 78.7$ hp, $Q\text{dot}_s = 264.3$ Btu/sec, $\eta = 21.04\%$.

5. The following are operating characteristics of a split-shaft gas turbine:

Atmospheric conditions—$p = 14.7$ psia, $T = 60°$F

Compressor—inlet pressure $= 14.5$ psia, inlet temperature $= 60°$F, $m\text{dot} = 1$ lbm/sec, $\eta = 0.8$, exit pressure $= 101.5$ psia

Combustion chamber—exit temperature $= 1800°$F, exit pressure $= 99$ psia

Turbine #1—$\eta = 0.85$

Power turbine #2—$\eta = 0.85$, exit pressure $= 14.9$ psia

Find the temperatures of all states of the cycle, power required by the compressor, power produced by turbine #1, power produced by power turbine #2, rate of heat transfer supplied in the combustion chamber, cycle efficiency, and indicated horse power (IHP).

ANSWERS: $T_{\text{compressor exit}} = 543.1°$F, $T_{\text{turbine #1 exit}} = 1317°$F, $T_{\text{powerturbine #2 exit}} = 981.3°$F, $\text{Power}_{\text{compressor}} = -163.8$ hp, $\text{Power}_{\text{turbine #1}} = 163.8$ hp, $\text{Power}_{\text{power turbine #2}} = -113.8$ hp, $Q\text{dot}_{\text{combustion chamber}} = 301.2$ Btu/s, $\eta = 26.71\%$, IHP $= 113.8$ hp.

6. A split-shaft Brayton gas turbine operates with the following information:

Inlet compressor temperature $= 70°$F $= T_1$

Inlet compressor pressure $= 14.5$ psia $= p_1$

Inlet combustion chamber pressure $= 145$ psia $= p_2$

Inlet ggt (gas generator turbine) temperature $= 2000°$F $= T_3$

Exit pt (power turbine) pressure $= 14.8$ psia $= p_{5a}$

Air mass rate of flow $= 2$ lbm/sec

Power turbine efficiency $= 85\%$

Gas generator turbine efficiency $= 100\%$

Compressor efficiency $= 100\%$

Draw the T–s diagram of the cycle.

Find the temperature at the exit of the compressor (T_2), the temperature at the exit of the gas generator turbine (T_4), the temperature at the exit of the power turbine (T_{5a}), compressor work required per unit mass, gas generator turbine work produced per unit mass, power turbine work produced per unit mass, heat supplied per unit mass, net work produced per unit mass, total power produced by the cycle, and cycle efficiency.

ANSWERS: $T_2 = 563.0°F$, $T_4 = 1516°F$, $T_{4a} = 1507°F$, $w_c = -118.1$ Btu/lbm, $q_s = 344.0$ Btu/lbm, $w_{t1} = 118.1$ Btu/lbm, $w_{t2} = 143.7$ Btu/lbm, $w_{net} = 143.7$ Btu/lbm, $P_{net} = 287.4$ Btu/s $= 740.9$ hp, $\eta = 41.72\%$.

7. A split-shaft Brayton gas turbine operates with the following information:

Inlet compressor temperature $= 60°F = T_1$
Inlet compressor pressure $= 14.5$ psia $= p_1$
Inlet combustion chamber pressure $= 145$ psia $= p_2$
Inlet ggt temperature $= 2000°F = T_3$
Exit pt pressure $= 14.8$ psia $= p_{5a}$
Air mass rate of flow $= 2$ lbm/sec
Power turbine efficiency $= 85\%$
Gas generator turbine efficiency $= 100\%$
Compressor efficiency $= 100\%$
Draw the T–s diagram of the cycle.

Find the temperature at the exit of the compressor (T_2), the temperature at the exit of the gas generator turbine (T_4), the temperature at the exit of the power turbine (T_{5a}), compressor work required per unit mass, gas generator turbine work produced per unit mass, power turbine work produced per unit mass, heat supplied per unit mass, net work produced per unit mass, total power produced by the cycle, and cycle efficiency.

ANSWERS: $T_2 = 543.7°F$, $T_4 = 1516°F$, $T_{5a} = 926°F$, $w_c = -115.9$ Btu/lbm, $q_s = 349.0$ Btu/lbm, $w_{t1} = 115.9$ Btu/lbm, $w_{t2} = 141.5$ Btu/lbm, $w_{net} = 141.5$ Btu/lbm, $P_{net} = 283.0$ Btu/sec $= 400.4$ hp, $\eta = 40.54\%$.

4.3 IMPROVEMENTS TO BRAYTON CYCLE

The thermal efficiency or net work of the Brayton cycle can be improved by several modifications to the basic cycle. These modifications include increasing the turbine inlet temperature, reheating, intercooling, regeneration, etc.

Increasing the turbine inlet temperature increases the thermal efficiency of the Brayton cycle. It is limited by the metallurgical material problem in the turbine blade.

Increasing the average temperature during the heat-addition process with a reheater without increasing the compressor pressure ratio increases the net work of the Brayton cycle. A multistage turbine is used. Gas is reheated between stages.

Using an intercooler without increasing the compressor pressure ratio increases the net work of the Brayton cycle. A multi-stage compressor is used, and gas is cooled between stages.

Increasing the average temperature during the heat-addition process can also be done by regenerating the gas. A multistage turbine is used. The exhaust gas is used to preheat the air before it is heated in the combustion chamber. In this way, the amount of heat added at the low temperature is reduced. So the average temperature during the heat-addition process is increased.

4.4 REHEAT AND INTERCOOL BRAYTON CYCLE

Methods for improving the gas turbine cycle performance are available.

Two ways to improve the cycle net work are to reduce the compressor work and to increase the turbine work. The intercool may be accomplished by compressing in stages with an intercooler, cooling the air as it passes from one stage to another. Similarly, the reheat may be accomplished by the expansion in stages with a reheater. Since there is more than sufficient air for combustion, some more can be injected. The reheated products of combustion return to the turbine. The products of combustion re-entering the turbine are usually at the same temperature as those entering the turbine. The schematic diagram of a reheat open Brayton cycle is illustrated in Fig. 4.11. The schematic diagram and *T–s* diagram for a *reheat and inter-cool gas turbine* cycle are illustrated in Figs. 4.12 and 4.13, respectively.

Notice that reheat and inter-cool increase the net work of the gas turbine cycle, but not necessarily the efficiency, unless a regenerator is also added.

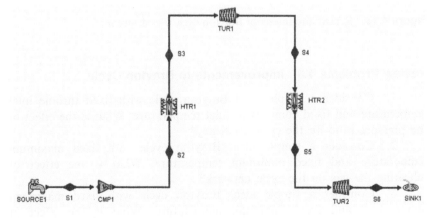

Figure 4.11 Reheat Brayton cycle.

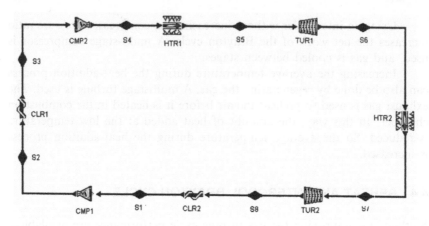

Figure 4.12   Reheat and intercool Brayton cycle.

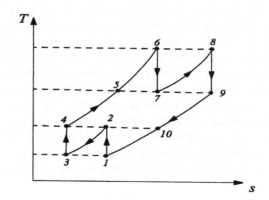

Figure 4.13   Reheat and intercool Brayton cycle *T–s* diagram.

Review Problems 4.3   Improvements to Brayton Cycle

1.   Consider a simple ideal Brayton cycle with fixed turbine inlet temperature and fixed compressor inlet temperature. What is the effect of the pressure ratio on the cycle efficiency?

2.   Consider a simple ideal Brayton cycle with fixed maximum temperature and fixed minimum temperature. What is the effect of reheating the gas on the cycle network?

3.   Consider a simple ideal Brayton cycle with fixed maximum temperature and fixed minimum temperature. What is the effect of intercooling the gas on the cycle network?

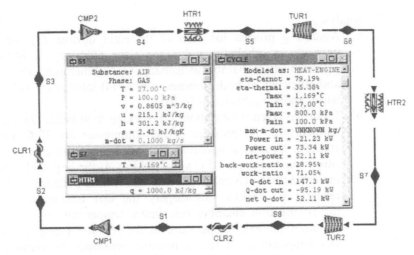

Figure 4.14   Ideal reheat and intercool Brayton cycle.

Example 4.5

An engine operates on an ideal reheat and intercooling Brayton cycle (Fig. 4.14). The low-pressure compressor has a compression ratio of 2, and the high-pressure compressor has a compression ratio of 4. The air enters the engine at 27°C and 100 kPa. The air is cooled to 27°C at the inlet of the high-pressure compressor. The heat added to the combustion chamber is 1 MJ/kg and air is heated to the maximum temperature of the cycle. The mass flow rate of air is 0.1 kg/sec. Air expands to 200 kPa through the first turbine. Air is heated again by the reheater to the maximum temperature of the cycle and then expanded through the second turbine to 100 kPa. Determine the power required for the first compressor, the power required for the second compressor, the maximum temperature of the cycle (at the exit of the combustion chamber), the power produced by the first turbine, the rate of heat added in the reheater, the power produced by the second turbine, the net power produced, back-work ratio, and the efficiency of the cycle. Show the cycle on a *T–s* diagram.

To solve this problem by CyclePad, we take the following steps:

1. Build:
   a. Take two compressors, two coolers (one is the intercooler), two combustion chambers (one is the reheater), and two turbines from the open-system inventory shop and connect the devices to form the reheat and intercooling Brayton cycle.
   b. Switch to analysis mode.

2. Analysis:
   a. Assume a process for each of the eight devices: (1) compressors as adiabatic and compressor efficiency is 86%, (2) combustion chamber as isobaric, (3) turbines as adiabatic, and efficiency of both turbines is 89%, (4) intercooler as isobaric, and (5) reheater as isobaric.
   b. Input the given information: (1) working fluid is air, (2) inlet temperature and pressure of the compressor are 27°C and 100 kPa, (3) inlet temperature and pressure of the high-pressure compressor are 27°C and 200 kPa, (4) the heat added to the combustion chamber is 1 MJ/kg, (5) inlet pressure of the reheater is 200 kPa, (6) display the exit temperature of the combustion chamber (maximum temperature of cycle), (7) input the exit temperature of the reheater (same as the exit temperature of the combustion chamber as found in point 6), (8) mass flow rate of air is 0.1 kg/sec, and (9) exit pressure of the low-pressure turbine is 100 kPa.
3. Display the T–s diagram and cycle properties' results. The cycle is a heat engine. The answers are power required for the first compressor $= -6.60$ kW, power required for the second compressor $= -14.64$ kW, maximum temperature of the cycle $= 1169°C$, power produced by the first turbine $= 47.34$ kW, rate of heat added in the reheater $= 47.29$ kW, power produced by the second turbine $= 26.00$ kW, net power produced $= 52.11$ kW, back-work ratio $= 28.95\%$, and efficiency of the cycle $\eta = 35.38\%$.

COMMENTS: Comparing with Example 4.1, we see that (1) the efficiency of the reheat and intercooler cycle does not increase, and (2) the net power of the reheat and intercooler cycle does increase.

Example 4.6

An engine operates on an actual reheat open Brayton cycle (Fig. 4.15a)). The air enters the compressor at 60°F and 14.7 psia, and exits at 120 psia. The maximum cycle temperature (at the exit of the combustion chamber) allowed due to material limitation is 2000°F. The exit pressure of the high-pressure turbine is 50 psia. The air is reheated to 2000°F, and the mass flow rate of air is 1 lbm/sec. The exit pressure of the low-pressure turbine is 14.7 psia. The compressor efficiency is 86% and the turbine efficiency is 89%. Determine the power required for the compressor, the power produced by the first turbine, the rate of heat added in the reheater, the power produced by the second turbine, the net power produced, back-work ratio, and the

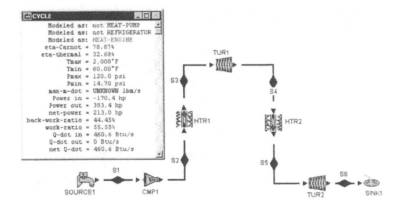

Figure 4.15a Actual reheat open Brayton cycle.

efficiency of the cycle. Show the cycle on a *T–s* diagram. Plot the sensitivity diagram of cycle efficiency versus inlet pressure of the low-pressure turbine. To solve this problem by CyclePad, we take the following steps:

1. Build:
   a. Take a source, a compressor, a combustion chamber (heater), a reheater, two turbines, and a sink from the open-system inventory shop and connect the devices to form the Brayton cycle.
   b. Switch to analysis mode.
2. Analysis:
   a. Assume a process for each of the five devices: (1) compressor as adiabatic with 85% efficiency, (2) combustion chamber as isobaric, (3) turbines as adiabatic with 89% efficiency, and (4) reheater as isobaric.
   b. Input the given information: (1) working fluid is air, (2) inlet pressure and temperature of the compression device are 15 psia and 60°F, (3) exit pressure of the compressor is 120 psia, (4) exit temperature of the combustion chamber is 2000°F, (5) exit pressure of the high-pressure turbine is 50 psia, (6) inlet temperature of the low-pressure turbine is 2000°F, (7) mass flow rate of air is 1 lbm/sec, and (8) exit pressure of the turbine is 15 psia.
3. Display results:
   a. Display the cycle properties' results. The cycle is a heat engine. The answers are power required for the compressor = −170.4 hp, power produced by the first turbine = 164.3 hp, rate of heat added to the combustion chamber = 334.5 Btu/ sec, rate of heat added in the reheater = 116.1 Btu/sec, power

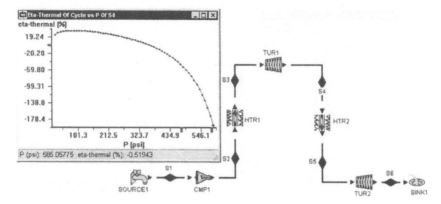

Figure 4.15b   Actual reheat open Brayton cycle sensitivity analysis.

produced by the second turbine = 219.1 hp, net power produced = 213.0 hp, back-work ratio = 44.45%, and efficiency of the cycle $\eta = 32.68\%$.

b.   Display the sensitivity diagram of cycle efficiency versus inlet pressure of the low-pressure turbine. It is seen that the optimal pressure is about 77 psia, which gives a maximum cycle efficiency of about 32%. (see Fig. 4.15b.)

Review Problems 4.4   Reheat and Intercool Brayton Cycle

1.   Does reheat or intercooling increase the net work of the Brayton cycle?

2.   Does reheat or intercooling increase the efficiency of the Brayton cycle?

3.   An ideal Brayton cycle is modified to incorporate multistage compression with intercooling, and multistage expansion with reheating. As a result of these modifications, does the efficiency increase?

4.   An ideal Brayton cycle is modified to incorporate multistage compression with intercooling, and multistage expansion with reheating. As a result of these modifications, does the net work increase?

5.   Atmospheric air is at 100 kPa and 300 K. 1 kg/sec of air at 800 kPa and 1200 K enters an actual two-stage (high-pressure stage and low-pressure stage) adiabatic turbine at steady state and exits to 100 kPa. Air is reheated to 1200 K and enters the low-pressure turbine. Air pressure at 300.1 kPa is measured at the exit of the high-pressure stage turbine. The high-pressure stage turbine, low-pressure stage turbine, and the compressor all are known to have an isentropic efficiency of 85%. Determine (a) the actual temperature

at the exit of the high-pressure stage turbine, (b) the actual temperature at the exit of the low-pressure stage turbine, (c) the cycle efficiency, (d) net power produced by the cycle, (e) power required by the compressor, (f) power produced by the high-pressure turbine, (g) power produced by the low-pressure turbine, (h) the back-work ratio, (i) the rate of heat added in the combustion chamber, and (j) the rate of heat added in the reheater.

6.   An ideal Brayton cycle with a one-stage compressor and a two-stage turbine has an overall pressure ratio of 10. Atmospheric air is at 101 kPa and 292 K. The high-pressure stage turbine, low-pressure stage turbine, and the compressor all are known to have an isentropic efficiency of 85%. Air enters each stage of the turbine at 1350 K. The mass rate of air flow is 0.56 kg/sec. The air pressure at the inlet of the second stage turbine is 307 kPa. Determine (a) the power required by the compressor, (b) power produced by the turbine, (c) rate of heat added, (d) back-work ratio, (e) net power produced, and (f) the cycle efficiency.

7.   An ideal Brayton cycle with a one-stage compressor and a two-stage turbine has an overall pressure ratio of 10. Atmospheric air is at 101 kPa and 292 K. The high-pressure stage turbine, low-pressure stage turbine, and the compressor all are known to have an isentropic efficiency of 85%. Air enters each stage of the turbine at 1350 K. The mass rate of air flow is 0.56 kg/sec. Use air pressure at the inlet of the second-stage turbine as a parameter (from 400 to 800 kPa). Determine (a) the maximum cycle efficiency by showing the sensitivity diagram, (b) the optimum air pressure at the inlet of the second-stage turbine at the maximum cycle efficiency condition, (c) power required by the compressor, (d) power produced by the turbine, (e) rate of heat added, (f) back-work ratio, and (g) net power produced.

8.   An ideal Brayton cycle with a one-stage compressor and a two-stage turbine has an overall pressure ratio of 10. Atmospheric air is at 14.6 psia and 65 F. The high-pressure stage turbine, low-pressure stage turbine, and the compressor all are known to have an isentropic efficiency of 85%. Air enters each stage of the turbine at 2000 F. The mass rate of air flow is 1.5 lbm/sec. The air pressure at the inlet of the second-stage turbine is 50 psia. Determine (a) the power required by the compressor, (b) power produced by the turbine, (c) rate of heat added, (d) back-work ratio, (e) net power produced, and (f) the cycle efficiency.

4.5 REGENERATIVE BRAYTON CYCLE

The thermal efficiency of the gas turbine cycle is not high. It is observed that the exhaust temperature of the turbine is quite high, indicating that

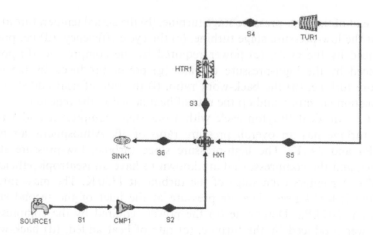

Figure 4.16   Regenerative Brayton cycle.

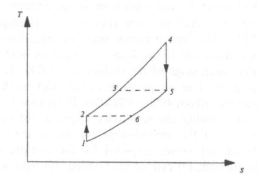

Figure 4.17   Regenerative Brayton cycle *T–s* diagram.

a large portion of available energy is wasted. One way to put this high-temperature available energy to use is to preheat the combustion air before it enters the combustion chamber. This increases the overall efficiency by decreasing the fuel required, hence heat added. The schematic diagram and *T–s* diagram for an ideal *regenerative gas turbine cycle* are illustrated in Fig. 4.16 and 4.17, respectively.

Example 4.7

An engine operates on the actual regenerative Brayton cycle (Fig. 4.18a). Air enters the engine at 60°F and 14.7 psia. The maximum cycle temperature and the maximum pressure are 2000°F and 120 psia. The compressor efficiency is 85% and the turbine efficiency is 89%. The mass flow rate of air is 1 lbm/sec. Determine the power required for the

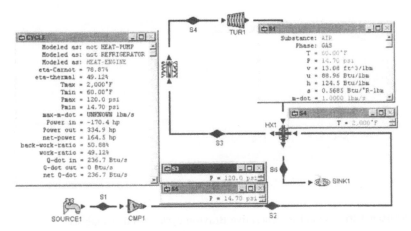

Figure 4.18a Actual regenerative Brayton cycle.

compressor, the power produced by the turbine, the rate of heat added in the combustion chamber, the net power produced, the back-work ratio, and the efficiency of the cycle. Show the cycle on a *T–s* diagram. Plot the sensitivity diagram of cycle efficiency versus inlet pressure of the low-pressure turbine. Plot the sensitivity diagram of cycle efficiency versus exit temperature of the turbine exhaust stream in the heat exchanger.

To solve this problem by CyclePad, we take the following steps:

1. Build:
   a. Take a source, a compressor, a combustion chamber (heater), a heat exchanger, a turbine, and a sink from the open-system inventory shop and connect the devices to form the actual regenerative Brayton cycle.
   b. Switch to analysis mode.
2. Analysis:
   a. Assume a process for each of the four devices: (1) compressor as adiabatic with efficiency of 85%, (2) combustion chamber as isobaric, (3) turbine as adiabatic with efficiency of 89%, and (4) heat exchanger as isobaric on both hot and cold sides.
   b. Input the given information: (1) working fluid is air, (2) inlet pressure and temperature of the compression device are 14.7 psia and 60°F, (3) inlet pressure and temperature of the turbine are 120 psia and 2000°F, (4) mass flow rate of air is 1 lbm/sec, (5) exit pressure of the turbine is 14.7 psia, (6) display the exit temperature of the compressor (it is 562.5°F), and (7) input the exit temperature of the exhaust turbine gas

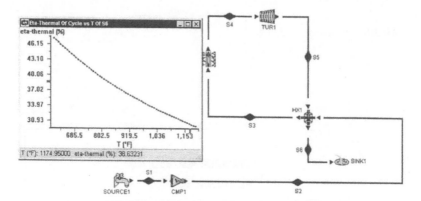

Figure 4.18b Actual Regenerative Brayton cycle sensitive analysis.

to be the same as the compressor exit temperature, 562.5°F, by assuming perfect regeneration.

3. Display the *T–s* diagram and cycle properties results. The cycle is a heat engine. The answers are power required for the compressor = −170.4 hp, the power produced by the turbine = 334.9 hp, the rate of heat added in the combustion chamber = 236.7 Btu/sec, the net power produced = 164.5 hp, back-work ratio = 50.88%, and the efficiency of the cycle $\eta = 49.12\%$. (See Fig. 4.18b.)

COMMENT: The cycle efficiency is increased because the rate of heat added is decreased by making use of the waste energy. The higher the exit temperature of the heat exchanger in the exhaust turbine gas stream, the less waste energy is used, and therefore the less the efficiency as shown by the sensitivity diagram.

Review Problems 4.5   Regenerative Brayton Cycle

1. What is regeneration?
2. Why is regeneration added to a Brayton cycle?
3. How does regeneration affect the efficiency of the Brayton cycle?
4. An ideal Brayton cycle is modified to incorporate multistage compression with intercooling, multistage expansion with reheating, and regeneration. As a result of these modifications, does the efficiency increase?
5. An ideal Brayton cycle with regeneration has a pressure ratio of 10. Air enters the compressor at 101 kPa and 290 K. Air leaves the regenerator and enters the combustion chamber at 580 K. Air enters the turbine at 1220 K. The air mass flow rate is 0.46 kg/sec. Determine (a) the power required by the compressor, (b) power produced by the turbine, (c) rate of

heat added, (d) back-work ratio, (e) net power produced, and (f) the cycle efficiency.

6. An ideal Brayton cycle with regeneration has a pressure ratio of 10. Air enters the compressor at 14.7 psia and 29°F. Air enters the combustion chamber at 610°F. Air enters the turbine at 1520°F. The turbine exit air pressure is 15.0 psia. The air mass flow rate is 0.41 lbm/sec. The turbine efficiency is 85%, and the compressor efficiency is 82%. Determine (a) the exit air temperature of the compressor, (b) the inlet air temperature of the combustion chamber, (c) the power required by the compressor, (d) power produced by the turbine, (e) rate of heat added, (f) back-work ratio, (g) net power produced, and (h) the cycle efficiency.

7. A regenerative gas-turbine power plant is to be designed according to the following specifications:

> Maximum cycle temperature = 1200 K.
> Turbine efficiency = 85%.
> Compressor efficiency = 82%.
> Inlet air temperature of the combustion chamber is 20 K higher than the exit air temperature of the compressor.
> Inlet air pressure to the compressor = 100 kPa.
> Turbine exit air pressure = 110 kPa.
> Exit air pressure of the compressor = 800 kPa.
> Mass flow rate of air = 1.4 kg/sec.

Determine (a) the exit air temperature of the compressor, (b) the inlet air temperature of the combustion chamber, (c) the power required by the compressor, (d) power produced by the turbine, (e) rate of heat added, (f) back-work ratio, (g) net power produced, and (h) the cycle efficiency.

4.6 BLEED AIR BRAYTON CYCLE

Real gas turbine engines send a portion of the air supplied by the compressor through alternative flow paths to provide cooling to the outside of the engine, to protect nearby components, to be remixed with the combustion products, and to drive ancillary equipments such as air conditioning and ventilation. The rate of the bleed air can be controlled. The schematic diagram for the *bleed air Brayton cycle* is illustrated in Fig. 4.19. It can be seen from the diagram that the engine's entire flow passes through the compressor, but only a fraction of the flow passes through the combustion chamber and the turbine.

Bleed air is a necessary feature of practical gas turbine engines. For example, when one enters a commercial aircraft, the cabin temperature is

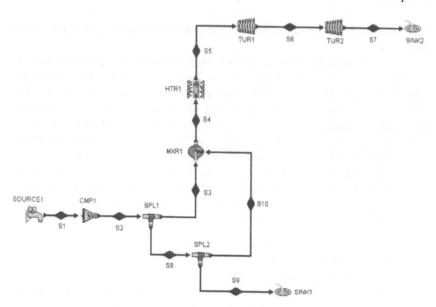

Figure 4.19 Bleed air Brayton cycle.

normally pleasant and the small vents above the seats are providing plenty of fresh air. The ventilation reduces dramatically as the aircraft prepares to take off, because the pilot has temporarily reduced the bleed air so that more air will flow through the combustion chamber to give the pilot more power available to get airborne. Bleed air does not improve the efficiency of the Brayton cycle.

Example 4.8

An open, split-shaft, bleed air Brayton cycle as illustrated in Fig. 4.19 has the following information:

Compressor efficiency = 80%, turbine efficiency = 80%, compressor inlet pressure = 14.5 psia, compressor inlet temperature = 60°F, compressor exit pressure = 145 psia, combustion chamber exit temperature = 1800°F, power turbine exit pressure = 14.9 psia, air mass flow rate through compressor = 1 lbm/sec, and air mass flow rate through combustion chamber = 0.9 lbm/sec.

Find the temperature of all states, power required by the compressor, power produced by turbine #1, which drives the compressor, power produced by the power turbine, rate of heat supplied by the combustion chamber, and cycle efficiency.

To solve this problem by CyclePad, we take the following steps:

1. Build the cycle as shown by Fig. 4.19.
2. Assume compressor as adiabatic and 80% efficient, combustion chamber and mixing chamber as isobaric, and turbines as adiabatic and 80% efficient.
3. Input working fluid is air, compressor inlet pressure = 14.5 psia, compressor inlet temperature = 60°F, compressor exit pressure = 145 psia, combustion chamber exit temperature = 1800°F, power turbine exit pressure = 14.9 psia, air mass flow rate through compressor = 1 lbm/sec, air mass flow rate through state 9 = 0 lbm/sec, and air mass flow rate through combustion chamber = 0.9 lbm/sec.
4. Display compressor power (− 205 hp); input turbine #1 power = 205 hp.
5. Display the cycle properties results and state results.

The answers are $T_2 = 664.6°F$, $T_3 = T_4 = T_8 = T_9 = T_{10} = 664.6°F$, $T_6 = 1128°F$, $T_7 = 913.1°F$, compressor power = − 205 hp, turbine #1 power = 205 hp, turbine #2 power = 65.66 hp, rate of heat added in the combustion chamber = 244.9 Btu/sec, net cycle power = 65.66 hp, and cycle efficiency = 18.95%. (See Fig. 4.20.)

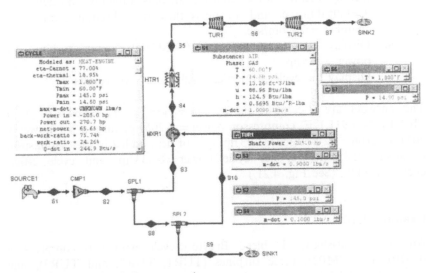

Figure 4.20 Bleed air Brayton cycle.

Example 4.9

An open, split-shaft, air-bleed Brayton cycle as illustrated in Fig. 4.19 has the following information:

Compressor efficiency = 80%, turbine efficiency = 80%, compressor inlet pressure = 14.5 psia, compressor inlet temperature = 60°F, compressor exit pressure = 145 psia, combustion chamber exit temperature = 1800°F, power turbine exit pressure = 14.9 psia, air mass flow rate through compressor = 1 lbm/sec, and air mass flow rate through combustion chamber = 0.9 lbm/sec.

Find the temperature of all states, power required by the compressor, power produced by turbine #1, which drives the compressor, power produced by the power turbine, rate of heat supplied by the combustion chamber, and cycle efficiency.

To solve this problem by CyclePad, we take the following steps:

1. Build the cycle as shown by Fig. 9.19.
2. Assume compressor as adiabatic and 80% efficient, combustion chamber and mixing chamber as isobaric, and turbines as adiabatic and 80% efficient.
3. Input working fluid is air, compressor inlet pressure = 14.5 psia, compressor inlet temperature = 60°F, compressor exit pressure = 145 psia, combustion chamber exit temperature = 1800°F, power turbine exit pressure = 14.9 psia, air mass flow rate through compressor = 1 lbm/sec, air mass flow rate through state 9 = 0.05 lbm/sec, and air mass flow rate through combustion chamber = 0.9 lbm/sec.
4. Display compressor power (-205 hp); input turbine #1 power = 205 hp.
5. Display the cycle properties results and state results.

The answers are $T_2 = 664.6°F$, $T_3 = T_4 = T_8 = T_9 = T_{10} = 664.6°F$, $T_6 = 1164°F$, $T_7 = 911.1°F$, compressor power = -205 hp, turbine #1 power = 205 hp, turbine #2 power = 81.33 hp, rate of heat added in the combustion chamber = 258.5 Btu/sec, net cycle power = 81.33 hp, and cycle efficiency = 22.24%. (See Fig. 4.21.)

Example 4.10

An open, split-shaft, air bleed Brayton cycle with two compressors (CMP1 and CMP2), three turbines (TUR1, TUR2, and TUR3), one intercooler (CLR1), one combustion chamber (HTR1), one reheater

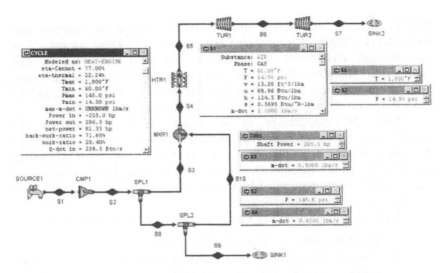

Figure 4.21 Bleed air Brayton cycle.

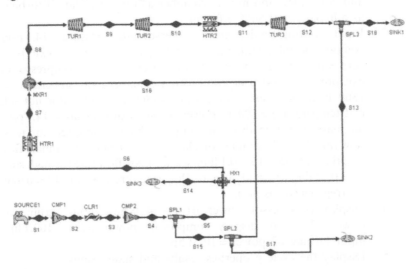

Figure 4.22a Bleed air Brayton cycle.

(HTR2), and one regenerator (HX1), illustrated in Fig. 4.22a, has the following information:

Compressor efficiency $= 80\%$, turbine efficiency $= 80\%$, compressor inlet pressure $= 14.5\,\text{psia}$, compressor inlet temperature $= 60°\text{F}$, compressor #1 exit pressure $= 40\,\text{psia}$, intercooler exit temperature $= 100°\text{F}$, compressor #2 exit pressure $= 140\,\text{psia}$, combustion chamber exit temperature $=$

1800°F, reheater exit temperature = 1800°F, regenerator hot-side exit temperature = 500°F, power turbine exit pressure = 14.9 psia, air mass flow rate through compressor #2 = 1 lbm/sec, air mass flow rate through combustion chamber = 0.9 lbm/sec, air mass flow rate through state 15 = 0.1 lbm/sec, air mass flow rate through state 17 = 0.01 lbm/sec, and air mass flow rate through state 13 = 0 lbm/sec (regenerator is off).

Find the temperature of all states, power required by the compressors, power produced by turbine #1, which drives compressor #1, power produced by turbine #2, which drives compressor #2, power produced by power turbine #3, rate of heat supplied by the combustion chamber, rate of heat supplied by the reheater, rate of heat removed from the intercooler, and cycle efficiency.

To solve this problem by CyclePad, we take the following steps:

1. Build the cycle as shown by Fig. 4.27a.
2. Assume all compressors as adiabatic and 80% efficient, combustion chamber, reheater, intercooler, heat exchanger (both hot and cold side), and mixing chamber as isobaric, and all turbines as adiabatic and 80% efficient.
3. Input working fluid is air, compressor inlet pressure = 14.5 psia, compressor inlet temperature = 60°F, compressor #1 exit pressure = 40 psia, intercooler exit temperature = 100°F, compressor #2 exit pressure = 140 psia, combustion chamber exit temperature = 1800°F, reheater exit temperature = 1800°F, regenerator hot-side exit temperature = 500°F, power turbine exit pressure = 14.9 psia, air mass flow rate through compressor #2 = 1 lbm/sec, air mass flow rate through combustion chamber = 0.9 lbm/sec, air mass flow rate through state 15 = 0.1 lbm/sec, air mass flow rate through state 17 = 0.01 lbm/sec, and air mass flow rate through state 13 = 0 lbm/ sec (regenerator is off).
4. Display compressor power #1 (-74.08 hp), input turbine #1 power = 74.08 hp; display compressor power #2 (-102.1 hp), input turbine #2 power = 102.1 hp.
5. Display the cycle properties results and state results.

The answers are $T_2 = 278.5°F$, $T_3 = 100°F$, $T_4 = 401.1°F = T_5 = T_{15} = T_{17} = 401.1°F$, $T_7 = T_{11} = 1800°F$, $T_8 = 1673°F$, $T_9 = 1452°F$, $T_{10} = 1148°F$, $T_{12} = 1359°F$, compressor #1 power $= -74.08$ hp, turbine #1 power = 74.08 hp, compressor #2 power $= -102.1$ hp, turbine #2 power = 102.1 hp, turbine #3 power = 148.1 hp, rate of heat added in the combustion chamber = 301.8 Btu/sec, rate of heat added in the reheater = 154.7 Btu/sec, rate of heat removed in the intercooler = -42.78 Btu/sec, net cycle power = 148.1 hp, and cycle efficiency = 22.94%. (See Fig. 4.22b.)

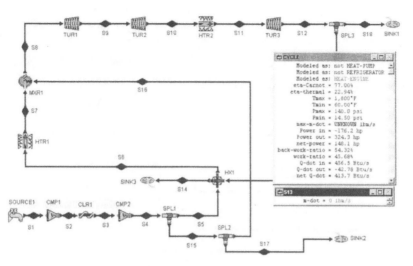

Figure 4.22b Bleed air Brayton cycle (regenerator off).

Example 4.11

An open, split-shaft, air-bleed Brayton cycle with two compressors (CMP1 and CMP2), three turbines (TUR1, TUR2, and TUR3), one intercooler (CLR1), one combustion chamber (HTR1), one reheater (HTR2), and one regenerator (HX1) is illustrated in Fig. 4.23 has the following information:

Compressor efficiency = 80%, turbine efficiency = 80%, compressor inlet pressure = 14.5 psia, compressor inlet temperature = 60°F, compressor #1 exit pressure = 40 psia, intercooler exit temperature = 100°F, compressor #2 exit pressure = 140 psia, combustion chamber exit temperature = 1800°F, reheater exit temperature = 1800°F, regenerator hot-side exit temperature = 500°F, power turbine exit pressure = 14.9 psia, air mass flow rate through compressor #2 = 1 lbm/sec, air mass flow rate through combustion chamber = 0.9 lbm/sec, air mass flow rate through state 15 = 0.1 lbm/sec, air mass flow rate through state 17 = 0.01 lbm/sec, and air mass flow rate through state 13 = 0.99 lbm/sec (regenerator is on).

Find the temperature of all states, power required by the compressors, power produced by turbine #1, which drives compressor #1, power produced by turbine #2, which drives compressor #2, power produced by the power turbine #3, rate of heat supplied by the combustion chamber, rate of heat supplied by the reheater, rate of heat removed from the intercooler, and cycle efficiency.

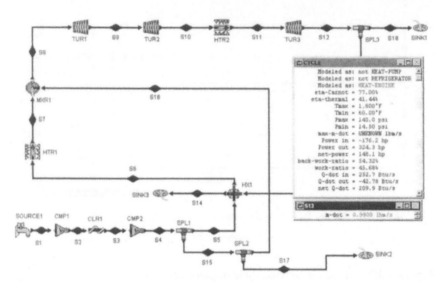

Figure 4.23 Bleed air Brayton cycle (regenerator on).

To solve this problem by CyclePad, we take the following steps:

1. Build the cycle as shown by Fig. 4.22.
2. Assume all compressors as adiabatic and 80% efficient, combustion chamber, reheater, intercooler, heat exchanger (both hot and cold side), and mixing chamber as isobaric, and all turbines as adiabatic and 80% efficient.
3. Input working fluid is air, compressor inlet pressure = 14.5 psia, compressor inlet temperature = 60°F, compressor #1 exit pressure = 40 psia, intercooler exit temperature = 100°F, compressor #2 exit pressure = 140 psia, combustion chamber exit temperature = 1800°F, reheater exit temperature = 1800°F, regenerator hot-side exit temperature = 500°F, power turbine exit pressure = 14.9 psia, air mass flow rate through compressor #2 = 1 lbm/sec, air mass flow rate through combustion chamber = 0.9 lbm/sec, air mass flow rate through state 15 = 0.1 lbm/sec, air mass flow rate through state 17 = 0.01 lbm/sec, and air mass flow rate through state 13 = 0.99 lbm/sec (regenerator is on).
4. Display compressor power #1 (−74.08 hp), input turbine #1 power = 74.08 hp; display compressor power #2 (−102.1 hp) and input turbine #2 power = 102.1 hp.
5. Display the cycle properties results and state results.

The answers are $T_2 = 278.5°F$, $T_3 = 100°F$, $T_4 = 401.1°F = T_5 = T_{15} = T_{17} = 401.1°F$, $T_7 = T_{11} = 1800°F$, $T_8 = 1673°F$, $T_9 = 1452°F$, $T_{10} = 1148°F$, $T_{12} = 1359°F$, compressor #1 power $= -74.08$ hp, turbine #1 power $= 74.08$ hp, compressor #2 power $= -102.1$ hp, turbine #2 power $= 102.1$ hp, turbine #3 power $= 148.1$ hp, rate of heat added in the combustion chamber $= 301.8$ Btu/sec, rate of heat added in the reheater $= 154.7$ Btu/sec, rate of heat removed in the intercooler $= -42.78$ Btu/sec, net cycle power $= 148.1$ hp, and cycle efficiency $= 22.94\%$.

Review Problems 4.6   Bleed Air Brayton Cycle

1. Why do we need bleed air in practical gas turbine cycles?
2. How do you control the power output of a real gas turbine cycle using bleed air?
3. Does the efficiency of a real gas turbine cycle increase using bleed air?
4. Bleed air is used in a split-shaft gas turbine cycle. The following information is provided:

> Compressor—air inlet temperature and pressure are 60°F and 14.5 psia, efficiency $= 85\%$, exit pressure is 145 psia, and $mdot = 1$ lbm/sec.
> Combustion chamber—$q_{supply} = 400$ Btu/lbm, and $mdot = 1$ lbm/sec (no bleed air).
> Turbine #1—efficiency $= 92.3\%$.
> Turbine #2—efficiency $= 92.3\%$, and air exit pressure $= 15$ psia.

Determine temperature of all states of the cycle, compressor power required, power produced by turbine #1, power produced by turbine #2, and cycle efficiency.

ANSWERS: $T_2 = 629°F$, $T_3 = 629°F$, $T_4 = 1881°F$, $T_5 = 1881°F$, $T_6 = 1312°F$, $T_7 = 837.6°F$, $Wdot_{compressor} = -192.9$ hp, $Wdot_{turbine \#1} = 192.9$ hp, $Wdot_{turbine \#2} = 160.8$ hp, $\eta = 37.88\%$.

5. Bleed air is used in a split-shaft gas turbine cycle. The following information is provided:

> Compressor—air inlet temperature and pressure are 60°F and 14.5 psia, efficiency $= 85\%$, exit pressure is 145 psia, and $mdot = 1$ lbm/sec.
> Combustion chamber—$q_{supply} = 400$ Btu/lbm, and $mdot = 0.95$ lbm/sec.

Turbine #1—efficiency $= 92.3\%$.

Turbine #2—efficiency $= 92.3\%$, and air exit pressure $= 15\,\text{psia}$.

Determine the temperature of all states of the cycle, compressor power required, power produced by turbine #1, power produced by turbine #2, and cycle efficiency.

ANSWERS: $T_2 = 629°\text{F}$, $T_3 = 629°\text{F}$, $T_4 = 1881°\text{F}$, $T_5 = 1881°\text{F}$, $T_6 = 1282°\text{F}$, $T_7 = 837.6°\text{F}$, $W\text{dot}_{\text{compressor}} = -192.9\,\text{hp}$, $W\text{dot}_{\text{turbine \#1}} = 192.9\,\text{hp}$, $W\text{dot}_{\text{turbine \#2}} = 143.2\,\text{hp}$, $\eta = 35.49\%$.

6. Bleed air is used in a split-shaft gas turbine cycle. The following information is provided:

> Compressor—air inlet temperature and pressure are 60°F and 14.5 psia, efficiency $= 85\%$, exit pressure is 145 psia, and $m\text{dot} = 1\,\text{lbm/sec}$.
>
> Combustion chamber—$q_{\text{supply}} = 400\,\text{Btu/lbm}$, and $m\text{dot} = 0.9\,\text{lbm/sec}$.
>
> Turbine #1—efficiency $= 92.3\%$.
>
> Turbine #2—efficiency $= 92.3\%$, and air exit pressure $= 15\,\text{psia}$.

Determine temperature of all states of the cycle, compressor power required, power produced by turbine #1, power produced by turbine #2, and cycle efficiency.

ANSWERS: $T_2 = 629°\text{F}$, $T_3 = 629°\text{F}$, $T_4 = 1881°\text{F}$, $T_5 = 1881°\text{F}$, $T_6 = 1249°\text{F}$, $T_7 = 837.6°\text{F}$, $W\text{dot}_{\text{compressor}} = -192.9\,\text{hp}$, $W\text{dot}_{\text{turbine \#1}} = 192.9\,\text{hp}$, $W\text{dot}_{\text{turbine \#2}} = 125.4\,\text{hp}$, $\eta = 32.83\%$.

7. Bleed air is used in a split-shaft gas turbine cycle. The following information is provided:

> Compressor—air inlet temperature and pressure are 60°F and 14.5 psia, efficiency $= 85\%$, exit pressure is 145 psia, and $m\text{dot} = 1\,\text{lbm/sec}$.
>
> Combustion chamber—$q_{\text{supply}} = 400\,\text{Btu/lbm}$, and $m\text{dot} = 0.9\,\text{lbm/sec}$.
>
> Turbine #1—efficiency $= 92.3\%$, and $m\text{dot} = 1\,\text{lbm/sec}$.
>
> Turbine #2—efficiency $= 92.3\%$, air exit pressure $= 15\,\text{psia}$, and $m\text{dot} = 1\,\text{lbm/sec}$.

Determine the temperature of all states of the cycle, compressor power required, power produced by turbine #1, power produced by turbine #2, and cycle efficiency.

ANSWERS: $T_2 = 629°F$, $T_3 = 629°F$, $T_4 = 1881°F$, $T_5 = 1756°F$, $T_6 = 1187°F$, $T_7 = 768.2°F$, $W dot_{compressor} = -192.9\,hp$, $W dot_{turbine\ \#1} = 192.9\,hp$, $W dot_{turbine\ \#2} = 141.9\,hp$, $\eta = 37.14\%$.

4.7 FEHER CYCLE

The *Feher cycle* as shown in Fig. 4.24 is a single-phase cycle operating above the critical point of the working fluid. It incorporates the efficient pumping of the Rankine cycle with the regenerative heating features of the Brayton cycle to achieve higher theoretical efficiencies. The components of the cycle are identical to those of the Brayton cycle except that the higher pressures permit a substantial reduction in the size of all components. The *T–s* diagram of the cycle is shown in Fig. 4.7.2. The four processes of the Feher cycle are isentropic compression process 1-2, isobaric heat-addition process 2-3, isentropic expansion process 3-4, and isobaric heat removal process 4-1. Carbon dioxide with a critical pressure of 73.9 bars (1072 psia) and critical temperature of 304 K (88°F) appears to be the most suitable working fluid for such a cycle.

Applying the first and second laws of thermodynamics of the closed system to each of the four processes of the cycle yields:

$$Q_{12} - W_{12} = m(h_2 - h_1), \quad Q_{12} = 0 \tag{4.13}$$

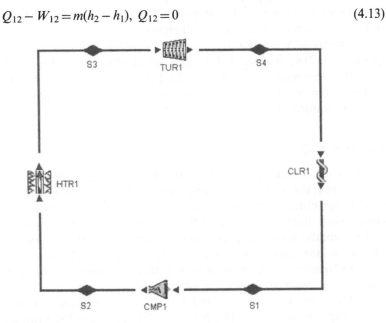

Figure 4.24 Feher cycle.

$$Q_{23} - W_{23} = m(h_3 - h_2), W_{23} = 0 \qquad (4.14)$$
$$Q_{34} - W_{34} = m(h_4 - h_3), Q_{34} = 0 \qquad (4.15)$$

and

$$Q_{41} - W_{41} = m(h_1 - h_4), \ W_{41} = 0 \qquad (4.16)$$

The net work (W_{net}), which is also equal to net heat (Q_{net}), is

$$W_{net} = W_{12} + W_{34} = Q_{net} = Q_{23} + Q_{41} \qquad (4.17)$$

The thermal efficiency of the cycle is

$$\eta = W_{net}/Q_{23} \qquad (4.18)$$

Example 4.12

A proposed Feher cycle using carbon dioxide has the following design information: turbine efficiency = 0.88, compressor efficiency = 0.88, mass rate flow of carbon dioxide = 1 lbm/sec, compressor inlet pressure = 1950 psia, compressor inlet temperature = 100°F, turbine inlet pressure = 4000 psia, and turbine inlet temperature = 1300°F.

Determine the compressor power, turbine power, rate of heat added, rate of heat removed, and cycle efficiency.

The answers are: $W{dot}_{compressor} = -31.66$ hp, $W{dot}_{turbine} = -65.59$ hp, $Q{dot}_{heater} = 218.5$ Btu/sec, $Q{dot}_{cooler} = -194.5$ Btu/sec, and $\eta = 10.98\%$.

If a regenerator is added to the Feher cycle, as shown in Fig. 4.25 and the following example, the cycle would have an efficiency of 39.82%, which is comparable to today's best steam power plant.

Example 4.13

A proposed Feher cycle with a regenerator (Fig. 4.26) using carbon dioxide has the following design information: turbine efficiency = 0.88, compressor efficiency = 0.88, mass rate flow of carbon dioxide = 1 lbm/sec, compressor inlet pressure = 1950 psia, compressor inlet temperature = 100°F, combustion chamber inlet temperature = 1000°F, turbine inlet pressure = 4000 psia, and turbine inlet temperature = 1300°F.

Determine the compressor power, turbine power, rate of heat added, rate of heat removed, and cycle efficiency.

The answers are: $W{dot}_{compressor} = -31.66$ hp, $W{dot}_{turbine} = -65.59$ hp, $Q{dot}_{heater} = 218.5$ Btu/sec, $Q{dot}_{cooler} = -194.5$ Btu/sec, and $\eta = 39.82\%$.

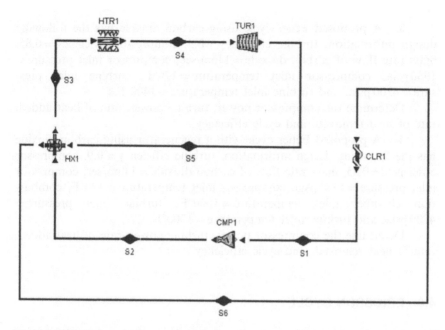

Figure 4.25 Feher cycle with regenerator.

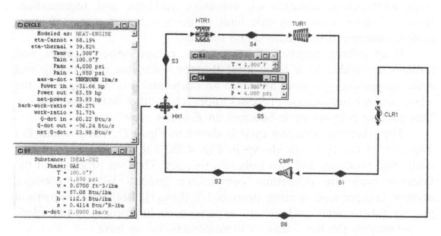

Figure 4.26 Feher cycle with regenerator.

Review Problems 4.7   Feher Cycle

1. What are the four processes of the Feher cycle?
2. What is the difference between the Feher and the Brayton cycle?

3. A proposed Feher cycle using carbon dioxide has the following design information: turbine efficiency = 0.9, compressor efficiency = 0.85, mass rate flow of carbon dioxide = 1 lbm/sec, compressor inlet pressure = 1950 psia, compressor inlet temperature = 120°F, turbine inlet pressure = 4200 psia, and turbine inlet temperature = 1400°F.s

Determine the compressor power, turbine power, rate of heat added, rate of heat removed, and cycle efficiency.

4. A proposed Feher cycle with a regenerator using carbon dioxide has the following design information: turbine efficiency = 0.9, compressor efficiency = 0.85, mass rate flow of carbon dioxide = 1 lbm/sec, compressor inlet pressure = 1950 psia, compressor inlet temperature = 120°F, combustion chamber inlet temperature = 1000°F, turbine inlet pressure = 4200 psia, and turbine inlet temperature = 1300°F.

Determine the compressor power, turbine power, rate of heat added, rate of heat removed, and cycle efficiency.

4.8 ERICSSON CYCLE

The thermal cycle efficiency of a Brayton cycle can be increased by adding more intercoolers, compressors, reheaters, and turbines, and regeneration. However, there is an economic limit to the number of stages of intercoolers, compressors, reheaters, and turbines.

If an infinite number of intercoolers, compressors, reheaters, and turbines are added to a basic ideal Brayton cycle, the intercooling and multicompression processes approach an isothermal process. Similarly, the reheat and multiexpansion processes approach another isothermal process. This limiting Brayton cycle becomes an *Ericsson cycle*.

The schematic Ericsson cycle is shown in Fig. 4.27. The p–v and T–s diagrams of the cycle are shown in Fig. 4.28. The cycle consists of two isothermal processes and two isobaric processes. The four processes of the Ericsson cycle are isothermal compression process 1-2 (compressor), isobaric compression heating process 2-3 (heater), isothermal expansion process 3-4 (turbine), and isobaric expansion cooling process 4-1 (cooler).

Applying the basic laws of thermodynamics, we have

$$q_{12} - w_{12} = h_2 - h_1 \tag{4.19}$$

$$q_{23} - w_{23} = h_3 - h_2, \quad w_{23} = 0 \tag{4.20}$$

$$q_{34} - w_{34} = h_4 - h_3 \tag{4.21}$$

$$q_{41} - w_{41} = h_1 - h_4, \quad w_{41} = 0 \tag{4.22}$$

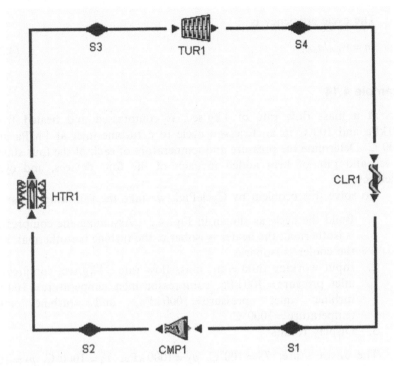

Figure 4.27 Schematic Ericsson cycle.

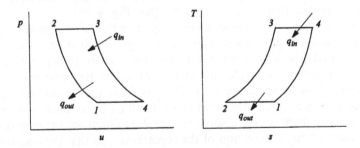

Figure 4.28 Ericsson cycle *p–v* and *T–s* diagram.

The net work produced by the cycle is

$$w_{net} = w_{12} + w_{34} \tag{4.23}$$

The heat added to the cycle is

$$q_{add} = q_{23} + q_{34} \tag{4.24}$$

The cycle efficiency is

$$\eta = w_{net}/q_{add} \tag{4.25}$$

Example 4.14

Air, at a mass flow rate of 1 kg/sec, is compressed and heated from 100 kPa and 100°C in an Ericsson cycle to a turbine inlet at 1 MPa and 1000°C. Determine the pressure and temperature of each of the four states, power and rate of heat added in each of the four devices, and cycle efficiency.

To solve this problem by CyclePad, we take the following steps:

1. Build the cycle as shown in Fig. 4.27. Assuming the compressor is isothermal, the heater is isobaric, the turbine is isothermal, and the cooler is isobaric.
2. Input working fluid = air, mass flow rate = 1 kg/sec, compressor inlet pressure = 100 kPa, compressor inlet temperature = 100°C, turbine inlet pressure = 1000 kPa, and turbine inlet temperature = 1000°C.
3. Display results

The answers are: $T_2 = 100°C$, $p_2 = 1000$ kPa, $T_4 = 1000°C$, $p_2 = 100$ kPa, $Q\text{dot}_{htr} = 903.1$ kW, $Q\text{dot}_{clr} = -903.1$ kW, $Q\text{dot}_{cmp} = -246.3$ kW, $W\text{dot}_{cmp} = -246.3$ kW, $Q\text{dot}_{tur} = 840.4$ kW, $W\text{dot}_{cmp} = 840.4$ kW, and $\eta = 34.08$. Notice that $\eta_{Carnot} = 70.69\%$. (See Fig. 4.29.)

An attempt to achieve Carnot cycle efficiency is made by the Ericsson cycle using an ideal regenerator. Figure 4.30 shows a schematic Ericsson cycle with a regenerator. In the regenerator, gas from the compressor enters as a cold-side stream at a low temperature (T_2) and leaves at a high temperature (T_1). The gas from the turbine enters as a hot-side stream at a high temperature (T_5) and leaves at a low temperature (T_6). Suppose there is only a small temperature difference between the two gas streams at any one section of the regenerator, so that the operation of the regenerator is almost ideal. Then the heat loss by the hot-side stream gas (Q_{56}) equals to the heat gain by the cold-side stream gas (Q_{23}). The cycle efficiency is close to that of the Carnot cycle operating between the same two temperatures.

Example 4.15

Air, at a mass flow rate of 1 kg/sec, is compressed and heated from 100 kPa and 100°C in an Ericsson cycle to a turbine inlet at 1 MPa and

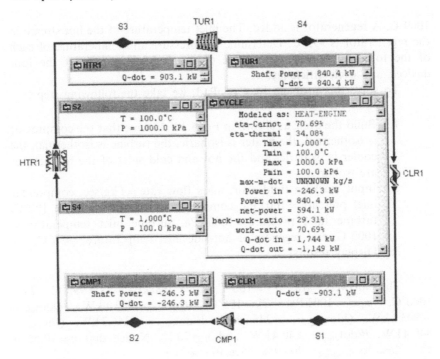

Figure 4.29 Ericsson cycle.

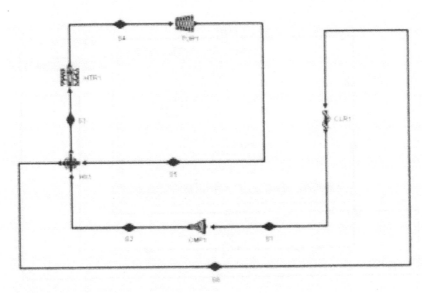

Figure 4.30 Schematic Ericsson cycle with a regenerator.

1000°C. A regenerator is added. The inlet temperature of the hot stream in the regenerator is 995°C. Determine the pressure and temperature of each of the four states, power and rate of heat added in each of the four devices, and cycle efficiency.

To solve this problem by CyclePad, we take the following steps:

1. Build the cycle as shown in Fig.4.30. Assume that the compressor is isothermal, the heater is isobaric, the turbine is isothermal, the cooler is isobaric, and the hot and cold sides of the regenerator are isobaric.
2. Input working fluid = air, mass flow rate = 1 kg/sec, compressor inlet pressure = 100 kPa, compressor inlet temperature = 100°C, turbine inlet pressure = 1000 kPa, turbine inlet temperature = 1000°C, and regenerator hot-side inlet temperature = 995°C.
3. Display results.

The answers are: $T_2 = 100°C$, $p_2 = 1000$ kPa, $T_3 = 995°C$, $T_4 = 1000°C$, $p_4 = .100$ kPa, $T_6 = 105°C$, $Qdot_{htr} = 903.1$ kW, $Qdot_{clr} = -903.1$ kW, $Qdot_{cmp} = -246.3$ kW, $Wdot_{cmp} = -246.3$ kW, $Qdot_{tur} = 840.4$ kW, $Wdot_{cmp} = 840.4$ kW, and $\eta = 70.52$. Notice that $\eta = 70.52$ is very close to $\eta_{Carnot} = 70.69\%$. (See Fig. 4.31.)

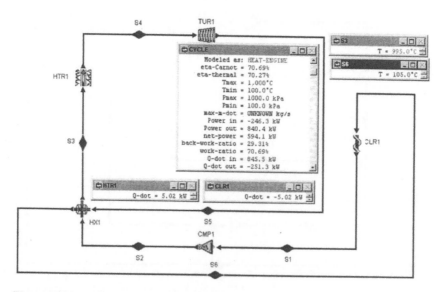

Figure 4.31 Ericsson cycle with regenerator.

Review Problems 4.8　Ericsson Cycle

1. What are the four processes of the Ericsson cycle?
2. What is the function of a regenerator?
3. Does the regenerator improve the efficiency of the Ericsson cycle?
4. Suppose an ideal regenerator is added to an Ericsson cycle. The regenerator would absorb heat from the system during part of the cycle and return exactly the same amount of heat to the system during another part of the cycle. What would be the difference between the Ericsson cycle efficiency and the Carnot cycle efficiency?
5. Air, at a mass flow rate of 1.2 kg/sec, is compressed and heated from 100 kPa and 300 K in an Ericsson cycle to a turbine inlet at 1200 kPa and 1500 K. The turbine efficiency is 85% and the compressor efficiency is 88%. Determine the pressure and temperature of each of the four states, power and rate of heat added in each of the four devices, and cycle efficiency.
6. Air, at a mass flow rate of 1.2 kg/sec, is compressed and heated from 100 kPa and 300 K in an Ericsson cycle to a turbine inlet at 1200 kPa and 1500 K. The turbine efficiency is 85% and the compressor efficiency is 88%. A regenerator is added. The inlet temperature of the hot stream in the regenerator is 1495 K. Determine the pressure and temperature of each of the states, power and rate of heat added in each of the four devices, and cycle efficiency. What would be the efficiency of the Carnot cycle operating between 300 and 1500 K?

4.9　BRAYSSON CYCLE

A *Braysson* cycle, proposed by Frost et al. (Frost, T.H., Anderson, A., and Agnew, B., A hybrid gas turbine cycle (Brayton/Ericsson): an alternative to conventional combined gas and steam turbine power plant. *Proceedings of the Institution of Mechanical Engineers, Part A, Journal of Power and Energy*, vol. 211, no. A2, pp. 121–131, 1997), is an alternative to the Brayton/Rankine combined gas and steam turbine power plant. The Braysson cycle is a combination of a single Brayton cycle and an Ericsson cycle. The cycle takes advantage of the high-temperature heat-addition process of the Brayton cycle and the low-temperature heat-rejection process of the Ericsson cycle. It employs one working fluid in the two cycles in such a way that the full waste heat from the top Brayton cycle serves as the heat source for the bottom Ericsson cycle. The total power output of the Braysson cycle is the summation of the power produced by the top and bottom cycles.

A design of such a novel Braysson cycle (Wu, C., Intelligent computer aided optimization of power and energy systems. *Proceedings of the Institution of Mechanical Engineers, Part A, Journal of Power and Energy*, vol. 213, no. A1, pp. 1–6, 1999), consisting of four compressors, one combustion chamber, two turbines, and two coolers, is shown in Fig. 4.32. The *T–s* diagram of the Braysson cycle is shown in Fig. 4.33. Another arrangement of the Braysson cycle, consisting of four compressors, one combustion chamber, two turbines, and two coolers, is shown in Fig. 4.34.

Neglecting kinetic and potential energy changes, a steady state and steady flow mass and energy balance on the components of the Braysson cycle have the general forms:

$$\sum m\text{dot}_e = \sum m\text{dot}_i \tag{4.26}$$

and

$$Q\text{dot} - W\text{dot} = \sum m\text{dot}_e h_e - \sum m\text{dot}_i h_i \tag{4.27}$$

Figure 4.32 Braysson cycle.

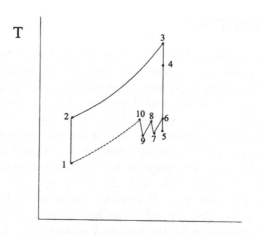

Figure 4.33 Braysson cycle *T–s* diagram.

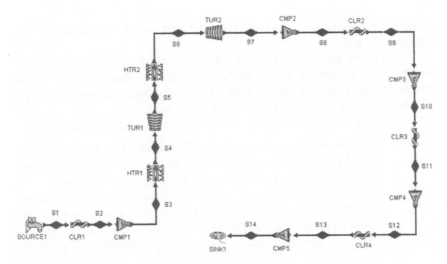

Figure 4.34 Braysson cycle.

The energy input of the cycle is the heat added by the heater. The net energy output of the cycle is the sum of the work added to the individual compressors and work produced by the turbines:

$$W\text{dot}_{net} = \sum W\text{dot}_{compressor} + \sum W\text{dot}_{turbine} \qquad (4.28)$$

and the efficiency of the cycle is

$$\eta = W\text{dot}/Q\text{dot}_{heater} \qquad (4.29)$$

The following examples illustrate the analysis of the Braysson cycle.

Example 4.16

A Braysson cycle (Fig. 4.32) uses air as the working fluid with 1 kg/sec of mass flow rate through the cycle. In the cycle, air enters from the atmospheric source to an isentropic compressor at 20°C and 1 bar (state 1) and leaves at 8 bars (state 2); air enters an isobaric heater (combustion chamber) and leaves at 1100°C (state 3); air enters a high-pressure isentropic turbine and leaves at 1 bar (state 4). Air enters a low-pressure isentropic turbine and leaves at 0.04 bar (state 5); air enters a first-stage isentropic compressor and leaves at 0.2 bar (state 6); air enters an isobaric intercooler and leaves at 20°C (state 7); air enters a second-stage isentropic compressor and leaves at 1 bar (state 8); and air is discharged to the atmospheric sink.

Determine the thermodynamic efficiency and the net power output of the Braysson combined plant. Plot the sensitivity diagram of η (cycle efficiency) versus p_6 (pressure at state 6) and sensitivity diagram of η (cycle efficiency) versus p_8 (pressure at state 8).

To solve this problem by CyclePad, we take the following steps:

1. Build the cycle as shown in Fig. 4.32. Assume that the compressors are adiabatic and isentropic, the heater is isobaric, the turbines are adiabatic and isentropic, and the coolers are isobaric.

2. Input working fluid = air, $p_1 = 1$ bar, $T_1 = 20°C$, $mdot = 1$ kg/sec, $p_2 = 8$ bar, $T_3 = 1100°C$, $p_4 = 1$ bar, $p_5 = 0.04$ bar, $p_6 = 0.2$ bar, compressor inlet temperature = 100°C, turbine inlet pressure = 1 MPa, turbine inlet temperature = 1000°C, $T_7 = 20°C$, $p_8 = 0.6$ bar, $T_9 = 20°C$, and $p_{10} = 1$ bar.

3. Display results. The answers are $\eta = 59.67\%$, power input = -570.4 kW, power output = 1075 kW, net power output = 504.2 kW, and $Qdot_{in} = 845.0$ kW. (see Fig. 4.35a.)

4. Display sensitivity diagrams of η (cycle efficiency) versus p_6 (pressure at state 6) and of η (cycle efficiency) versus p_8 (pressure at state 8). (see Figs. 4.35b and 4.35c.)

The design of differently arranged Braysson cycles such as a simple Braysson cycle with a two-stage compressor, and a Braysson cycle with a four-stage compressor can be made. The four-stage compressor Braysson

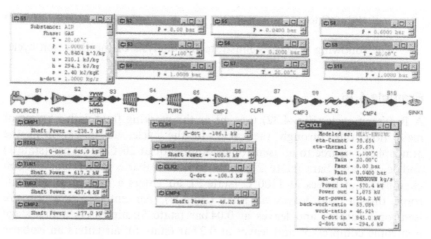

Figure 4.35a Braysson cycle.

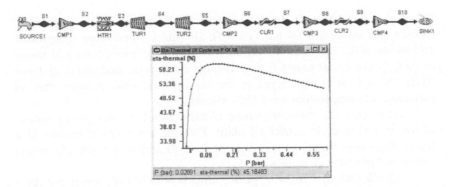

Figure 4.35b Braysson cycle sensitivity diagram.

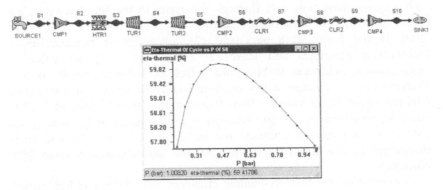

Figure 4.35c Braysson cycle sensitivity diagram.

cycle is a refinement of the three-stage compressor Braysson cycle. Other improvement include (1) a reheater added in the Brayton cycle of the combined cycle, (2) more reheaters and more turbines added in the Brayton cycle, (3) more compressors and more intercoolers added in the Ericsson cycle, etc.

Review Problems 4.9   Braysson Cycle

1. What is a Braysson cycle?
2. Why is the efficiency of the Braysson cycle high?
3. A Braysson cycle (Fig. 4.32) uses air as the working fluid with 1 kg/ sec of mass flow rate through the cycle. In the Brayton cycle, air enters from the atmospheric source to a compressor at 20°C and 1 bar (state 1) and leaves at 8 bars (state 2); air enters an isobaric heater (combustion chamber) and leaves at 1100°C (state 3); and air enters a high-pressure turbine and

leaves at 1 bar (state 4). In the Ericsson cycle, air enters a low-pressure turbine and leaves at 0.04 bar (state 5); air enters a first-stage compressor and leaves at 0.2 bar (state 6); air enters an isobaric inter-cooler and leaves at 20°C (state 7); air enters a second-stage compressor and leaves at 1 bar (state 8); and air is discharged to the atmospheric sink. Assume that all turbines and compressors have 85% efficiency.

Determine the thermodynamic efficiency and the net power output of the Braysson cycle combined plant. Plot the sensitivity diagrams of η (cycle efficiency) versus p_6 (pressure at state 6) and of η (cycle efficiency) versus p_8 (pressure at state 8).

ANSWERS: $\eta = 26.34\%$, power input $= -749.3$ kW, power output $= 960.7$ kW, Qdot$_{in} = 802.9$ kW.

4. A Braysson cycle (Fig. 4.32) uses air as the working fluid with 1 kg/sec mass flow rate through the cycle. In the Brayton cycle, air enters from the atmospheric source to a compressor at 20°C and 1 bar (state 1) and leaves at 8 bars (state 2); air enters an isobaric heater (combustion chamber) and leaves at 1100°C (state 3); air enters a high-pressure isentropic turbine and leaves at 1 bar (state 4). In the Ericsson cycle, air enters a low-pressure isentropic turbine and leaves at 0.04 bar (state 5); air enters a first-stage compressor and leaves at 0.2 bar (state 6); air enters an isobaric intercooler and leaves at 20°C (state 7); air enters a second-stage compressor and leaves at 1 bar (state 8); and air is discharged to the atmospheric sink. Assume all compressors have 85% efficiency.

Determine the thermodynamic efficiency and the net power output of the Braysson combined plant. Plot the sensitivity diagrams of η (cycle efficiency) versus p_6 (pressure at state 6) and of η (cycle efficiency) versus p_8 (pressure at state 8).

ANSWERS: $\eta = 50.26\%$, power input $= -671.1$ kW, Power output $= 1075$ kW, Qdot$_{in} = 802.9$ kW.

4.10 STEAM INJECTION GAS TURBINE CYCLE

The injection of water or steam in gas turbines has been known (Nicolin, C., A gas turbine with steam injection. Swedish Patent application No.8112/51, Stockholm, Sweden, 1951) as an efficient method for NO_x abatement and power boosting. Several cycle configurations are possible with respect to water/steam injection. Figure 4.36 is the schematic diagram of the *Steam-injection gas turbine cycle*. Air is compressed from state 1 to state 2. Water is pumped from state 7 to state 8. Steam at state 9 is generated in a recovery boiler (heat exchanger) from state 8 by the hot exhaust gas. Steam at state 8

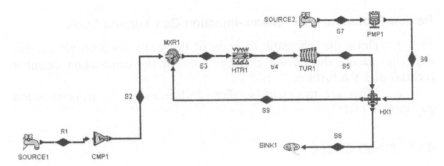

Figure 4.36 Steam-injection gas turbine cycle.

is injected into air at state 2 in a mixing chamber. Air and steam is then heated in the combustion chamber from state 3 to state 4, expanded in the gas turbine from state 4 to state 5, and exhausted to the recovery boiler from state 5 to state 6. The mixing chamber can be located either between the compressor and the combustion chamber (heater), or between the combustion chamber (heater) and the turbine. The mass flow rate of injected steam is of the order of 15% of the mass flow rate of air supplied to the gas turbine. Comparing the cycle with a gas turbine cycle without steam injection, we can see that the compressor work is not effected, but the turbine work increases considerably due to increase of gas mass and increase of substance specific heat (c_p). Therefore, the net output work of the steam-injection gas turbine cycle increases.

The beneficial influences of the steam-injection gas turbine cycle include:

1. It provides an increase in both power output and overall efficiency. For a given temperature at inlet to the gas turbine, extra fuel has to be supplied in order to heat the injected steam to that temperature, but the additional power arising from the expansion of the injected steam as it passes through the gas turbine more than offsets the otherwise adverse effect on the overall efficiency of the cycle of the increase in fuel supply.

2. As a result of the cooling effect of the steam in the primary flame zone of the combustion chamber, it results in a reduction in the emission of noxious oxides of nitrogen, NO_x, from the cycle.

CyclePad is not able to perform the steam-injection gas turbine cycle, because there is no binary working fluid in the substance menu of the software.

Review Problems 4.10   Steam-Injection Gas Turbine Cycle

1. Draw the schematic diagram of the steam-injection gas turbine cycle by placing the mixing chamber between the combustion chamber (heater) and the turbine.

2. What are the main beneficial influences of the steam-injection gas turbine cycle?

4.11   FIELD CYCLE

The *Field cycle* is a supergenerative cycle that makes use of the high-temperature heat addition of the Brayton cycle and the low-temperature heat removal of the Rankine cycle (Field, J.F., The application of gas turbine technique to steam power, *Proc. of the Inst. of Mech. Eng.*, 1950). Therefore, it is able to achieve a high mean temperature of heat addition. The gain due to high-temperature heat addition, however, is offset by the reduction in cycle efficiency resulting from the irreversibility of the mixing process. The schematic diagram of the Field cycle is shown in Fig. 4.37. The arrangement includes one compressor, five turbines, three pumps, one boiler and one reheater (heaters), one regenerator (heat exchanger), one condenser (cooler), three mixing chambers, and two splitters. Processes 1-2, 2-3, 3-4, 4-5, 5-6, and 6-7 take advantage of the high-temperature heat addition of Brayton cycle, and the other processes take advantage of the low-temperature heat removal and regenerative condensing of the Rankine cycle.

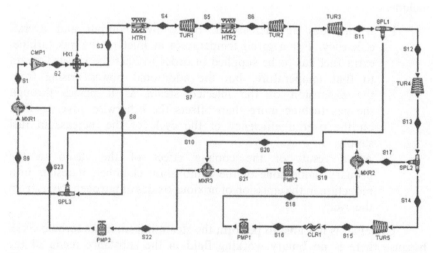

Figure 4.37   Field cycle schematic diagram.

Example 4.17

An ideal Field cycle with perfect regeneration, as shown in Fig. 4.37, is designed according to the following data: $p_{16} = 10\,\text{kPa}$, $x_{16} = 0$, $p_{19} = 200\,\text{kPa}$, $x_{19} = 0$, $p_{22} = 1\,\text{MPa}$, $x_{22} = 0$, $p_{23} = 2\,\text{MPa}$, $p_2 = 6\,\text{MPa}$, $T_4 = 500°\text{C}$, $mdot_4 = 1\,\text{kg/sec}$, $p_5 = 4\,\text{MPa}$, $T_6 = 500°\text{C}$, $T_8 = 300°\text{C}$, $mdot_{10} = 0.9\,\text{kg/sec}$, and $mdot_{17} = 0.1\,\text{kg/sec}$.

Determine (1) the pressure and temperature of each state of the cycle, (2) power produced by each of the five turbines, rate of heat added by each of the two heaters, power reuired by the compressor and each of the three pumps, rate of heat removed by the condenser, and (3) net power produced by the cycle, and cycle efficiency.

To solve this problem by CyclePad, we take the following steps:

1. Build the cycle as shown in Fig. 4.37. Assume that the compressor, turbines, and pump are adiabatic and isentropic, the heaters, mixing chambers, cooler, and regenerator are isobaric, and the splitters are isoparametric.

2. Input working fluid = water, $p_{16} = 10\,\text{kPa}$, $x_{16} = 0$, $p_{19} = 200\,\text{kPa}$, $x_{19} = 0$, $p_{22} = 1\,\text{MPa}$, $x_{22} = 0$, $p_{23} = 2\,\text{MPa}$, $p_2 = 6\,\text{MPa}$, $T_4 = 500°\text{C}$, $mdot_4 = 1\,\text{kg/sec}$, $p_5 = 4\,\text{MPa}$, $T_6 = 500°\text{C}$, $T_8 = 300°\text{C}$, $mdot_{10} = 0.9\,\text{kg/sec}$, and $mdot_{17} = 0.9\,\text{kg/sec}$.

3. Display results as shown in Fig. 4.38.

The answers are: (1) $p_1 = 2\,\text{MPa}$, $T_1 = 212.4°\text{C}$, $p_2 = 6\,\text{MPa}$, $T_2 = 237.3°\text{C}$, $p_3 = 6\,\text{MPa}$, $T_3 = 275.6°\text{C}$, $p_4 = 6\,\text{MPa}$, $T_4 = 500°\text{C}$, $p_5 =$

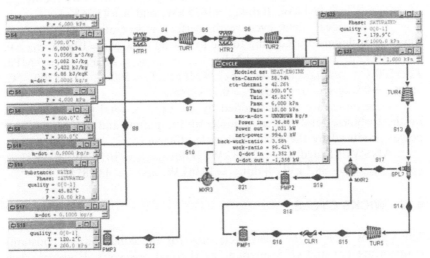

Figure 4.38 Field cycle.

4 MPa, $T_5 = 432.8°C$, $p_6 = 4$ MPa, $T_6 = 500°C$, $p_7 = 2$ MPa, $T_7 = 388.9°C$, $p_8 = 2$ MPa, $T_8 = 300°C$, $p_9 = 2$ MPa, $T_9 = 300°C$, $p_{10} = 2$ MPa, $T_{10} = 300°C$, $p_{11} = 1$ MPa, $T_{11} = 214.6°C$, $p_{12} = 1$ MPa, $T_{12} = 214.6°C$, $p_{13} = 200$ kPa, $T_{13} = 120.2°C$, $p_{14} = 200$ kPa, $T_{14} = 120.2°C$, $p_{15} = 10$ kPa, $T_{15} = 45.82°C$, $p_{16} = 10$ kPa, $T_{16} = 45.82°C$, $p_{17} = 200$ kPa, $T_{17} = 120.2°C$, $p_{18} = 200$ kPa, $T_{18} = 45.83°C$, $p_{19} = 200$ kPa, $T_{19} = 120.2°C$, $p_{20} = 1$ MPa, $T_{20} = 214.6°C$, $p_{21} = 1$ MPa, $T_{21} = 120.3°C$, $p_{22} = 2$ MPa, $T_{17} = 179.9°C$, $p_{23} = 2$ MPa, and $T_{17} = 185.7°C$;   (2) $W\text{dot}_{T\#1} = 131.9$ kW, $W\text{dot}_{T\#2} = 222.4$ kW, $W\text{dot}_{T\#3} = 144.7$ kW, $W\text{dot}_{T\#4} = 238.9$ kW, $W\text{dot}_{T\#5} = 293.2$ kW, $W\text{dot}_{P\#1} = -0.1598$ kW, $W\text{dot}_{P\#2} = -0.6973$ kW, $W\text{dot}_{P\#3} = -28.45$ kW, $W\text{dot}_{Compressor} = -7.57$ kW, $Q\text{dot}_{Htr\#1} = 2197$ kW, $Q\text{dot}_{Htr\#2} = 155$ kW, and $Q\text{dot}_{Condenser} = -1358$ kW, (3) $W\text{dot}_{net} = 994.0$ kW, and $\eta = 42.26\%$.

Review Problems 4.11   Field Cycle

1.   What is the concept of the Field cycle?

2.   An ideal Field cycle with perfect regeneration, as shown in Fig. 4.37, is designed according to the following data: $p_{16} = $ 10 kPa, $x_{16} = 0$, $p_{19} = 200$ kPa, $x_{19} = 0$, $p_{22} = 1$ MPa, $x_{22} = 0$, $p_{23} = $ 2 MPa, $p_2 = 7$ MPa, $T_4 = 500°C$, $m\text{dot}_4 = 1$ kg/sec, $p_5 = 4$ MPa, $T_6 = 500°C$, $T_8 = 300°C$, $m\text{dot}_{10} = 0.9$ kg/sec, and $m\text{dot}_{17} = $ 0.1 kg/sec.

Determine rate of heat added by the heaters, total power produced by the turbines, total power required by the pumps and compressor, net power produced by the cycle, and cycle efficiency.

ANSWERS: $Q\text{dot}_{add} = 2393$ kW, $W\text{dot}_{Turbines} = 1073$ kW, $W\text{dot}_{Pumps}$ and Compressor $= -38.1$ kW, $W\text{dot}_{net} = 1035$ kW, and $\eta = 43.26\%$.

3.   An ideal field cycle with perfect regeneration as shown in Fig. 4.37 is designed according to the following data: $p_{16} = 10$ kPa, $x_{16} = 0$, $p_{19} = 200$ kPa, $x_{19} = 0$, $p_{22} = 1$ MPa, $x_{22} = 0$, $p_{23} = 2$ MPa, $p_2 = 7$ MPa, $T_4 = 500°C$, $m\text{dot}_4 = 1$ kg/sec, $p_5 = 4$ MPa, $T_6 = 500°C$, $T_8 = 300$ °C, $m\text{dot}_{10} = 0.9$ kg/sec, and $m\text{dot}_{17} = 0.12$ kg/sec.

Determine rate of heat added by the heaters, total power produced by the turbines, total power required by the pumps and compressor, net power produced by the cycle, and cycle efficiency.

ANSWERS: $Q\text{dot}_{add} = 2393$ kW, $W\text{dot}_{Turbines} = 1074$ kW, $W\text{dot}_{Pumps}$ and Compressor $= -38.1$ kW, $W\text{dot}_{net} = 1036$ kW, and $\eta = 43.28\%$.

4.12   WICKS CYCLE

The Carnot cycle is the ideal cycle only for the conditions of constant-temperature hot and cold surrounding thermal reservoirs. However, such conditions do not exist for fuel-burning engines. For these engines, the combustion products are artificially created as a finite-size hot reservoir that releases heat over the entire temperature range from its maximum to ambient temperature. The natural environment in terms of air or water bodies is the cold reservoir and can be considered as an infinite reservoir relative to the engine. Thus, an ideal fuel-burning engine should operate reversibly between a finite-size hot reservoir and an infinite-size cold reservoir. Wicks (Wicks, F., The thermodynamic theory and design of an ideal fuel burning engine. *Proceedings of the Intersociety Engineering Conference of Energy Conversion*, vol. 2, pp. 474–481, 1991) proposed a three-process ideal fuel-burning engine consisting of an isothermal compression, an isobaric heat addition, and an adiabatic expansion process. The schematic *Wicks cycle* is shown in Fig. 4.39, and an example of the cycle is given in Example 4.18.

Example 4.18

A Wicks cycle as shown in Fig. 4.39 is designed according to the following data: $p_1 = 101$ kPa, $T_1 = 5°C$, $p_2 = 28$ MPa, $T_3 = 1100°C$, $mdot_1 = 1$ kg/sec, and $T_4 = 5°C$.

Determine the power produced, rate of heat added, power input, net power produced by the cycle, and cycle efficiency.

To solve this problem by CyclePad, we take the following steps:

1. Build the cycle as shown in Fig. 4.39. Assume that the compressor, heater, and turbine are isothermal, isobaric, and isentropic.
2. Input working fluid = air, $p_1 = 101$ kPa, $T_1 = 5°C$, $mdot_1 = 1$ kg/sec, $p_2 = 28$ MPa, $T_3 = 1100°C$, and $T_4 = 5°C$.
3. Display results.

The answers are: $Wdot_{in} = -448.5$ kW, $Wdot_{out} = 1099$ kW, $Wdot_{net} = 650.2$ kW, $Qdot_{in} = 1099$ kW, $Qdot_{out} = -448.5$ kW, and $\eta = 59.18\%$, as shown in Fig. 4.40.

Review Problems 4.12   Wicks Cycle

1. What is the concept of the Wicks cycle?
2. A Wicks cycle as shown in Fig. 4.39 is designed according to the following data: $p_1 = 101$ kPa, $T_1 = 10°C$, $p_2 = 28$ MPa, $T_3 = 1050°C$, $mdot_1 = 1$ kg/sec, and $T_4 = 10°C$.

Figure 4.39 Wicks cycle.

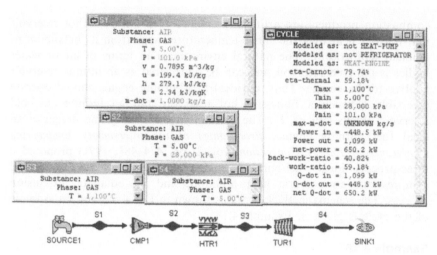

Figure 4.40 Wicks cycle.

Determine power produced, rate of heat added, power input, net power produced by the cycle, and cycle efficiency.

4.13 ICE CYCLE

Silverstein (Silverstein, C.C., The Ice Cycle: High gas turbine efficiency at moderate temperature. *Proceedings of the Intersociety Energy Conversion Engineering Conference*, paper number 889341, pp. 285–289, 1988) proposed an ice cycle, which consists of isothermal compression, isentropic compression, isothermal expansion, and isentropic expansion processes as shown in Fig. 4.41. Its efficiency is the same as that of the Carnot cycle. An actual ice cycle is characterized by efficiencies of 35–40%, peak temperatures below 1080 K, and overall pressure ratios of 300–500. Isothermal compression and isothermal expansion are approximated by the use of a heat exchanger after each stage, which is an integral part of the rotating equipment. A heat-pipe heat exchanger appears to be particularly well adapted to integral intercooling and reheat. The *T–s* diagram of the cycle is shown in Fig. 4.42. The use of pressure rather than temperature and/or regenerative heat exchange to achieve high cycle efficiency can lead to major design and economic benefits for gas turbine cycles. Compressors and turbines can be fabricated from materials that will retain good strength characteristics at peak operating temperatures.

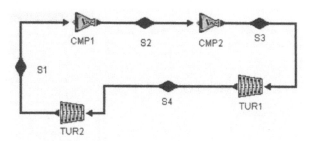

Figure 4.41 Ice cycle.

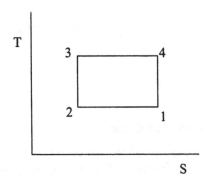

Figure 4.42 Ice cycle T–s diagram.

Example 4.19

An ice cycle as shown in Fig. 4.41 is designed according to the following data: $p_1 = 100$ kPa, $T_1 = 300$ K, $p_2 = 200$ kPa, and $T_3 = 1200$ K, and $mdot_1 = 1$ kg/sec.

Determine power produced, rate of heat added, power input, net power produced by the cycle, and cycle efficiency.

To solve this problem by CyclePad, we take the following steps:

1. Build the ice cycle as shown in Fig. 4.41. Assume that the compressors and turbines are isothermal and isentropic.
2. Input working fluid = air, $p_1 = 100$ kPa, $T_1 = 300$ K, $mdot_1 = 1$ kg/sec, $p_2 = 200$ kPa, and $T_3 = 1200$ K.
3. Display results.

The answers are: $Wdot_{in} = -962.7$ kW, $Wdot_{out} = 1142$ kW, $Wdot_{net} = 178.8$ kW, $Qdot_{in} = 238.5$ kW, $Qdot_{out} = -59.62$ kW, and $\eta = 75\%$, as shown in Fig. 4.43.

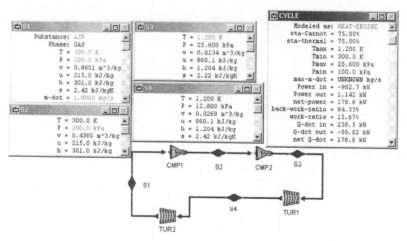

Figure 4.43   Ice cycle.

Review Problems 4.13   Ice Cycle

1.   What is the concept of the ice cycle?

2.   An ice cycle as shown in Fig. 4.13 is designed according to the following data: $p_1 = 100\,\text{kPa}$, $T_1 = 290\,\text{K}$, $p_2 = 200\,\text{kPa}$, $T_3 = 1400\,\text{K}$, and $m\text{dot}_1 = 1\,\text{kg/sec}$.

Determine power produced, rate of heat added, power input, net power produced by the cycle, and cycle efficiency.

4.14   DESIGN EXAMPLES

Typically, CyclePad's "build" mode allows the designer to select parts from one of its two inventory shops (closed system and open system) and connect them. Each component is clearly labeled with its input and output denoted by arrows. After connecting the parts in a complete cycle, the software will prompt the designer to stay in the "build" mode or move on to "analysis" mode. In the "analysis" mode, CyclePad combines user input and thermodynamic principles to solve cycles numerically. Here, the designer selects the working fluid, component properties, and boundary conditions. CyclePad solves all possible variables as the designer adds conditions to the cycle. When all necessary entries have been made, the software will have solved for properties such as total heat input, total heat output, net power, thermal efficiency, etc. In the event that the designer enters conflicting conditions, CyclePad enters the "contradiction" mode.

In this mode, the designer is told of a conflict and the program will not proceed until the contradiction is resolved. This is done in a pop-up window which shows all inputs that contribute to the error. In this way, the designer does not necessarily have to remove the entry that forced the contradiction, a very useful function indeed. If the designer wishes, CyclePad will explain what the contradiction is and how the current assumptions cannot be correct. This editing technique is very similar to the senior design engineer monitoring junior engineers who are not yet experienced enough to see the error in advance.

The intelligent computer software CyclePad is a very effective tool in design cycles. Any complicated gas cycle can be easily designed and analyzed using CyclePad. Optimization of design parameters of the cycle is demonstrated by the following examples.

Example 4.20

A four-stage reheat and four-stage intercool Brayton air cycle as shown in Fig. 4.44a has been designed by a junior engineer with the following design input information:

$p_1 = p_{21} = p_{22} = 100$ kPa, $p_2 = p_3 = p_{20} = p_{19} = 200$ kPa, $p_4 = p_5 = p_{17} = p_{18} = 400$ kPa, $p_6 = p_7 = p_{15} = p_{16} = 600$ kPa, $p_8 = p_9 = p_{13} = p_{14} = 800$ kPa, $p_{10} = p_{11} = p_{12} = 1$ MPa, $T_1 = T_3 = T_5 = T_7 = T_9 = 20°C$, $T_{12} = T_{14} = T_{16} = T_{18} = T_{20} = 1200°C$, $T_{22} = 400°C$, $mdot_1 = 1$ kg/sec, $\eta_{tur1} = \eta_{tur2} = \eta_{tur3} = \eta_{tur4} = \eta_{tur5} = 85\%$, and $\eta_{cmpr1} = \eta_{cmpr2} = \eta_{cmpr3} = \eta_{cmpr4} = \eta_{cmpr5} = 85\%$. (See Fig. 4.44b.)

The following output results as shown in Fig. 4.44c are obtained from his design:

$\eta_{cycle} = 55.16\%$, $Wdot_{input} = -589.8$ kW, $Wdot_{output} = -1334$ kW, $Wdot_{net\ output} = 744.1$ kW, $Qdot_{add} = 1349$ kW, $Qdot_{remove} = -605.0$ kW, $Wdot_{cmp1} = -75.79$ kW, $Wdot_{cmp2} = -75.79$ kW, $Wdot_{cmp3} = -42.50$ kW, $Wdot_{cmp4} = -29.65$ kW, $Wdot_{cmp5} = -366.1$ kW, $Wdot_{tur1} = 645.9$ kW, $Wdot_{tur2} = 99.14$ kW, $Wdot_{tur3} = 137.4$ kW, $Wdot_{tur4} = 225.7$ kW, $Wdot_{tur5} = 225.7$ kW, $Qdot_{htr1} = 241.0$ kW, $Qdot_{htr2} = 645.9$ kW, $Qdot_{htr3} = 99.14$ kW, $Qdot_{htr4} = 137.4$ kW, $Qdot_{htr5} = 225.7$ kW, $Qdot_{clr1} = -381.4$ kW, $Qdot_{clr2} = -75.79$ kW, $Qdot_{clr3} = -75.79$ kW, $Qdot_{clr4} = -42.50$ kW, $Qdot_{clr5} = -29.65$ kW, $T_{10} = 384.8°C$, $T_{11} = 959.8°C$, and $T_{21} = 400°C$.

The T–s diagram of the cycle is shown in Fig. 4.44d.

Try to modify his design (use p_{16}, p_{18}, p_6 and p_8 as design parameters only) to get a better cycle thermal efficiency than his $\eta_{cycle} = 55.16\%$.

The sensitivity analyses of η_{cycle} versus p_{16}, and η_{cycle} versus p_{18} are shown in Figs. 4.44e and 4.44f. The optimization design values of p_{16} and p_{18} can be easily identified.

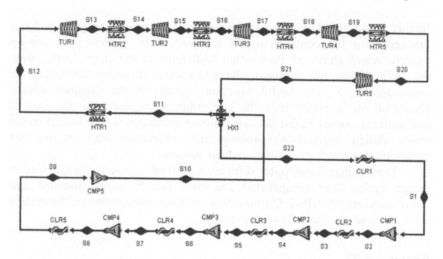

Figure 4.44a Four-stage reheat and four-stage intercool Brayton air cycle.

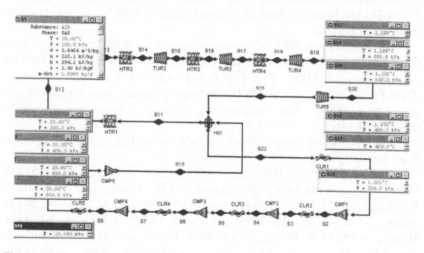

Figure 4.44b Brayton air cycle design input.

Review Problems 4.14   Design

A four-stage reheat and four-stage intercool Brayton air cycle as shown in Fig. 4.44a has been designed by a junior engineer with the following design input information:

$$p_1 = p_{21} = p_{22} = 100 \, \text{kPa}, \quad p_2 = p_3 = p_{20} = p_{19} = 200 \, \text{kPa}, \quad p_4 = p_5 = p_{17} = p_{18} = 500 \, \text{kPa}, \quad p_6 = p_7 = p_{15} = p_{16} = 700 \, \text{kPa}, \quad p_8 = p_9 = p_{13} = p_{14} = 900 \, \text{kPa},$$

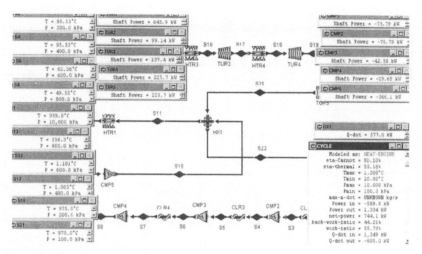

Figure 4.44c Brayton air cycle design output.

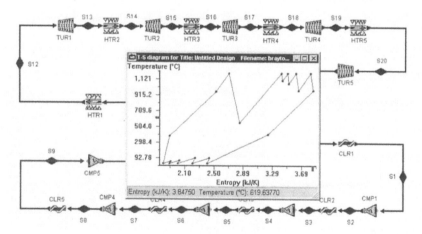

Figure 4.44d Brayton air cycle T–s diagram.

$p_{10} = p_{11} = p_{12} = 1100$ kPa, $T_1 = T_3 = T_5 = T_7 = T_9 = 20°C$, $T_{12} = T_{14} = T_{16} = T_{18} = T_{20} = 1200°C$, $T_{22} = 400°C$, $\dot{m}dot_1 = 1$ kg/sec, $\eta_{tur1} = \eta_{tur2} = \eta_{tur3} = \eta_{tur4} = \eta_{tur5} = 83\%$, and $\eta_{cmpr1} = \eta_{cmpr2} = \eta_{cmpr3} = \eta_{cmpr4} = \eta_{cmpr5} = 80\%$.

Try to modify his design (use p_{16}, p_{18}, p_6, and p_8 as design parameters only) to get a better cycle thermal efficiency than his cycle efficiency.

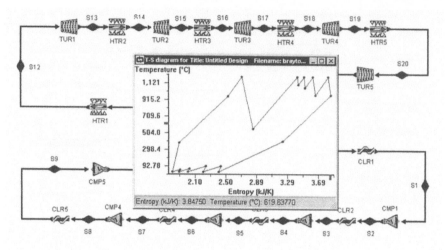

Figure 4.44e   Brayton air cycle design parameter optimization.

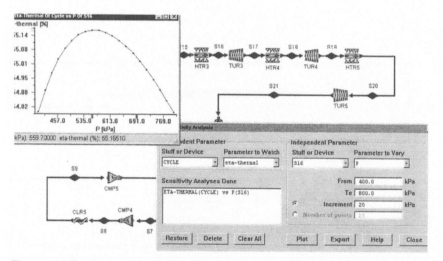

Figure 4.44f   Brayton air cycle design parameter optimization.

4.15 SUMMARY

Heat engines that use gases as the working fluid in an open system model
are treated in this chapter. The modern gas turbine engine operates on the
Brayton cycle. The basic Brayton cycle consists of an isentropic
compression process, an isobaric combustion process, an isentropic

expansion process, and an isobaric cooling process. The thermal efficiency of the basic Brayton cycle depends on the compression ratio across the compressor. The compression ratio is defined as $r_p = p_{compressor\ exit}/p_{compressor\ inlet}$. The thermal efficiency of the basic Brayton cycle can be improved by a split shaft and regeneration. The net work of the basic Brayton cycle can be improved by intercooling and reheating. The power of the basic Brayton cycle can be controlled by air-bleed.

The Ericsson, Wicks, and ice cycles are modified Brayton cycles with many stages of intercooling and reheat. It has the same efficiency of the Carnot cycle operating between the same temperature limits. The Feher cycle is a cycle operating above the critical point of the working fluid.

5

Combined Cycle and Cogeneration

5.1 COMBINED CYCLE

There are situations where it is desirable to combine several cycles in series in order to take advantage of a very wide temperature range or to utilize what would otherwise be waste heat to improve efficiency. Such a cycle is called a *cascaded cycle*. A cascaded cycle made of three Rankine cycles in series is shown schematically in Fig. 5.1. The cascaded cycle is made of three subcycles. Subcycle A, 1-2-3-4-1, is the topping cycle; subcycle B, 5-6-7-8-5, is the middle cycle; and subcycle C, 9-10-11-12-9, is the bottom cycle. The waste heat of the upstream cycle is the heat input of the downstream cycle. Since the net work output is equal to the sum of the three outputs and the heat input is that of the topping cycle alone, a substantial efficiency increase is possible.

A cascaded cycle made of two cycles in series, called a *combined cycle*, is shown schematically in Fig. 5.2. The combined cycle is made of two subcycles. Subcycle A, 1-2-3-4-1, is the upstream topping cycle; subcycle B, 5-6-7-8-5, is the downstream bottom cycle. The waste heat of the upstream topping cycle is the heat added to the downstream bottom cycle. The power output is the sum of the output of the upstream topping cycle and the output of the downstream bottom cycle.

The energy flow of the combined cycle is shown in Fig. 5.3.

The overall efficiency of the combined cycle is the total output work $(W_1 + W_2)$ divided by the heat input, Q_1. Referring to Fig. 5.3, we have

$$\eta = (W_1 + W_2)/Q_1 \tag{5.1}$$

$$W_1 = \eta_A Q_1 \tag{5.2}$$

and

$$W_2 = \eta_B Q_2 \tag{5.3}$$

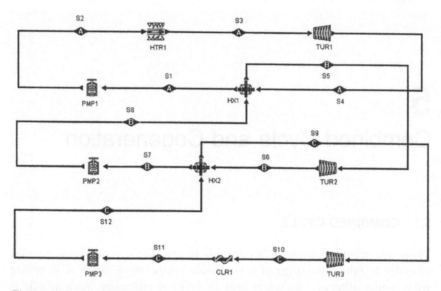

Figure 5.1 Cascaded cycle.

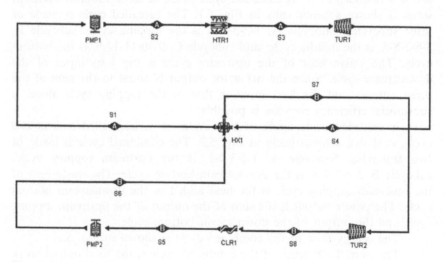

Figure 5.2 Combined cycle.

By substituting W_1 and W_2 into Eq. (5.1), the following efficiency expression is obtained:

$$\eta = 1 - (1 - \eta_A)(1 - \eta_B) \tag{5.4}$$

The combined cycle efficiency may, therefore, be substantially greater than the cycle efficiency of any of its components operating alone.

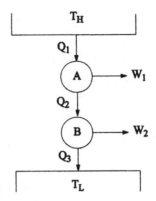

Figure 5.3 Combined cycle energy flow diagram.

A numerical example is given in the following to illustrate the cycle analysis of the combined cycle.

Example 5.1

A combined cycle made of two cycles is shown in Fig. 5.2. The upstream topping cycle is a steam Rankine cycle and the downstream bottom cycle is an ammonia Rankine cycle. The following information is provided: steam boiler pressure $= 2\,MPa$, steam superheater temperature $= 400°C$, steam condenser (heat exchanger) pressure $= 20\,kPa$, ammonia boiler (heat exchanger) pressure $= 1200\,kPa$, ammonia condenser pressure $= 800\,kPa$, and mass flow rate of steam $= 1\,kg/sec$.

Determine the total pump power input, total turbine power output, rate of heat added, rate of heat removed, cycle efficiency, and mass flow rate of ammonia.

To solve this problem with CyclePad, we take the following steps:

1. Build as shown in Fig. 5.2.
2. Analysis:
   a. Assume a process each for the seven devices: (1) pumps as adiabatic with 100% efficiency, (2) turbines as adiabatic with 100% efficiency, (3) heat exchanger as isobaric on both cold and hot sides, (4) ammonia condenser as isobaric, and (5) steam boiler as isobaric.
   b. Input the given information: (1) working fluid of cycle B is ammonia, and working fluid of cycle A is water, (2) inlet pressure and quality of the ammonia pump are 800 kPa and 0, (3) inlet temperature and pressure of the steam turbine are

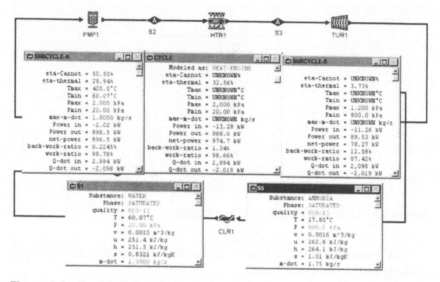

Figure 5.4 Combined Rankine cycle.

400°C and 2 MPa, (4) inlet quality and pressure of the ammonia turbine are 1 and 1200 kPa, (5) inlet pressure and quality
of the water pump are 20 kPa and 0, and (6) steam mass flow
rate is 1 kg/sec.
3. Display results.

The answers are: combined cycle—power input = −13.28 kW, power
output = 988.0 kW, net power output = 974.7 kW, rate of heat
added = 2994 kW, rate of heat removed = − 2019 kW, and η = 32.56%;
topping steam cycle—power input = − 2.02 kW, power output = 898.5 kW,
net power output = 896.5 kW, rate of heat added = 2994 kW, rate of heat
removed = − 2098 kW, and η = 29.94%; bottom ammonia cycle—power
input = − 11.26 kW, power output = 89.53 kW, net power output =
78.27 kW, rate of heat added = 2098 kW, rate of heat removed =
− 2019 kW, η = 3.73%, and mass rate flow = 1.75 kg/sec. (See Fig. 5.4.)

Example 5.2

Figure 5.5a depicts a combined plant in which a closed Brayton helium
nuclear plant releases heat to a recovery steam generator, which supplies
heat to a Rankine steam plant. The generator is provided with a gas
burner for supplementary additional heat when the demand of steam
power is high. The Rankine plant is a regenerative cycle.

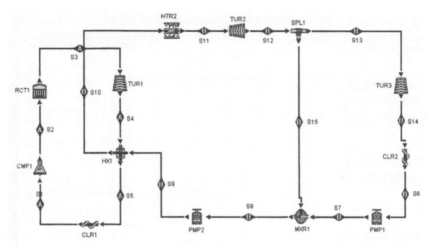

Figure 5.5a Combined Rankine cycle.

The data given below correspond approximately to the design conditions for the combined plant: $p_1 = 100\,\text{kPa}$, $T_1 = 30°\text{C}$, $p_2 = 800\,\text{kPa}$, $T_3 = 1400°\text{C}$, $T_5 = 200°\text{C}$, $p_6 = 5\,\text{kPa}$, $x_6 = 0$, $p_8 = 1\,\text{MPa}$, $x_8 = 0$, $p_9 = 6\,\text{MPa}$, $T_{10} = 400°\text{C}$, $mdot_{10} = 1\,\text{kg/sec}$, $T_{11} = 500°\text{C}$, and $\eta_{\text{compressor}} = \eta_{\text{turbine}} = \eta_{\text{pump}} = 100\%$.

Determine the power required by the compressor, power required by pumps #1 and #2, power produced by turbines #1, #2, and #3, rate of heat added by the nuclear reactor, net power produced by the Brayton gas turbine plant, net power produced by the Rankine plant, rate of heat removed by coolers #1 and #2, rate of heat exchanged in the heat exchanger, rate of heat added in the gas burner, mass rate flow of helium in the Brayton cycle, mass rate flow of steam extracted to the feed-water heater (mixing chamber), cycle efficiency of the Brayton plant, cycle efficiency of the Rankine plant, and cycle efficiency of the combined Brayton–Rankine plant.

To solve this problem with CyclePad, we take the following steps:

1. Build as shown in Fig. 5.5a.
2. Analysis:
   a. Assume a process each for the twelve devices: (1) pumps as adiabatic with 100% efficiency, (2) turbines as adiabatic with 100% efficiency, (3) heat exchanger as isobaric on both cold and hot sides, (4) nuclear reactor, mixing chamber, heater, and coolers as isobaric, and (5) splitter as isoparametric.
   b. Input the given information: (1) working fluid of cycle A is helium, and working fluid of cycle B is water, (2) $p_1 = 100\,\text{kPa}$, $T_1 = 30°\text{C}$, $p_2 = 800\,\text{kPa}$, $T_3 = 1400°\text{C}$, $T_5 = 200°\text{C}$, $p_6 = 5\,\text{kPa}$,

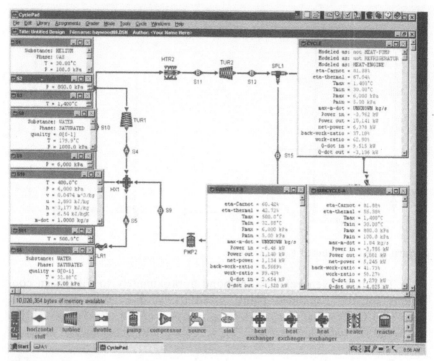

Figure 5.5b   Combined Rankine cycle.

$x_6 = 0$, $p_8 = 1$ MPa, $x_8 = 0$, $p_9 = 6$ MPa, $T_{10} = 400°C$, $mdot_{10} = 1$ kg/sec, and $T_{11} = 500°C$.

3. Display results.

The answers are: $Wdot_{compressor} = -3756$ kW, $Wdot_{pump\ \#1\ and\ \#2} = -6.46$ kW, $Wdot_{turbine\#1} = 9001$ kW, $Wdot_{turbine\#2} = 512.7$ kW, $Wdot_{turbine\#3} = 637.5$ kW, $Qdot_{reactor} = 9270$ kW, $Wdot_{net\ Brayton} = 5245$ kW, $Wdot_{nee\ Rankine} = 1134$ kW, $Qdot_{cooler\#1} = -1616$ kW, $Qdot_{cooler\#2} = -1520$ kW, $Qdot_{HX} = 2408$ kW, $Qdot_{gas\ burner} = 245.3$ kW, $mdot_{helium} = 1.84$ kg/sec, $mdot_{15} = 0.2245$ kg/sec, $\eta_{Brayton} = 5245/9270 = 56.58\%$, $\eta_{Rankine} = 1134/2654 = 42.73\%$, and $\eta_{combined} = (5245 + 1134)/(9270 + 245.3) = 67.04\%$. (See Fig. 5.5b.)

The combined cycle is designed to gain maximum efficiency from the primary heat source. In most cases, both cycles are used for the same purpose—usually to generate electricity. The major combined-cycle options currently under development include open-cycle gas turbines, closed-cycle turbines, fuel cells, and magnetohydrodynamics with vapor cycles. Other combined cycles include the Diesel/Rankine cycle (Boretz, J.E.,

Rankine engine compounding of Diesel engines. *Proceedings of the Intersociety Energy Conversion Engineering Conference*, vol. 2, pp. 193–197, 1990) and dual gas turbine combined cycle (Weston, K.C., *Proceedings of the Intersociety Energy Conversion Engineering Conference*, vol. 1, pp. 955–958, 1993), etc.

Review Problems 5.1   Combined Cycle

1. What is a combined cycle?
2. What is the heat input to the whole combined cycle?
3. What is the total work output of the whole combined cycle?
4. Is the efficiency of the combined cycle better than that of any of the individual cycles that made the combined cycle? Why?
5. Redo Example 5.2 without the gas burner.

5.2   TRIPLE CYCLE IN SERIES

A cascaded cycle made of three cycles in series, called a *triple cycle*, is shown schematically in Fig. 5.6. The combined cycle is made of three subcycles. Subcycle A, 1-2-3-4-5, is the upstream topping cycle; subcycle C, 6-7-8-9-6, is the midcycle, and subcycle B, 10-11-12-13-10, is the downstream bottom cycle. A part of the waste heat of the upstream topping cycle is the heat added to the midcycle, and the waste heat of the

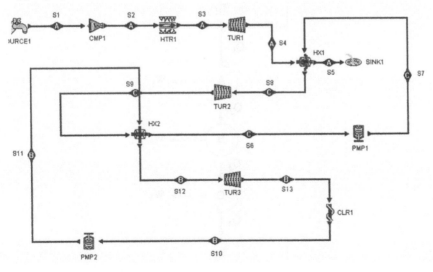

Figure 5.6   Triple cycle.

midcycle is the heat added to the downstream bottom cycle. The power output is the sum of the output of the upstream topping cycle, the output of the midcycle, and the output of the downstream bottom cycle.

The energy flow of the combined cycle is shown in Fig. 5.7.

The overall efficiency of the triple cycle is the total output work $(W_1 + W_2 + W_3)$ divided by the heat input, Q_1. Referring to Fig. 5.7, we have

$$\eta = (W_1 + W_2 + W_{Co})/Q_1 \tag{5.5}$$

$$W_1 = \eta_A Q_1 \tag{5.6}$$

$$W_2 = \eta_B Q_2 \tag{5.7}$$

and

$$W_3 = \eta_C Q_3 \tag{5.8}$$

Substituting W_1, W_2, and W_3 into Eq. (5.5), the following efficiency expression is obtained:

$$\eta = 1 - (1 - \eta_A)(1 - \eta_B)(1 - \eta_C) \tag{5.9}$$

The triple cycle efficiency may, therefore, be substantially greater than the cycle efficiency of any of its components operating alone.

A numerical example is given in the following to illustrate the cycle analysis of the triple cycle in series.

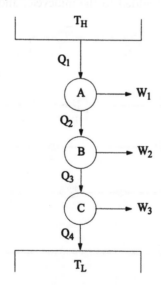

Figure 5.7 Triple cycle in series energy flow diagram.

Example 5.3

Figure 5.6 depicts a triple cycle in which an open Brayton plant releases heat to a recovery steam generator, which supplies heat to a Rankine steam plant. The Rankine steam plant releases heat to a recovery R-12 generator, which supplies heat to a Rankine R-12 plant. The data given below correspond approximately to the design conditions for the triple plant: $p_1 = 101.3\,\text{kPa}$, $T_1 = 15°C$, $p_2 = 8\,p_1$, $T_3 = 1200°C$, $p_4 = 101.3\,\text{kPa}$, $T_5 = 450°C$, $p_6 = 20\,\text{kPa}$, $x_6 = 0$, $T_7 = 60.1°C$, $p_8 = 2\,\text{MPa}$, $T_8 = 400°C$, $x_{10} = 0$, $p_{10} = 500\,\text{kPa}$, $T_{10} = 400°C$, $p_{12} = 900\,\text{kPa}$, $x_{12} = 1$, $m\text{dot}_1 = 1\,\text{kg/sec}$, and $\eta_{\text{compressor}} = \eta_{\text{turbine}} = \eta_{\text{pump}} = 100\%$.

Determine the power required by the compressor, power required by pumps #1 and #2, power produced by turbine #1, #2, and #3, rate of heat added to the Brayton cycle, net power produced by the Brayton gas turbine plant, net power produced by the steam Rankine plant, rate of heat exchanged in the heat exchanger #1, rate of heat added to the R-12 Rankine plant, mass rate flow of air in the Brayton cycle, mass rate flow of steam in the Rankine steam plant, mass rate flow of R-12 in the Rankine R-12 plant, cycle efficiency of the Brayton plant, cycle efficiency of the steam Rankine plant, cycle efficiency of the R-12 Rankine plant, and cycle efficiency of the triple plant.

To solve this problem with CyclePad, we take the following steps:

1. Build as shown in Fig. 5.6.
2. Analysis:
   a. Assume a process each for the devices: (1) pumps as adiabatic with 100% efficiency, (2) turbines as adiabatic with 100% efficiency, (3) heat exchanger as isobaric on both cold and hot sides, and (4) heater and coolers as isobaric.
   b. Input the given information as shown in Fig. 5.8a: (1) working fluid of cycle A is air, working fluid of cycle B is R-12, and working fluid of cycle C is water, (2) $p_1 = 101.3\,\text{kPa}$, $T_1 = 15°C$, $p_2 = 8\,p_1$, $T_3 = 1200°C$, $p_4 = 101.3\,\text{kPa}$, $T_5 = 450°C$, $p_6 = 20\,\text{kPa}$, $x_6 = 0$, $T_7 = 60.1°C$, $p_8 = 2\,\text{MPa}$, $T_8 = 400°C$, $x_{10} = 0$, $p_{10} = 500\,\text{kPa}$, $T_{10} = 400°C$, $p_{12} = 900\,\text{kPa}$, $x_{12} = 1$, and $m\text{dot}_1 = 1\,\text{kg/sec}$.
3. Display results.

The answers shown in Fig. 5.8b are: $W\text{dot}_{\text{compressor} \#1} = -234.6\,\text{kW}$, $W\text{dot}_{\text{pump}\#1} = -0.0609\,\text{kW}$, $W\text{dot}_{\text{pump}\#2} = -0.9171\,\text{kW}$, $W\text{dot}_{\text{turbine}\#1} = 662.2\,\text{kW}$, $W\text{dot}_{\text{turbine}\#2} = 27.13\,\text{kW}$, $W\text{dot}_{\text{turbine}\#3} = 4.32\,\text{kW}$, $Q\text{dot}_{\text{HTR}\#1} = 954.4\,\text{kW}$, $W\text{dot}_{\text{net Brayton}} = 427.5\,\text{kW}$, $W\text{dot}_{\text{net steam Rankine}} = 27.07\,\text{kW}$, $W\text{dot}_{\text{net R12 Rankine}} = 3.40\,\text{kW}$, $W\text{dot}_{\text{net triple}} =$

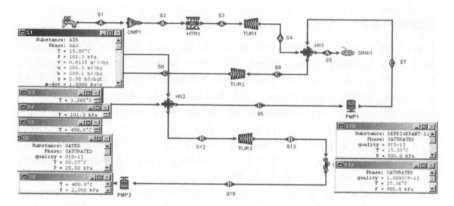

Figure 5.8a Input information;

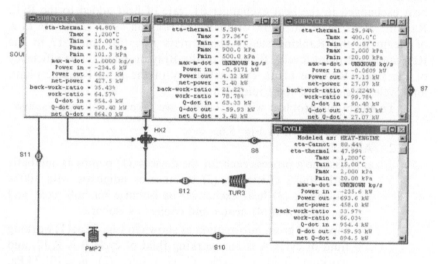

Figure 5.8b Output information.

458.0 kW, $\dot{Q}$dot$_{cooler\,\#1}$ = $-$ 59.93 kW, $\dot{m}$dot$_{steam}$ = 0.0302 kg/sec, $\dot{m}$dot$_{R12}$ = 0.4237 kg/sec, $\eta_{Brayton}$ = 44.8%, $\eta_{steam\;Rankine}$ = 29.94%, $\eta_{R12\;Rankine}$ = 5.38%, and η_{triple} = 47.99%.

Review Problems 5.2   Triple Cycle in Series

Redo Example 5.3 with $\eta_{compressor}$ = $\eta_{turbine}$ = 83% and η_{pump} = 100%.

5.3   TRIPLE CYCLE IN PARALLEL

A cycle made of three cycles in parallel is shown schematically in Fig. 5.9. The triple cycle is made of three subcycles: Subcycle A, 1-2-3-4-5-6, is the upstream topping open gas turbine cycle; subcycle C, 7-8-9-10-7, and subcycle B, 11-12-13-14-11, are the downstream parallel bottom cycles. A part of the waste heat of the upstream topping cycle is the heat added to subcycle C, and another part of the waste heat of the upstream topping cycle is the heat added to subcycle B. The power output is the sum of the output of subcycles A, B, and C. The overall efficiency of the triple cycle in parallel is the total output work $(W_{net,A} + W_{net,B} + W_{Net,C})$ divided by the heat input added to the heater in the gas turbine cycle, Q_1.

$$\eta = (W_{net,A} + W_{net,B} + W_{Net,C})/Q_1 \tag{5.10}$$

The triple cycle efficiency may, therefore, be substantially greater than the cycle efficiency of any of its components operating alone.

A numerical example is given in the following to illustrate the cycle analysis of the triple cycle in parallel.

Example 5.4

An open Brayton plant releases a part of its waste heat to a recovery steam generator, which supplies heat to a Rankine steam plant, and another part of its waste heat to a recovery R-12 generator, which supplies

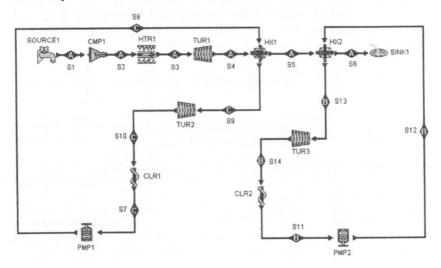

Figure 5.9   Triple cycle in parallel.

heat to a Rankine R-12 plant. The data given below correspond approximately to the design conditions for a triple plant in parallel: $p_1 = 101.3\,\text{kPa}$,　$T_1 = 15°\text{C}$,　$p_2 = 8\,p_1$,　$T_3 = 1200°\text{C}$,　$p_4 = 101.3\,\text{kPa}$, $T_5 = 450°\text{C}$,　$T_6 = 40°\text{C}$,　$x_7 = 0$,　$p_7 = 20\,\text{kPa}$,　$p_9 = 2\,\text{MPa}$,　$T_9 = 400°\text{C}$, $x_{11} = 0$,　$p_{11} = 500\,\text{kPa}$,　$p_{13} = 900\,\text{kPa}$,　$x_{13} = 1$,　$m\text{dot}_1 = 1\,\text{kg/sec}$,　and $\eta_{\text{compressor}} = \eta_{\text{turbine}} = \eta_{\text{pump}} = 100\%$.

Determine the power required by the compressor, power required by pumps #1 and #2, power produced by turbines #1, #2, and #3, rate of heat added to the Brayton cycle, net power produced by the Brayton gas turbine plant, net power produced by the steam Rankine plant, rate of heat exchanged in the heat exchanger #1, rate of heat added to the R-12 Rankine plant, mass rate flow of air in the Brayton cycle, mass rate flow of steam in the Rankine steam plant, mass rate flow of R-12 in the Rankine R-12 plant, cycle efficiency of the Brayton plant, cycle efficiency of the steam Rankine plant, cycle efficiency of the R-12 Rankine plant, and cycle efficiency of the triple plant.

To solve this problem with CyclePad, we take the following steps:

1. Build as shown in Fig. 5.9.
2. Analysis:
   a. Assume a process each for the devices: (1) pumps as adiabatic with 100% efficiency, (2) turbines as adiabatic with 100% efficiency, (3) heat exchanger as isobaric on both cold and hot sides, and (4) heater and coolers as isobaric.
   b. Input the given information as shown in Fig. 5.9a: (a) working fluid of cycle A is air, working fluid of cycle B is R-12, and working fluid of cycle C is water, (2) $p_1 = 101.3\,\text{kPa}$,

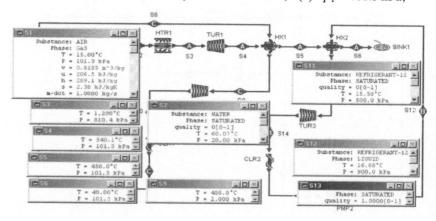

Figure 5.9a　Input information.

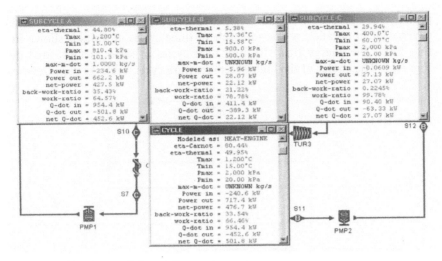

Figure 5.9b   Output information.

$T_1 = 15°C$, $p_2 = 8\,p_1$, $T_3 = 1200°C$, $p_4 = 101.3\,\text{kPa}$, $T_5 = 450°C$,
$T_6 = 40°C$, $x_7 = 0$, $p_7 = 20\,\text{kPa}$, $p_9 = 2\,\text{MPa}$, $T_9 = 400°C$,
$x_{11} = 0$, $p_{11} = 500\,\text{kPa}$, $p_{13} = 900\,\text{kPa}$, $x_{13} = 1$, $mdot_1 = 1\,\text{kg}/$
sec, and $mdot_1 = 1\,\text{kg/sec}$.

3.   Display results.

The answers shown in Fig. 5.9b are: $W\text{dot}_{\text{compressor}\#1} = -234.6\,\text{kW}$, $W\text{dot}_{\text{pump}\#1} = -0.0609\,\text{kW}$, $W\text{dot}_{\text{pump}\#2} = -5.96\,\text{kW}$, $W\text{dot}_{\text{turbine}\#1} = 662.2\,\text{kW}$, $W\text{dot}_{\text{turbine}\#2} = 27.13\,\text{kW}$, $W\text{dot}_{\text{turbine}\#3} = 28.07\,\text{kW}$, $Q\text{dot}_{\text{HTR}\#1} = 954.4\,\text{kW}$, $W\text{dot}_{\text{net Brayton}} = 427.5\,\text{kW}$, $W\text{dot}_{\text{net steam Rankine}} = 27.07\,\text{kW}$, $W\text{dot}_{\text{net R12 Rankine}} = 22.12\,\text{kW}$, $W\text{dot}_{\text{net triple}} = 476.7\,\text{kW}$, $Q\text{dot}_{\text{cooler}\#1} = -63.33\,\text{kW}$, $mdot_{\text{steam}} = 0.0302\,\text{kg/sec}$, $mdot_{\text{R12}} = 2.75\,\text{kg/sec}$, $\eta_{\text{Brayton}} = 44.8\%$, $\eta_{\text{steam Rankine}} = 29.94\%$, $\eta_{\text{R12 Rankine}} = 5.38\%$, and $\eta_{\text{triple}} = 49.95\%$.

Review Problems 5.3   Triple Cycle in Parallel

Redo Example 5.4 with $\eta_{\text{compressor}} = \eta_{\text{turbine}} = 81\%$ and $\eta_{\text{pump}} = 100\%$.

5.4   CASCADED CYCLE

There are applications when the temperature difference between the heat source and the heat sink is quite large. A single power cycle usually cannot be used to utilize the full range of the available temperature difference. A *cascade cycle* must be used to gain maximum possible efficiency from the primary heat source. A cascade cycle is several (n) power cycles connecting in series or in parallel. A cascade cycle of three cycles in series is shown in Fig. 5.10. The energy flow diagram of n cycles in series is illustrated in Fig. 5.11. The cooler of the highest temperature cycle (cycle A) provides the

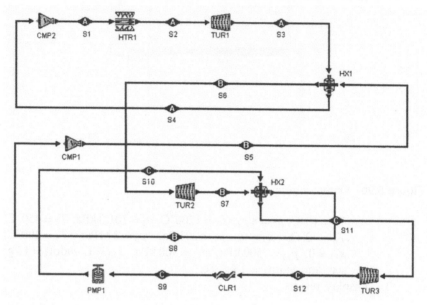

Figure 5.10   Cascade cycle with $n = 3$.

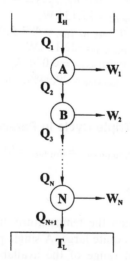

Figure 5.11   Cascade cycle energy flow diagram.

heat input to the heater of the second-highest temperature cycle (cycle B), ...,
and the cooler of the next-to-the-lowest temperature cycle provides the heat
input to the heater of the lowest temperature cycle (cycle N).

The overall efficiency of the cascaded cycle is the total output
work $(W_1 + W_2 + \cdots + W_N)$ divided by the heat input, Q_1. Referring to
Fig. 5.11, we have

$$\eta = (W_1 + W_2 + \cdots + W_N)/Q_1 \tag{5.11}$$

$$W_1 = \eta_A Q_1 \tag{5.12}$$

$$W_2 = \eta_B Q_2 \tag{5.13}$$

....

and
$$W_N = \eta_N Q_n \tag{5.14}$$

Substituting $W_1, W_2, \ldots, W_N$ into Eq. (5.11), the following efficiency
expression is obtained:

$$\eta_{\text{cascaded}} = 1 - (1 - \eta_1)(1 - \eta_2)\ldots(1 - \eta_N) \tag{5.15}$$

where η_{cascaded} is the efficiency of a cascaded cycle with n-component cycles.

The cascaded cycle efficiency may, therefore, be substantially greater
than the cycle efficiency of any of its components operating alone.

Review Problems 5.4   Cascaded Cycle

1. Build a cascaded cycle in series with $n = 5$ using CyclePad.
2. Build a cascaded cycle in parallel with $n = 5$ using CyclePad. The
   topping cycle is an open gas turbine cycle.

5.5   BRAYTON/RANKINE COMBINED CYCLE

Advances in the combined cycle power plant focus on a high-temperature
gas turbine cycle combined with a steam vapor cycle. Improvement in
cycle efficiency can be achieved by using the hot exhaust waste heat of a
high-temperature cycle to power, either partially or totally, a low-
temperature cycle. For example, since the boiler temperature of the basic
Rankine cycle is about 500°C, the exhaust gases of the gas turbine cycle with a
temperature of 500°C could be used for the boiler heat input. One arrangement of the *Brayton/Rankine cycle*, which is a combination of a two-stage
reheat Brayton cycle and a two-stage reheat Rankine cycle, is shown in

Fig. 5.12. In general, modifications of both the Brayton and Rankine cycles could also be included. Since the net work output is equal to the sum of the two outputs and the heat input is that of the topping cycle alone, a substantial increase in cycle efficiency is possible. Another arrangement of the Brayton/Rankine cycle, which is a combination of a two-stage reheat Brayton cycle and a two-stage reheat Rankine cycle, is shown in Fig. 5.13.

Example 5.5

A Brayton/Rankine cycle (Fig. 5.12) uses water as the working fluid with 1 kg/sec mass flow rate through the Rankine cycle, and air as the working

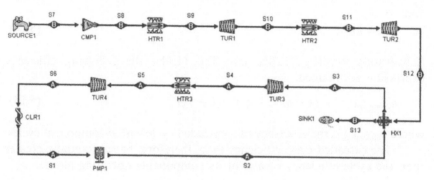

Figure 5.12   Brayton/Rankine combined cycle.

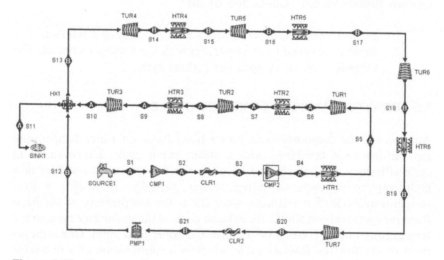

Figure 5.13   Two-stage Brayton/two-stage Rankine combined cycle.

fluid in the Brayton cycle. In the Rankine cycle, the condenser pressure is 15 kPa (p_1), the boiler pressure is 8 MPa (p_2), the reheater pressure is 5 MPa (p_4), and the superheater and reheater temperatures (T_3 and T_5) are both 400°C. In the Brayton cycle, air enters from the atmospheric source to an isentropic compressor at 20°C and 100 kPa (T_7 and p_7) and leaves at 1 MPa (p_8); air enters an isobaric heater (combustion chamber) and leaves at 1800°C (T_9); and air enters a high-pressure isentropic turbine and leaves at 600 kPa (p_{11}). Air enters a low-pressure isentropic turbine and leaves at 100 kPa (p_{12}); air enters an isobaric regenerator and leaves at 500°C (T_{13}); and air is discharged to the atmospheric sink.

Determine the mass rate flow of air through the Brayton cycle, and the thermodynamic efficiency and net power output of the Brayton/Rankine combined plant. Plot the sensitivity diagram of η (cycle efficiency) versus p_{11} (pressure at state 11).

To solve this problem by CyclePad, we take the following steps:

1. Build the cycle as shown in Fig. 5.12. Assume that the compressor, turbines, and pump are adiabatic and isentropic, and the heater, cooler, and regenerator are isobaric.
2. Input working fluid = air, $p_1 = 15$ kPa, $x_1 = 0$, mdot = 1 kg/sec, $p_3 = 8$ MPa, $T_3 = 400$°C, $p_5 = 5$ MPa, $T_5 = 400$°C, $p_7 = 100$ kPa, $T_7 = 20$°C, $p_9 = 1$ MPa, $T_9 = 1800$°C, $p_{11} = 600$ kPa, $T_{11} = 1600$°C, $p_{12} = 100$ kPa, and $T_{13} = 500$°C.
3. Display results. The answers shown in Fig. 5.14a are: (1) cycle A: $\eta = 37.52\%$, power input = −8.12 kW, power output = 1165 kW, net power output = 1157 kW, $Q\mathrm{dot_{in}} = 3084$ kW; (2) cycle B: $\eta = 47.79\%$, power input = −2267 kW, power output = 8575 kW, net power output = 6308 kW, $Q\mathrm{dot_{in}} = 13{,}200$ kW, mdot = 8.28 kg/sec; and (3) combined cycle: $\eta = 55.79\%$, power input = −2275 kW, power output = 9740 kW, net power output = 7465 kW, $Q\mathrm{dot_{in}} = 13{,}380$ kW.
4. Display sensitivity diagram of η (cycle efficiency) versus p_{11} (pressure at state 11) as shown in Fig. 5.14b.

Review Problems 5.5   Brayton/Rankine Combined Cycle

1. What is a combined Brayton/Rankine cycle? What is its purpose?
2. What is the heat source for the Rankine cycle in the combined Brayton/Rankine cycle?
3. Why is the combined Brayton/Rankine cycle more efficient than either of the cycles operating alone?

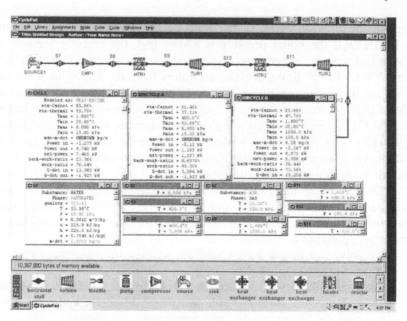

Figure 5.14a   Brayton/Rankine combined cycle.

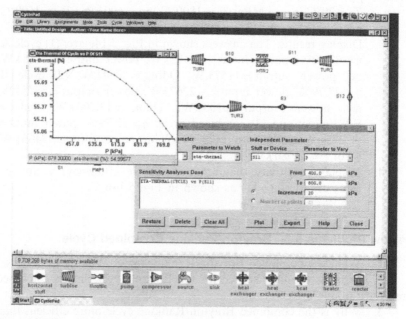

Figure 5.14b   Brayton/Rankine combined cycle.

4. A Brayton/Rankine cycle (Fig. 5.12) uses water as the working fluid with 1 kg/sec mass flow rate through the Rankine cycle, and air as the working fluid in the Brayton cycle. In the Rankine cycle, the condenser pressure is 15 kPa (p_1), the boiler pressure is 8 MPa (p_2), the reheater pressure is 5 MPa (p_4), and the superheater and reheater temperatures $(T_3$ and $T_5)$ are both 400°C. In the Brayton cycle, air enters from the atmospheric source to an isentropic compressor at 20°C and 100 kPa $(T_7$ and $p_7)$ and leaves at 1 MPa (p_8); air enters an isobaric heater (combustion chamber) and leaves at 1800°C (T_9); air enters a high-pressure isentropic turbine and leaves at 600 kPa (p_{11}). Air enters a low-pressure isentropic turbine and leaves at 100 kPa (p_{12}); air enters an isobaric regenerator and leaves at 500°C (T_{13}); and air is discharged to the atmospheric sink. Assume that the compressor efficiency is 85%.

Determine the mass flow rate of air through the Brayton cycle, and the thermodynamic efficiency and net power output of the Brayton/Rankine combined plant.

ANSWERS: (1) cycle A: $\eta = 37.52\%$, power input $= -8.12$ kW, power output $= 1165$ kW, net power output $= 1157$ kW, $Q\text{dot}_{in} = 3084$ kW; (2) cycle B: $\eta = 46.16\%$, power input $= -2667$ kW, power output $= 8575$ kW, net power output $= 5908$ kW, $Q\text{dot}_{in} = 12,800$ kW, $m\text{dot} = 8.28$ kg/sec; and (3) combined cycle: $\eta = 54.43\%$, power input $= -2675$ kW, power output $= 9740$ kW, net power output $= 7065$ kW, $Q\text{dot}_{in} = 12,980$ kW.

5. A Brayton/Rankine cycle (Fig. 5.12) uses water as the working fluid with 1 kg/sec of mass flow rate through the Rankine cycle, and air as the working fluid in the Brayton cycle. In the Rankine cycle, the condenser pressure is 15 kPa (p_1), the boiler pressure is 8 MPa (p_2), the reheater pressure is 5 MPa (p_4), and the superheater and reheater temperatures $(T_3$ and $T_5)$ are both 400°C. In the Brayton cycle, air enters from the atmospheric source to an isentropic compressor at 20°C and 100 kPa $(T_7$ and $p_7)$ and leaves at 1 MPa (p_8); air enters an isobaric heater (combustion chamber) and leaves at 1800°C (T_9); air enters a high-pressure isentropic turbine and leaves at 600 kPa (p_{11}). Air enters a low-pressure isentropic turbine and leaves at 100 kPa (p_{12}); air enters an isobaric regenerator and leaves at 500°C (T_{13}); and air is discharged to the atmospheric sink. Assume that the efficiency of the gas turbines and compressor is 85%.

Determine the mass rate flow of air through the Brayton cycle, and the thermodynamic efficiency and net power output of the Brayton/Rankine combined plant.

ANSWERS: (1) cycle A: $\eta = 37.52\%$, power input $= -8.12$ kW, power output $= 1165$ kW, net power output $= 1157$ kW, $Q\text{dot}_{in} = 3084$ kW;

(2)   cycle   B: $\eta = 37.12\%$,   power   input $= -2017$ kW,   power output $= 5513$ kW,   net   power   output $= 3496$ kW, Qdot$_{in} = 9416$ kW, $m$dot $= 6.26$ kg/sec;   and   (3)   combined   cycle: $\eta = 48.48\%$,   power input $= -2025$ kW,   power   output $= 6678$ kW,   net   power   output $=$ 4653 kW, Qdot$_{in} = 9596$ kW.

   6.   A Brayton/Rankine cycle (Fig. 5.12) uses water as the working fluid with 1 kg/sec mass rate of flow through the Rankine cycle, and air as the working fluid in the Brayton cycle. In the Rankine cycle, the condenser pressure is 15 kPa (p_1), the boiler pressure is 8 MPa (p_2), the reheater pressure is 5 MPa (p_4), and the superheater and reheater temperatures $(T_3$ and $T_5)$ are both 400°C. In the Brayton cycle, air enters from the atmospheric source to an isentropic compressor at 20°C and 100 kPa $(T_7$ and $p_7)$, and leaves at 1 MPa (p_8); air enters an isobaric heater (combustion chamber) and leaves at 1800°C (T_9); air enters a high-pressure isentropic turbine and leaves at 600 kPa (p_{11}). Air enters a low-pressure isentropic turbine and leaves at 100 kPa (p_{12}); air enters an isobaric regenerator and leaves at 500°C (T_{13}); and air is discharged to the atmospheric sink. Assume that all turbine and compressor efficiencies are 85%.

   Determine the mass flow rate of air through the Brayton cycle, and the thermodynamic efficiency and net power output of the Brayton/ Rankine combined plant.

   ANSWERS: (1) cycle A: $\eta = 32.04\%$, power input $= -8.12$ kW, power   output $= 990.3$ kW,   net   power   output $= 982.2$ kW, Qdot$_{in} = 3066$ kW; (2) cycle B: $\eta = 37.12\%$, power input $= -2017$ kW, power output $= 5513$ kW, net power output $= 3496$ kW, Qdot$_{in} = 9416$ kW, $m$dot $= 6.26$ kg/sec;   and   (3)   combined   cycle: $\eta = 46.75\%$,   power input $= -2025$ kW,   power   output $= 6503$ kW,   net   power   output $=$ 4478 kW, Qdot$_{in} = 9578$ kW.

5.6   BRAYTON/BRAYTON COMBINED CYCLE

The Brayton gas turbine engine has low capital cost compared with steam power plants. It has environmental advantages and short construction lead time. However, conventional industrial Brayton gas turbine engines have lower efficiencies. One of the technologies adopted nowadays for efficiency improvement is the utilization of Brayton/Brayton combined cycles (Najjar, Y.S.H. and Zaamout, M.S., Performance analysis of gas turbine air-bottoming combined system. *Energy Conversion and Management*, vol. 37, no. 4, 399–403, 1996). An air-bottoming cycle instead of steam bottoming reduces the cost of hardware installations and could achieve a thermal efficiency of about 49%, which does not deteriorate at part load as happens with the basic Brayton gas turbine engine. A Brayton/Brayton combined cycle is shown in Fig. 5.15. In this system, an air turbine is used to convert the split-shaft turbine exhaust heat from the top cycle. Three intercooled compressor stages are used to reduce the compressor work and the temperature of the air delivered to an air-to-gas heat exchanger. Air, heated by the exhaust gas, is then delivered to the air turbine which, in turn, produces power. The combined system is expected to be simpler and much less expensive to build, operate, and maintain than the Brayton/ Rankine combined system.

Example 5.6

A Brayton/Brayton cycle (Fig. 5.15) uses air as the working fluid with 1 kg/sec mass flow rate through the top Brayton split-shaft turbine cycle, and air as the working fluid in the bottom Brayton cycle.

In the top Brayton cycle, air at a mass flow rate of 1 kg/sec enters from the atmospheric source to an isentropic compressor at 290 K and 100 kPa (T_1 and p_1) and leaves at 1 MPa (p_2); air enters an isobaric heater (combustion chamber) and leaves at 1400 K (T_3); air goes through a high-pressure isentropic turbine (TUR1) and a low-pressure isentropic turbine (TUR2); air enters an isobaric heat exchanger and leaves at 700 K (T_6) and 100 kPa (p_6); and air is discharged to the atmospheric sink. In the bottom Brayton cycle, air at a mass flow rate of 0.12 kg/sec enters from the atmospheric source to an isentropic compressor (CMP2) at 290 K and 100 kPa (T_7 and p_7), and leaves at 200 kPa (p_2); air enters an isobaric inter-cooler (CLR1) and leaves at 290 K (T_9); air leaves an isentropic compressor (CMP3) at 400 kPa (p_{10}); air enters an isobaric intercooler (CLR2) and leaves at 290 K (T_{11}); air leaves another isentropic compressor

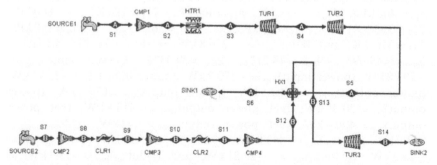

Figure 5.15 Brayton/Brayton combined cycle.

(CMP4) at 800 kPa (p_{12}); and air at 473 K and 100 kPa (T_{14} and p_{14}) is discharged to the atmospheric sink.

Determine the pressure and temperature of each state, power required by the top-cycle compressor, power produced by the top-cycle turbines, thermal efficiency of the combined cycle, thermal efficiency of the top cycle, thermal efficiency of the bottom cycle, power input to the combined cycle, power output by the combined cycle, power net output of the combined cycle, rate of heat added to the combined cycle, rate of heat removed from the combined cycle, power input to the top cycle, power output by the top cycle, power net output of the top cycle, rate of heat added to the top cycle, rate of heat removed from the top cycle, power input to the bottom cycle, power output by the bottom cycle, power net output of the bottom cycle, rate of heat added to the bottom cycle, and rate of heat removed from the bottom cycle.

To solve this problem by CyclePad, we take the following steps:

1. Build the cycle as shown in Fig. 5.15. Assume that the compressors and turbines are adiabatic and isentropic, and the heater, coolers, and heat exchanger are isobaric.
2. Input working fluid = air, $p_1 = 100$ kPa, $T_1 = 290$ K, $mdot = 1$ kg/sec, $p_2 = 1$ MPa, $T_3 = 1400$ K, $p_6 = 100$ kPa, $T_6 = 700$ K, $p_7 = 100$ kPa, $T_7 = 290$ K, $p_9 = 200$ kPa, $T_9 = 290$ K, $p_{11} = 400$ kPa, $T_{11} = 290$ K, $p_{12} = 800$ kPa, $p_{14} = 100$ kPa, and $T_{14} = 473$ K. Read $Wdot_{cmp1} = -270.8$ kW and input $Wdot_{tur1} = 270.8$ kW.
3. Display results.

The answers are: (1) $p_1 = 100$ kPa, $T_1 = 290$ K, $p_2 = 1$ MPa, $T_2 = 559.9$ K, $p_3 = 1$ MPa, $T_3 = 1400$ K, $p_4 = 472.6$ kPa, $T_4 = 1130$ K, $p_5 = 100$ kPa, $T_5 = 725.1$ K, $p_6 = 100$ kPa, $T_6 = 700$ K, $p_7 = 100$ kPa, $T_7 = 290$ K, $p_8 = 200$ kPa, $T_8 = 353.5$ K, $p_9 = 100$ kPa, $T_9 = 290$ K, $p_{10} = 400$ kPa, $T_{10} = 353.5$ K, $p_{11} = 400$ kPa, $T_{11} = 290$ K, $p_{12} = 800$ kPa, $T_{12} = 353.5$ K, $p_{13} = 800$ kPa, $T_{13} = 562.9$ K, $p_{14} = 100$ kPa, $T_{14} = 310.7$ K, and $Wdot_{cmp1} = -270.8$ kW as shown in Fig. 5.15a; (2) $\eta_{comb} = 49.09\%$, $\eta_{top} = 48.21\%$, $\eta_{bot} = 29.43\%$, (power input)$_{comb} = -293.8$ kW, (power input)$_{top} = -270.2$ kW, (power input)$_{bot} = -22.94$ kW, (power output)$_{comb} = 707.5$ kW, (power output)$_{top} = 677.2$ kW, (power output)$_{bot} = 30.36$ kW, (net power output)$_{comb} = 413.8$ kW, (net power output)$_{top} = 406.4$ kW, (net power output)$_{bot} = 7.42$ kW, ($Qdot_{in}$)$_{comb} = 843.0$ kW, ($Qdot_{in}$)$_{top} = 843.0$ kW, ($Qdot_{in}$)$_{bot} = 25.21$ kW; ($Qdot_{out}$)$_{comb} = -15.3$ kW, ($Qdot_{out}$)$_{top} = -25.21$ kW, and ($Qdot_{out}$)$_{bot} = -15.3$ kW as shown in Fig. 5.15b.

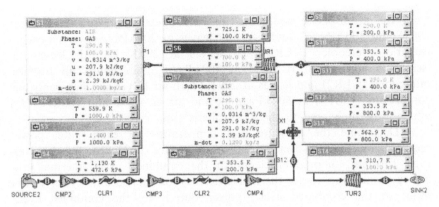

Figure 5.15a   Brayton/Brayton combined cycle result.

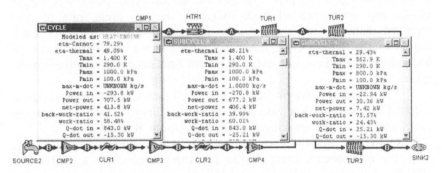

Figure 5.15b   Brayton/Brayton combined cycle result.

Review Problems 5.6   Brayton/Brayton Combined Cycle

1.   What are the advantages of a combined Brayton/Brayton cycle?

2.   A Brayton/Brayton cycle (Fig. 5.15) uses air as the working fluid with 1.2 kg/sec mass flow rate through the top Brayton split-shaft turbine cycle, and air as the working fluid in the bottom Brayton cycle.

In the top Brayton cycle, air at a mass flow rate of 1 kg/sec enters from the atmospheric source to an isentropic compressor at 290 K and 100 kPa (T_1 and p_1), and leaves at 1 MPa (p_2); air enters an isobaric heater (combustion chamber) and leaves at 1400 K (T_3); air goes through a high-pressure isentropic turbine (TUR1) and a low-pressure isentropic turbine (TUR2); air enters an isobaric heat exchanger and leaves at 700 K (T_6) and 100 kPa (p_6); and air is discharged to the atmospheric sink. In the bottom Brayton cycle, air at a mass flow rate of 0.12 kg/sec enters from the atmospheric source to an isentropic compressor (CMP2) at 290 K and 100 kPa (T_7 and p_7), and leaves at 200 kPa (p_2); air enters an isobaric intercooler (CLR1) and leaves at 290 K (T_9); air leaves an isentropic compressor (CMP3) at 400 kPa (p_{10}); air enters an isobaric intercooler (CLR2) and leaves at 290 K (T_{11}); air leaves another isentropic compressor (CMP4) at 800 kPa (p_{12}); and air at 473 K and 100 kPa (T_{14} and p_{14}) is discharged to the atmospheric sink.

Determine the pressure and temperature of each state, power required by the top-cycle compressor, power produced by the top-cycle turbines, thermal efficiency of the combined cycle, thermal efficiency of the top cycle, thermal efficiency of the bottom cycle, power input to the combined cycle, power output by the combined cycle, power net output of the combined cycle, rate of heat added to the combined cycle, rate of heat removed from the combined cycle, power input to the top cycle, power output by the top cycle, power net output of the top cycle, rate of heat added to the top cycle, rate of heat removed from the top cycle, power input to the bottom cycle, power output by the bottom cycle, power net output of the bottom cycle, rate of heat added to the bottom cycle, and rate of heat removed from the bottom cycle.

3. A Brayton/Brayton cycle (Fig. 5.15) uses air as the working fluid with 1.2 kg/sec mass flow rate through the top Brayton split-shaft turbine cycle, and air as the working fluid in the bottom Brayton cycle.

In the top Brayton cycle, air at a mass flow rate of 1 kg/sec enters from the atmospheric source to an isentropic compressor at 290 K and 100 kPa (T_1 and p_1), and leaves at 1 MPa (p_2); air enters an isobaric heater (combustion chamber) and leaves at 1500 K (T_3); air goes through a high-pressure isentropic turbine (TUR1) and a low-pressure isentropic turbine (TUR2); air enters an isobaric heat exchanger and leaves at 700 K (T_6) and 100 kPa (p_6); and air is discharged to the atmospheric sink. In the bottom Brayton cycle, air at a mass flow rate of 0.12 kg/sec enters from the atmospheric source to an isentropic compressor (CMP2) at 290 K and 100 kPa (T_7 and p_7), and leaves at 200 kPa (p_2); air enters an isobaric intercooler (CLR1) and leaves at 290 K (T_9); air leaves an isentropic compressor (CMP3) at 400 kPa (p_{10}); air enters an isobaric intercooler (CLR2) and leaves at 290 K (T_{11}); air leaves another isentropic compressor (CMP4) at 800 kPa (p_{12}); and air at 473 K and 100 kPa (T_{14} and p_{14}) is discharged to the atmospheric sink.

Determine the pressure and temperature of each state, power required by the top-cycle compressor, power produced by the top-cycle turbines, thermal efficiency of the combined cycle, thermal efficiency of the top cycle, thermal efficiency of the bottom cycle, power input to the combined cycle, power output by the combined cycle, power net output of the combined cycle, rate of heat added to the combined cycle, rate of heat removed from the combined cycle, power input to the top cycle, power output by the top cycle, power net output of the top cycle, rate of heat added to the top cycle, rate of heat removed from the top cycle, power input to the bottom cycle, power output by the bottom cycle, power net output of the bottom cycle, rate of heat added to the bottom cycle, and rate of heat removed from the bottom cycle.

5.7 RANKINE/RANKINE COMBINED CYCLE

A Rankine/Rankine combined cycle is shown in Fig. 5.16. The exhaust from the top steam turbine (TUR1) is hot enough to generate freon vapor in a waste-heat boiler. The freon vapor generated can power a freon turbine, thus increasing the total work produced. The Rankine/Rankine combined cycle has a thermal efficiency greater than either a steam or freon cycle may have by itself. The power plant occupies less area, and the fuel requirements are less.

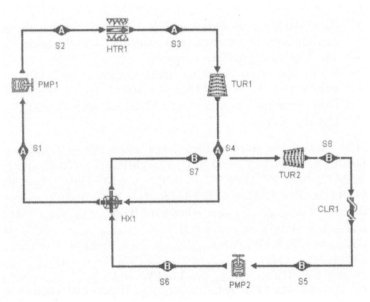

Figure 5.16 Rankine/Rankine combined cycle.

Example 5.7

A Rankine/Rankine cycle (Fig. 5.16) uses steam as the working fluid with 1 kg/sec mass flow rate through the top Rankine cycle, and Freon12 as the working fluid in the bottom Rankine cycle. The steam condenser (HX1) pressure is 20 kPa, the boiler pressure is 3 MPa, and the steam superheater temperature is 400°C. The steam mass flow rate is 1 kg/sec.

In the bottom cycle, the freon condenser (CLR1) temperature is 20°C, and the freon boiler temperature is 35°C. There is no superheater in the freon cycle.

Determine (1) the temperature and pressure of each state, and (2) the mass flow rate of the freon cycle, thermal efficiency of the combined cycle, thermal efficiency of the top cycle, thermal efficiency of the bottom cycle, power input to the combined cycle, power output by the combined cycle, power net output of the combined cycle, rate of heat added to the combined cycle, rate of heat removed from the combined cycle, power input to the top cycle, power output by the top cycle, power net output of the top cycle, rate of heat added to the top cycle, rate of heat removed from the top cycle, power input to the bottom cycle, power output by the bottom cycle, power net output of the bottom cycle, rate of heat added to the bottom cycle, and rate of heat removed from the bottom cycle.

To solve this problem by CyclePad, we take the following steps:

1. Build the cycle as shown in Fig. 5.16. Assume that the pumps and turbines are adiabatic and isentropic, and the heater, cooler, and heat exchanger are isobaric.
2. Input top cycle working fluid = steam, $p_1 = 20$ kPa, $x_1 = 0$, $mdot = 1$ kg/sec, $p_3 = 3$ MPa, $T_3 = 400$°C, bottom cycle working fluid = Freon12, $x_7 = 1$, $T_7 = 35$°C, $x_5 = 0$, and $T_5 = 20$°C.
3. Display results.

The answers are given in Figs. 5.16a and 5.16b as: (1) $p_1 = 20$ kPa, $T_1 = 60.7$°C, $p_2 = 3$ MPa, $T_2 = 60.20$°C, $p_3 = 3$ MPa, $T_3 = 400$°C, $p_4 = 20$ kPa, $T_4 = 60.07$°C, $p_5 = 567.3$ kPa, $T_5 = 20$°C, $p_6 = 847.9$ kPa, $T_6 = 20.96$°C, $p_7 = 847.9$ kPa, $T_7 = 35$°C, $p_8 = 567.3$ kPa, $T_8 = 20$°C; and (2) $mdot_{Freon} = 14.06$ kg/sec, $Wdot_{pump\ top\ cycle} = -3.04$ kW, $Wdot_{turb\ top\ cycle} = 941.1$ kW, $\eta_{combined\ cycle} = 34.06\%$, $\eta_{top\ cycle} = 31.52\%$, $\eta_{bottom\ cycle} = 3.70\%$, $Wdot_{in\ combined\ cycle} = -25.76$ kW, $Wdot_{out\ combined\ cycle} = 1039$ kW, $Wdot_{net\ combined\ cycle} = 1014$ kW, $Qdot_{in\ combined\ cycle} = 2976$ kW, $Qdot_{out\ combined\ cycle} = -1962$ kW, $Wdot_{in\ top\ cycle} = -3.04$ kW, $Wdot_{out\ top\ cycle} = 941.1$ kW, $Wdot_{net\ top\ cycle} = 938.1$ kW, $Qdot_{in\ combined\ cycle} = 2976$ kW, $Qdot_{out\ combined\ cycle} = -2038$ kW, $Wdot_{in\ bottom\ cycle} = -22.71$ kW, $Wdot_{out\ bottom\ cycle} =$

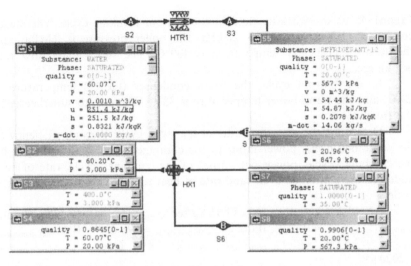

Figure 5.16a Rankine/Rankine combined cycle.

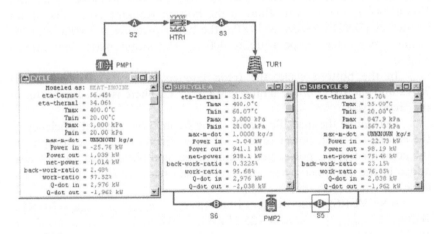

Figure 5.16b Rankine/Rankine combined cycle.

98.19 kW, $\dot{W}$dot$_{\text{net bottom cycle}} = 75.46$ kW, $\dot{Q}$dot$_{\text{in combined cycle}} = 2038$ kW, and $\dot{Q}$dot$_{\text{out combined cycle}} = -1962$ kW.

Review Problems 5.7 Rankine/Rankine Combined Cycle

1. A Rankine/Rankine cycle (Fig. 5.16) uses steam as the working fluid with 1 kg/sec mass flow rate through the top Rankine cycle, and

Freon134a as the working fluid in the bottom Rankine cycle. The steam condenser (HX1) pressure is 20 kPa, the boiler pressure is 2 MPa, and the steam superheater temperature is 400°C. The steam mass flow rate is 1 kg/sec.

In the bottom cycle, the freon condenser (CLR1) temperature is 20°C, and the freon boiler temperature is 35°C. There is no superheater in the freon cycle.

Determine the mass flow rate of the freon cycle, thermal efficiency of the combined cycle, power input to the combined cycle, power output by the combined cycle, power net output of the combined cycle, rate of heat added to the combined cycle, and rate of heat removed from the combined cycle.

ANSWERS: $mdot_{Freon} = 11.16$ kg/sec, $\eta_{combined\ cycle} = 32.52\%$, $Wdot_{in\ combined\ cycle} = -24.94$ kW, $Wdot_{out\ combined\ cycle} = 998.6$ kW, $Wdot_{net\ combined\ cycle} = 973.6$ kW, $Qdot_{in\ combined\ cycle} = 2994$ kW, $Qdot_{out\ combined\ cycle} = -2020$ kW.

2. A Rankine/Rankine cycle (Fig. 5.16) uses steam as the working fluid with 1 kg/sec mass flow rate through the top Rankine cycle, and Freon134a as the working fluid in the bottom Rankine cycle. The steam condenser (HX1) pressure is 20 kPa, the boiler pressure is 3 MPa, and the steam superheater temperature is 400°C. The steam mass flow rate is 1 kg/sec.

In the bottom cycle, the freon condenser (CLR1) temperature is 20°C, and the freon boiler temperature is 35°C. There is no superheater in the freon cycle.

Determine the mass flow rate of the freon cycle, thermal efficiency of the combined cycle, power input to the combined cycle, power output by the combined cycle, power net output of the combined cycle, rate of heat added to the combined cycle, and rate of heat removed from the combined cycle.

ANSWERS: $mdot_{Freon} = 10.84$ kg/sec, $\eta_{combined\ cycle} = 34.04\%$, $Wdot_{in\ combined\ cycle} = -25.31$ kW, $Wdot_{out\ combined\ cycle} = 1038$ kW, $Wdot_{net\ combined\ cycle} = 1013$ kW, $Qdot_{in\ combined\ cycle} = 2976$ kW, $Qdot_{out\ combined\ cycle} = -1963$ kW.

5.8 FIELD CYCLE

The *Field cycle* is a supergenerative cycle that makes use of the high-temperature heat addition of the Brayton cycle and the low-temperature heat removal of the Rankine cycle. Therefore, it is able to achieve a high mean temperature of heat addition. The gain due to high-temperature heat addition, however, is offset by the reduction in cycle efficiency resulting from the irreversibility of the mixing process. A schematic diagram of the Field cycle is shown in Fig. 5.17. The arrangement includes one compressor, five turbines, three pumps, one boiler and one reheater (heaters), one regenerator (heat exchanger), one condenser (cooler), three mixing chambers, and two splitters. Processes 1-2, 2-3, 3-4, 4-5, 5-6, and 6-7 take advantage of the high-temperature heat addition of Brayton cycle, and the other processes take advantage of the low-temperature heat removing and regenerative condensing of the Rankine cycle.

Example 5.8

An ideal field cycle with perfect regeneration as shown in Fig. 5.17 is designed according to the following data: $p_{16} = 10\,kPa$, $x_{16} = 0$, $p_{19} = 200\,kPa$, $x_{19} = 0$, $p_{22} = 1\,MPa$, $x_{22} = 0$, $p_{23} = 2\,MPa$, $p_2 = 6\,MPa$, $T_4 = 500°C$, $mdot_4 = 1\,kg/sec$, $p_5 = 4\,MPa$, $T_6 = 500°C$, $T_8 = 300°C$, $mdot_{10} = 0.9\,kg/sec$, and $mdot_{17} = 0.1\,kg/sec$.

Determine (1) the pressure and temperature of each state of the cycle, (2) power produced by each of the five turbines, rate of heat added by each of the two heaters, power required by the compressor and each of the three pumps, rate of heat removed by the condenser, and (3) net power produced by the cycle and cycle efficiency.

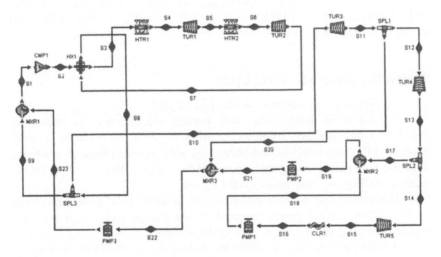

Figure 5.17 Field cycle schematic diagram.

To solve this problem by CyclePad, we take the following steps:

1. Build the cycle as shown in Fig. 5.17. Assume that the compressor, turbines, and pump are adiabatic and isentropic; the heaters, mixing chambers, cooler, and regenerator are isobaric; and the splitters are isoparametric.
2. Input working fluid = water, $p_{16} = 10\,\text{kPa}$, $x_{16} = 0$, $p_{19} = 200\,\text{kPa}$, $x_{19} = 0$, $p_{22} = 1\,\text{MPa}$, $x_{22} = 0$, $p_{23} = 2\,\text{MPa}$, $p_2 = 6\,\text{MPa}$, $T_4 = 500°\text{C}$, $mdot_4 = 1\,\text{kg/sec}$, $p_5 = 4\,\text{MPa}$, $T_6 = 500°\text{C}$, $T_8 = 300°\text{C}$, $mdot_{10} = 0.9\,\text{kg/sec}$, and $mdot_{17} = 0.9\,\text{kg/sec}$.
3. Display results.

The answers are: (1) $p_1 = 2\,\text{MPa}$, $T_1 = 212.4°\text{C}$, $p_2 = 6\,\text{MPa}$, $T_2 = 237.3°\text{C}$, $p_3 = 6\,\text{MPa}$, $T_3 = 275.6°\text{C}$, $p_4 = 6\,\text{MPa}$, $T_4 = 500°\text{C}$, $p_5 = 4\,\text{MPa}$, $T_5 = 432.8°\text{C}$, $p_6 = 4\,\text{MPa}$, $T_6 = 500°\text{C}$, $p_7 = 2\,\text{MPa}$, $T_7 = 388.9°\text{C}$, $p_8 = 2\,\text{MPa}$, $T_8 = 300°\text{C}$, $p_9 = 2\,\text{MPa}$, $T_9 = 300°\text{C}$, $p_{10} = 2\,\text{MPa}$, $T_{10} = 300°\text{C}$, $p_{11} = 1\,\text{MPa}$, $T_{11} = 214.6°\text{C}$, $p_{12} = 1\,\text{MPa}$, $T_{12} = 214.6°\text{C}$, $p_{13} = 200\,\text{kPa}$, $T_{13} = 120.2°\text{C}$, $p_{14} = 200\,\text{kPa}$, $T_{14} = 120.2°\text{C}$, $p_{15} = 10\,\text{kPa}$, $T_{15} = 45.82°\text{C}$, $p_{16} = 10\,\text{kPa}$, $T_{16} = 45.82°\text{C}$, $p_{17} = 200\,\text{kPa}$, $T_{17} = 120.2°\text{C}$, $p_{18} = 200\,\text{kPa}$, $T_{18} = 45.83°\text{C}$, $p_{19} = 200\,\text{kPa}$, $T_{19} = 120.2°\text{C}$, $p_{20} = 1\,\text{MPa}$, $T_{20} = 214.6°\text{C}$, $p_{21} = 1\,\text{MPa}$, $T_{21} = 120.3°\text{C}$, $p_{22} = 2\,\text{MPa}$, $T_{17} = 179.9°\text{C}$, $p_{23} = 2\,\text{MPa}$, and $T_{17} = 185.7°\text{C}$; (2) $Wdot_{T\#1} = 131.9\,\text{kW}$, $Wdot_{T\#2} = 222.4\,\text{kW}$, $Wdot_{T\#3} = 144.7\,\text{kW}$, $Wdot_{T\#4} = 238.9\,\text{kW}$, $Wdot_{T\#5} = 293.2\,\text{kW}$, $Wdot_{P\#1} = -0.1598\,\text{kW}$, $Wdot_{P\#2} = -0.6973\,\text{kW}$, $Wdot_{P\#3} = -28.45\,\text{kW}$, $Wdot_{Compressor} = -7.57\,\text{kW}$, $Qdot_{Htr\#1} = 2197\,\text{kW}$, $Qdot_{Htr\#2} = 155\,\text{kW}$, and $Qdot_{Condenser} = -1358\,\text{kW}$; and (3) $Wdot_{net} = 994.0\,\text{kW}$ and $\eta = 42.26\%$. (See Fig. 5.18.)

Review Problems 5.8   Field Cycle

1. What is the concept of the Field cycle?
2. An ideal field cycle with perfect regeneration as shown in Fig. 5.18 is designed according to the following data: $p_{16} = 10\,\text{kPa}$, $x_{16} = 0$, $p_{19} = 200\,\text{kPa}$, $x_{19} = 0$, $p_{22} = 1\,\text{MPa}$, $x_{22} = 0$, $p_{23} = 2\,\text{MPa}$, $p_2 = 7\,\text{MPa}$, $T_4 = 500°\text{C}$, $mdot_4 = 1\,\text{kg/sec}$, $p_5 = 4\,\text{MPa}$, $T_6 = 500°\text{C}$, $T_8 = 300°\text{C}$, $mdot_{10} = 0.9\,\text{kg/sec}$, and $mdot_{17} = 0.1\,\text{kg/sec}$.

Determine rate of heat added by the heaters, total power produced by the turbines, total power required by the pumps and compressor, net power produced by the cycle, and cycle efficiency.

ANSWERS: $Qdot_{add} = 2393\,\text{kW}$, $Wdot_{Turbines} = 1073\,\text{kW}$, $Wdot_{Pumps\ and\ Compressor} = -38.1\,\text{kW}$, $Wdot_{net} = 1035\,\text{kW}$, and $\eta = 43.26\%$.

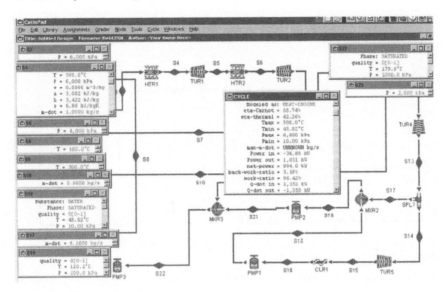

Figure 5.18 Field cycle.

3. An ideal field cycle with perfect regeneration as shown in Fig. 5.18 is designed according to the following data: $p_{16} = 10\,kPa$, $x_{16} = 0$, $p_{19} = 200\,kPa$, $x_{19} = 0$, $p_{22} = 1\,MPa$, $x_{22} = 0$, $p_{23} = 2\,MPa$, $p_2 = 7\,MPa$, $T_4 = 500°C$, $mdot_4 = 1\,kg/sec$, $p_5 = 4\,MPa$, $T_6 = 500°C$, $T_8 = 300°C$, $mdot_{10} = 0.9\,kg/sec$, and $mdot_{17} = 0.12\,kg/sec$.

Determine rate of heat added by the heaters, total power produced by the turbines, total power required by the pumps and compressor, net power produced by the cycle, and cycle efficiency.

ANSWERS: $Qdot_{add} = 2393\,kW$, $Wdot_{Turbines} = 1074\,kW$, $Wdot_{Pumps\ and\ Compressor} = -38.1\,kW$, $Wdot_{net} = 1036\,kW$, and $\eta = 43.28\%$.

5.9 COGENERATION

The cycles considered so far in this chapter are power cycles. However, there are applications in which Rankine cycles are used for the combined supply of power and process heat. The heat may be used as process steam for industrial processes, or steam to heat water for central or district heating. This type of combined heat and power plant is called *cogeneration*. A schematic cogeneration plant is illustrated in Fig. 5.19. A different schematic cogeneration plant is illustrated in Fig. 5.20.

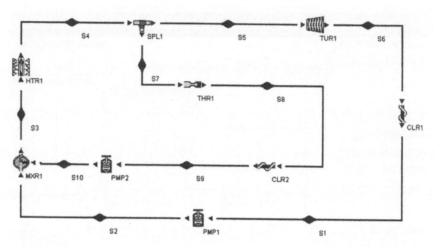

Figure 5.19 Cogeneration plant.

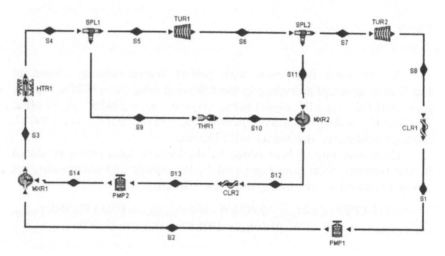

Figure 5.20 Cogeneration plant.

The one regenerative Rankine basic cycle is composed of the following seven processes:

    1-2 Isentropic compression
    3-4 Isobaric heat addition
    5-6 Isentropic expansion
    6-1 Isobaric heat removing
    7-8 Constant enthalpy throttling

8-9 Isobaric heat removing

9-10 Isentropic compression

Applying the mass balance and the first and second laws of thermodynamics of the open system to the mixing chamber and the splitter of the cogeneration Rankine cycle yields:

$$m_3 = m_2 + m_{10} \tag{5.16}$$

$$m_4 = m_7 + m_5 \tag{5.17}$$

$$Q_{89} = m_9(h_9 - h_8) \tag{5.18}$$

and

$$W_{56} = m_5(h_5 - h_6) \tag{5.19}$$

The net work (W_{net}) is

$$W_{net} = W_{56} + W_{9-10} + W_{12} \tag{5.20}$$

The thermal efficiency of the cycle is

$$\eta = W_{net}/Q_{34} \tag{5.21}$$

To take account of the desired heat output from process 8-9 (Q_{89}), the cogeneration ratio λ is

$$\lambda = Q_{89}/Q_{34} \tag{5.22}$$

Therefore, the combined power and heat cogeneration energy utility factor (EUF) is

$$EUF = \eta + \lambda \tag{5.23}$$

Example 5.9

A cogeneration cycle as shown in Fig. 5.19 is to be designed according to the following specifications: boiler temperature $= 500°C$, boiler pressure $= 7\,MPa$, condenser pressure $= 5\,kPa$, process steam (cooler #2) pressure $= 500\,kPa$, mass rate flow through the boiler $= 15\,kg/sec$, and mass rate flow through the turbine $= 14\,kg/sec$.

Determine the rate of heat supply, net power output, process heat output, cycle efficiency, cogeneration ratio, and energy utility factor of the cycle.

To solve this problem with CyclePad, we take the following steps:

1. Build as shown in Fig. 5.19.

2. Analysis:
    a. Assume a process each for the following devices: (1) pumps as adiabatic with 100% efficiency, (2) turbines as adiabatic with 100% efficiency, (3) splitter as isoparametric, (4) mixing chamber as isobaric, (5) condenser and cooler as isobaric, and (6) boiler as isobaric.
    b. Input the given information: (1) working fluid is water, (2) inlet pressure and quality of pump #1 are 5 kPa and 0, (3) inlet temperature and pressure of the turbine are 500°C and 7 MPa, (4) inlet quality and pressure of pump #2 are 0 and 500 kPa, (5) steam mass flow rate is 15 kg/sec at state 4, and (6) steam mass flow rate is 14 kg/sec at state 5.
3. Display results.

The answers are: rate of heat supply = 48,482 kW, net power output = 18,607 kW, process heat output ≈ 2770 kW, cycle efficiency = 0.3838, cogeneration ratio = 2770/48,482 = 0.0571, and energy utility factor of the cycle = 0.3838 + 0.0571 = 0.4409. (See Fig. 5.21.)

Example 5.10

The cogeneration cycle as shown in Fig. 5.19 is to produce power only according to the following specifications: boiler temperature = 500°C,

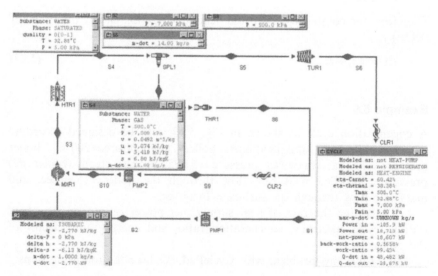

Figure 5.21 Cogeneration.

boiler pressure = 7 MPa, condenser pressure = 5 kPa, process steam (cooler #2) pressure = 500 kPa, mass rate flow through the boiler = 15 kg/sec, and mass rate flow through the turbine = 15 kg/sec.

Determine the net power output, rate of heat supply, and cycle efficiency of the cycle.

To solve this problem with CyclePad, we take the following steps:

1. Build as shown in Fig. 5.19.
2. Analysis:
   a. Assume a process each for the following devices: (1) pumps as adiabatic with 100% efficiency, (2) turbines as adiabatic with 100% efficiency, (3) splitter as isoparametric, (4) mixing chamber as isobaric, (5) condenser and cooler as isobaric, and (6) boiler as isobaric.
   b. Input the given information: (1) working fluid is water, (2) inlet pressure and quality of pump #1 are 5 kPa and 0, (3) inlet temperature and pressure of the turbine are 500°C and 7 MPa, (4) inlet quality and pressure of pump #2 are 0 and 500 kPa, (5) steam mass flow rate is 15 kg/sec at state 4, and (6) steam mass flow rate is 15 kg/sec at state 5.
3. Display results.

The answers are: rate of heat supply = 48,985 kW, net power output = 19,944 kW, and cycle efficiency = 0.4071. (See Fig. 5.22.)

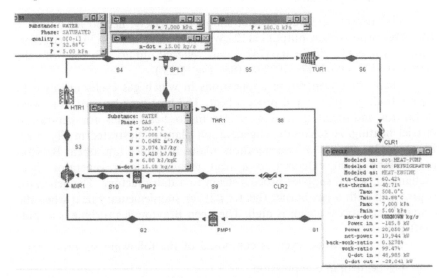

Figure 5.22 Cogeneration without heat output.

Example 5.11

A cogeneration cycle as shown in Fig. 5.20 is to be designed according to the following specifications: boiler temperature $= 400°C$, boiler pressure $= 40$ bars, inlet pressure of low-pressure turbine $= 10$ bars, condenser pressure $= 0.1$ bar, process steam (cooler #2) pressure $= 10$ bars, mass rate flow through the boiler $= 1$ kg/sec, mass rate flow through the turbine #1 $= 0.98$ kg/sec, and mass rate flow through the turbine #1 $= 0.95$ kg/sec.

Determine the rate of heat supply, net power output, process heat output, cycle efficiency, cogeneration ratio, and energy utility factor of the cycle.

To solve this problem with CyclePad, we take the following steps:

1. Build as shown in Fig. 5.20.
2. Analysis:
   a. Assume a process each for the following devices: (1) pumps as adiabatic with 100% efficiency, (2) turbines as adiabatic with 100% efficiency, (3) splitter as isoparametric, (4) mixing chamber as isobaric, (5) condenser and cooler as isobaric, and (6) boiler as isobaric.
   b. Input the given information: (1) working fluid is water, (2) $p_1 = 0.1$ bar, $x_1 = 0$, $p_2 = 40$ bars, $p_6 = 10$ bars, $mdot_4 = 1$ kg/sec, $mdot_9 = 0.02$ kg/sec, $mdot_{11} = 0.03$ kg/sec, $T_4 = 400°C$, and $x_{13} = 0$.
3. Display results.

The answers are: rate of heat supply $= 2989$ kW, net power output $= 1023$ kW, process heat output $= 112$ kW, cycle efficiency $= 0.3421$, cogeneration ratio $= 112/2989 = 0.03747$, and energy utility factor of the cycle $= 0.3421 + 0.03747 = 0.3796$. (See Fig. 5.23.)

There are cogeneration applications in which gas cycles are used to supply both gas power and process heat. The heat may be used as process steam for industrial processes, or steam to heat water for central or district heating. A schematic cogeneration plant is illustrated in Fig. 5.24. Figure 5.25 depicts a cogeneration plant in which an open Brayton gas turbine plant exhausts to a heat recovery steam generator (heat exchanger), which supplies process steam to a dairy factory. The generator is provided with a gas burner (heater #2) for supplementary heat when the demand for process steam is high. The open Brayton gas turbine is a split-shaft plant.

The cogeneration cycle is composed of the following six processes:

1-2 Isentropic compression
2-3 Isobaric heat addition

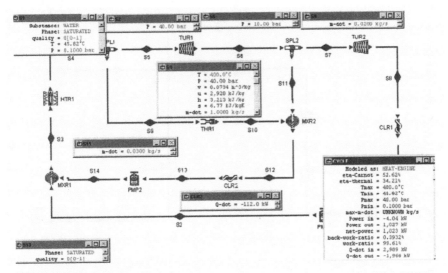

Figure 5.23 Cogeneration.

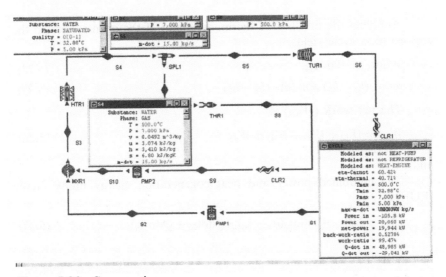

Figure 5.24 Cogeneration.

3-4 Isentropic expansion
4-5 Isentropic expansion
5-6 Isobaric heat removing
7-8 Isobaric heat addition
8-9 Isobaric heat addition

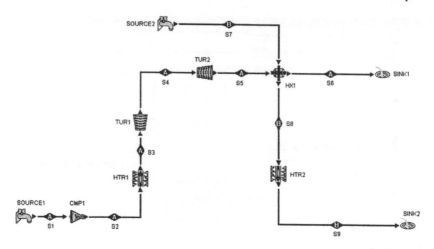

Figure 5.25 Cogeneration.

Applying the first law of thermodynamics of the open system to the cogeneration cycle yields:

$$W\text{dot}_{12} = W\text{dot}_{34}, \tag{5.24}$$

$$m\text{dot}_5(h_5 - h_6) = m\text{dot}_7\,(h_8 - h_7) \tag{5.25}$$

The net work (W_net) is

$$W_\text{net} = W_{12} + W_{34} + W_{45} = W_{45} \tag{5.26}$$

and

The combined power and heat cogeneration energy utility factor (EUF) is

$$\text{EUF} = [W_\text{net} + m\text{dot}_5(h_5 - h_6)]/(Q\text{dot}_{23} + Q\text{dot}_{89}) \tag{5.27}$$

Example 5.12

The data given below correspond approximately to the design conditions for a dairy factory cogeneration plant: $m\text{dot}_1 = 20.45\,\text{kg/sec}$, $p_1 = 1\,\text{bar}$, $T_1 = 25°C$, $p_2 = 7\,\text{bars}$, $T_3 = 850°C$, $p_5 = 1\,\text{bar}$, $T_6 = 138°C$, $\eta_\text{compressor} = \eta_\text{turbine \#1} = \eta_\text{turbine \#2} = 85\%$, $T_7 = 90°C$, $p_7 = 13\,\text{bars}$, and $x_9 = 1$.

Determine the power required by the compressor, power produced by turbines #1 and #2, rate of heat added to the combustion chamber, net power produced by the open Brayton gas turbine plant, cycle efficiency of the open Brayton gas turbine plant, rate of heat added to the process steam, rate of process steam, and energy utility of the cogeneration plant.

To solve this problem by CyclePad, we take the following steps:

1. Build the cycle as shown in Fig. 5.25. Assume that compressor and turbines are adiabatic and 85% efficient, and the heaters and heat exchanger are isobaric.

2. Input cycle A working fluid = air, $p_1 = 1$ bar, $T_1 = 25°C$, $mdot_1 = 20.45$ kg/sec, $p_2 = 7$ bars, $T_3 = 850°C$, $-Wdot_{12} = Wdot_{34}$, $T_6 = 138°C$, cycle B working fluid = water, $T_7 = 90°C$, $p_8 = 13$ bars, and $x_8 = 1$.

3. Display results.

The answers are: $Wdot_{compressor} = -5352$ kW, $Wdot_{turbine \#1} = 5352$ kW, $Wdot_{turbine \#2} = 3172$ kW, $Wdot_{net} = 3172$ kW, $Qdot_{comb\ chamber} = 11,576$ kW, $\eta = 3172/11576 = 27.40\%$, $Qdot_{HX} = -6086$ kW, $mdot_{steam} = 2.53$ kg/sec, and EUF $= (3172 + 6086)/11576 = 0.7998$. (See Fig. 5.26.)

Example 5.13

Referring to the dairy factory cogeneration design conditions where 5 kg/sec of process steam is needed, determine the rate of heat provided by the gas burner.

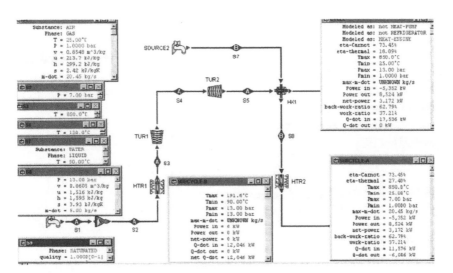

Figure 5.26 Cogeneration.

To solve this problem by CyclePad, we do the same thing as in Example 5.12, delete $x_8 = 1$, and let $x_9 = 1$ and $m\mathrm{dot}_8 = 5\,\mathrm{kg/sec}$.

The answers are $Q\mathrm{dot} = 5960\,\mathrm{kW}$ and $\mathrm{EUF} = (3172 + 6086)/(11576 + 5960) = 0.5279$. (See Fig. 5.27.)

Review Problems 5.9   Cogeneration

1.   What is cogeneration?
2.   How is the cogeneration ratio defined?
3.   How is the cogeneration energy utility factor defined?
4.   A cogeneration cycle as shown in Fig. 5.19 is to be designed according to the following specifications: turbine efficiency = 89%, boiler temperature = 500°C, boiler pressure = 7 MPa, condenser pressure = 5 kPa, process steam (cooler #2) pressure = 500 kPa, mass rate flow through the boiler = 15 kg/sec, and mass rate flow through the turbine = 13 kg/sec.

Determine the rate of heat supply, net power output, process heat output, cycle efficiency, cogeneration ratio, and energy utility factor of the cycle.

ANSWERS: cycle efficiency = 32.01%, rate of heat supply = 47,980 kW, rate of cogeneration heat = −5540 kW.

5.   A cogeneration cycle as shown in Fig. 5.19 is to be designed according to the following specifications: turbine efficiency = 89%, boiler temperature = 500°C, boiler pressure = 6 MPa, condenser pressure = 500 kPa,

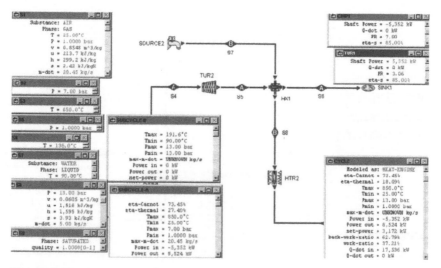

Figure 5.27   Cogeneration.

process steam (cooler #2) pressure = 500 kPa, mass rate flow through the boiler = 15 kg/sec, and mass rate flow through the turbine = 13 kg/sec.

Determine the rate of heat supply, net power output, process heat output, cycle efficiency, cogeneration ratio, and energy utility factor of the cycle.

ANSWERS: cycle efficiency = 31.63%, rate of heat supply = 48,173 kW, rate of cogeneration heat = −5564 kW.

6. A cogeneration cycle as shown in Fig. 5.19 is to be designed according to the following specifications: turbine efficiency = 89%, boiler temperature = 500°C, boiler pressure = 7 MPa, condenser pressure = 5 kPa, process steam (cooler #2) pressure = 500 kPa, mass rate flow through the boiler = 15 kg/sec, and mass rate flow through the turbine = 14 kg/sec.

Determine the rate of heat supply, net power output, process heat output, cycle efficiency, cogeneration ratio, and energy utility factor of the cycle.

ANSWERS: cycle efficiency = 38.38%, rate of heat supply = 48,482 kW, rate of cogeneration heat = −2770 kW.

7. A cogeneration cycle as shown in Fig. 5.20 is to be designed according to the following specifications: high-pressure turbine efficiency = 84%, low-pressure turbine efficiency = 100%, boiler temperature = 400°C, boiler pressure = 40 bars, inlet pressure of low-pressure turbine = 10 bars, condenser pressure = 0.1 bar, process steam (cooler #2) pressure = 10 bars, mass rate flow through the boiler = 1 kg/sec, mass rate flow through the turbine #1 = 0.98 kg/sec, and mass rate flow through the turbine #1 = 0.95 kg/sec.

Determine the rate of heat supply, net power output, process heat output, cycle efficiency, cogeneration ratio, and energy utility factor of the cycle.

ANSWERS: cycle efficiency = 33.02%, rate of heat supply = 2989 kW, rate of cogeneration heat = −113.7 kW.

8. Referring to the dairy factory design conditions, except that the compressor efficiency is 80% and the mass flow rate of process steam is 4 kg/sec, determine the power required by the compressor, power produced by turbines #1 and #2, rate of heat added to the combustion chamber, net power produced by the open Brayton gas turbine plant, cycle efficiency of the open Brayton gas turbine plant, rate of heat added to the process steam, rate of heat added in the gas burner, and energy utility of the cogeneration plant.

ANSWERS: $\dot{W}_{compressor} = -5352$ kW, $\dot{W}_{turbine\,\#1} = 5352$ kW, $\dot{W}_{turbine\,\#2} = 3172$ kW, $\dot{W}_{net} = 3172$ kW, $\dot{Q}_{comb\,chamber} = 11,576$ kW, $\eta = 3172/11,576 = 27.40\%$, $\dot{Q}_{HX} = -6086$ kW, $\dot{Q}_{gas\,burner} = 2.53$ kg/sec, and EUF = $(3172 + 6086)/11576 = 0.7998$.

5.10 DESIGN EXAMPLES

One of the purposes of this book is to illuminate elements of conceptual thermodynamic cycle design. In this endeavor, the intent is to build upon and extend information previously acquired in thermodynamics. Possible steps in an intelligent computer-aided design process, as shown in Example 5.14, involves the following steps:

1. The first step is to identify a need. For example, a power plant manager might decide that use of a combined cycle would increase cycle efficiency.
2. The second step is to develop several conceptual plants (e.g., cycles A, B, and C) to meet the identified need. One of the several plants is described in Example 5.14. In this example, a three-stage regenerative steam Rankine cycle and a four-stage intercool and four-stage reheat air Brayton cycle are combined to meet the need.
3. The third step is to model the components of the conceptual plant. For example, a steam turbine may be modeled as an adiabatic process with 85% isentropic efficiency.
4. The fourth step is to estimate input numerical values for the major parameters. For example, the inlet gas temperature to a gas turbine is 1200°C based on physical feasibility.
5. The fifth step is to analyze the conceptual thermodynamic cycle.
6. The sixth step is to refine and optimize the conceptual thermodynamic cycle with sensitivity analysis.
7. The seventh step is to compare the optimal cycles and choose the best cycle.

Example 5.14

A three-stage regenerative steam Rankine cycle and a four-stage intercool and four-stage reheat air Brayton cycle is combined by a heat exchanger as shown in Fig. 5.28a has been designed by a junior engineer with the following design input information, as shown in Fig. 5.28b.

The preliminary design information is:

Brayton cycle

$p_1 = 100\,\text{kPa}$, $T_1 = 20°C$, $p_3 = 200\,\text{kPa}$, $T_3 = 20°C$, $p_5 = 300\,\text{kPa}$, $T_5 = 20°C$, $p_7 = 500\,\text{kPa}$, $T_7 = 20°C$, $p_9 = 800\,\text{kPa}$, $T_9 = 20°C$, $p_{11} = 1200\,\text{kPa}$, $T_{11} =$

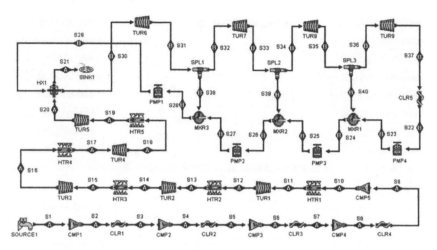

Figure 5.28a Combined Brayton–Rankine cycle.

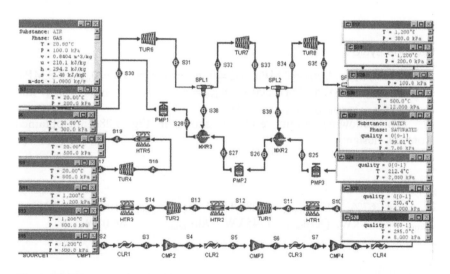

Figure 5.28b Combined Brayton–Rankine cycle input.

1200°C, $p_{13} = 800\,\text{kPa}$, $T_{13} = 1200°\text{C}$, $p_{15} = 500\,\text{kPa}$, $T_{15} = 1200°\text{C}$, $p_{17} = 300\,\text{kPa}$, $T_{17} = 1200°\text{C}$, $p_{19} = 200\,\text{kPa}$, $T_{19} = 1200°\text{C}$, $p_{20} = 100\,\text{kPa}$, $T_{21} = 550°\text{C}$, $m\text{dot}_1 = 1\,\text{kg/sec}$, $\eta_{\text{tur1}} = \eta_{\text{tur2}} = \eta_{\text{tur3}} = \eta_{\text{tur4}} = \eta_{\text{tur5}} = 85\%$, and $\eta_{\text{cmp1}} = \eta_{\text{cmp2}} = \eta_{\text{cmp3}} = \eta_{\text{cmp4}} = \eta_{\text{cmp5}} = 85\%$.

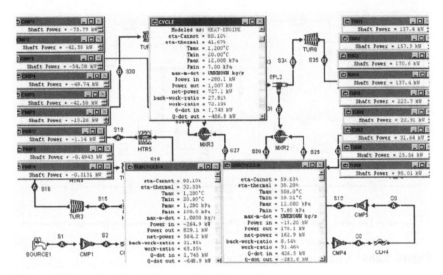

Figure 5.28c Combined Brayton–Rankine cycle output.

Rankine cycle

$p_{22} = 7\,\text{kPa}$, $x_{22} = 0$, $p_{24} = 2\,\text{MPa}$, $x_{24} = 0$, $p_{26} = 4\,\text{MPa}$, $x_{26} = 0$, $p_{28} = 8\,\text{MPa}$, $x_{28} = 0$, $p_{30} = 12\,\text{MPa}$, $T_{30} = 500°\text{C}$, $\eta_{\text{tur6}} = \eta_{\text{tur7}} = \eta_{\text{tur8}} = \eta_{\text{tur9}} = 85\%$, and $\eta_{\text{pmp1}} = \eta_{\text{pmp2}} = \eta_{\text{pmp3}} = \eta_{\text{pmp4}} = 85\%$.

The following output results as shown in Fig. 5.28c are obtained from his design:

Combined cycle

$\eta_{\text{cycle}} = 41.67\%$, $W\text{dot}_{\text{input}} = -280.1\,\text{kW}$, $W\text{dot}_{\text{output}} = 1007\,\text{kW}$, $W\text{dot}_{\text{net output}} = 727.1\,\text{kW}$, $Q\text{dot}_{\text{add}} = 1745\,\text{kW}$, $Q\text{dot}_{\text{remove}} = -486.0\,\text{kW}$.

Brayton cycle

$\eta_{\text{cycle}} = 32.33\%$, $W\text{dot}_{\text{input}} = -264.9\,\text{kW}$, $W\text{dot}_{\text{output}} = 829.1\,\text{kW}$, $W\text{dot}_{\text{net output}} = 564.2\,\text{kW}$, $Q\text{dot}_{\text{add}} = 1745\,\text{kW}$, $Q\text{dot}_{\text{remove}} = -648.9\,\text{kW}$, $W\text{dot}_{\text{cmp1}} = -75.79\,\text{kW}$, $W\text{dot}_{\text{cmp2}} = -42.50\,\text{kW}$, $W\text{dot}_{\text{cmp3}} = -54.38\,\text{kW}$, $W\text{dot}_{\text{cmp4}} = -49.74\,\text{kW}$, $W\text{dot}_{\text{cmp5}} = -42.50\,\text{kW}$, $W\text{dot}_{\text{tur1}} = 137.4\,\text{kW}$, $W\text{dot}_{\text{tur2}} = 157.9\,\text{kW}$, $W\text{dot}_{\text{tur3}} = 170.6\,\text{kW}$, $W\text{dot}_{\text{tur4}} = 137.4\,\text{kW}$, $W\text{dot}_{\text{tur5}} = 225.7\,\text{kW}$, $Q\text{dot}_{\text{clr1}} = -75.79\,\text{kW}$, $Q\text{dot}_{\text{clr2}} = -42.50\,\text{kW}$, $Q\text{dot}_{\text{clr3}} = -54.38\,\text{kW}$, $Q\text{dot}_{\text{clr4}} = -49.74\,\text{kW}$, $Q\text{dot}_{\text{htr1}} = 1142\,\text{kW}$,

$Q\text{dot}_{htr2} = 137.4\,\text{kW}, \qquad Q\text{dot}_{htr3} = 157.9\,\text{kW}, \qquad Q\text{dot}_{htr4} = 170.6\,\text{kW},$
$Q\text{dot}_{htr5} = 137.4\,\text{kW},$ and $Q\text{dot}_{\text{heat exch}} = -426.5\,\text{kW}.$

Rankine cycle

$\eta_{\text{cycle}} = 38.20\%, \qquad W\text{dot}_{\text{input}} = -15.20\,\text{kW}, \qquad W\text{dot}_{\text{output}} = 178.1\,\text{kW},$
$W\text{dot}_{\text{net output}} = 162.9\,\text{kW},\; Q\text{dot}_{\text{add}} = 426.5\,\text{kW},\; Q\text{dot}_{\text{remove}} = -263.6\,\text{kW},$
$W\text{dot}_{\text{pmp1}} = -13.26\,\text{kW},\; W\text{dot}_{\text{pmp2}} = -1.14\,\text{kW},\; W\text{dot}_{\text{pmp3}} = -0.4943\,\text{kW},$
$W\text{dot}_{\text{pmp4}} = -0.3131\,\text{kW}, \qquad W\text{dot}_{\text{tur6}} = 22.91\,\text{kW}, \qquad W\text{dot}_{\text{tur7}} = 31.64\,\text{kW},$
$W\text{dot}_{\text{tur8}} = 25.56\,\text{kW}, \qquad W\text{dot}_{\text{tur9}} = 98.01\,\text{kW}, \qquad Q\text{dot}_{\text{clr5}} = -263.6\,\text{kW},$
$Q\text{dot}_{\text{heat exch}} = 426.5\,\text{kW},\; m\text{dot}_{24} = 0.1782\,\text{kg/sec},\; m\text{dot}_{26} = 0.1939\,\text{kg/sec},$
$m\text{dot}_{28} = 0.2164\,\text{kg/sec}, \qquad m\text{dot}_{31} = 0.2164\,\text{kg/sec}, \qquad m\text{dot}_{32} = 0.1939\,\text{kg/sec},$
$m\text{dot}_{34} = 0.1782\,\text{kg/sec}, \qquad m\text{dot}_{36} = 0.1304\,\text{kg/sec}, \qquad m\text{dot}_{38} = 0.0225\,\text{kg/sec},$
$m\text{dot}_{39} = 0.0157\,\text{kg/sec},$ and $m\text{dot}_{40} = 0.0478\,\text{kg/sec}.$

Let us try to modify his design (use p_5, p_7, p_{15}, and p_{17} as design parameters only) to get a better cycle thermal efficiency than his $\eta_{\text{cycle}} = 41.67\%.$

The sensitivity analyses of η_{cycle} versus p_5, η_{cycle} versus p_7, η_{cycle} versus p_{15}, and η_{cycle} versus p_{17} are shown in Figs. 5.28d–5.28g.

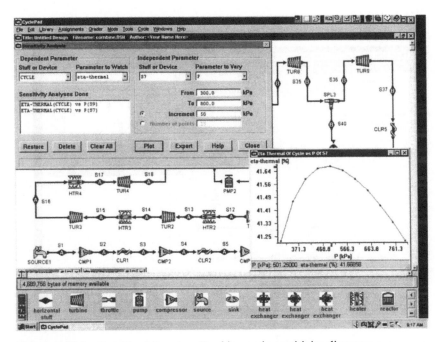

Figure 5.28d Combined Brayton–Rankine cycle sensitivity diagram.

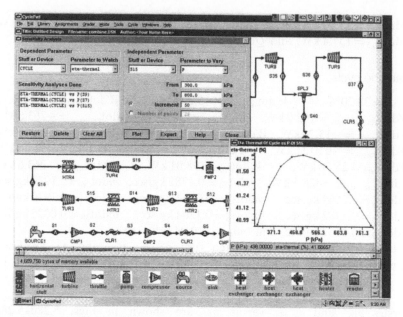

Figure 5.28e   Combined Brayton–Rankine cycle sensitivity diagram.

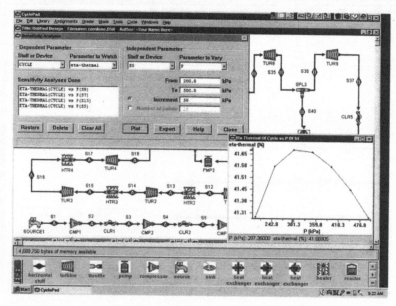

Figure 5.28f   Combined Brayton–Rankine cycle sensitivity diagram.

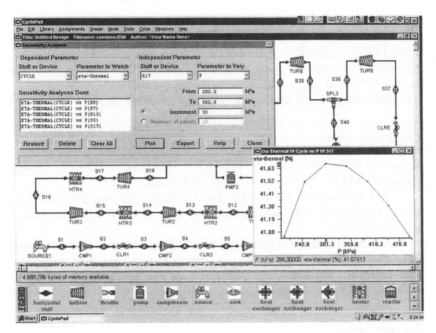

Figure 5.28g Combined Brayton–Rankine cycle sensitivity diagram; Review Problems 5.10 Design.

The optimization design values of p_5, p_7, p_{15}, and p_{18} can be easily identified.

Review Problem 5.10 Design

1. A three-stage regenerative steam Rankine cycle and a four-stage intercool and four-stage reheat air Brayton cycle combined with a heat exchanger, as shown in Fig. 5.28a, has been designed by a junior engineer with the following design input information:

Brayton cycle

$p_1 = 100\,\text{kPa}$, $T_1 = 20°\text{C}$, $p_3 = 200\,\text{kPa}$, $T_3 = 20°\text{C}$, $p_5 = 300\,\text{kPa}$, $T_5 = 20°\text{C}$, $p_7 = 500\,\text{kPa}$, $T_7 = 20°\text{C}$, $p_9 = 800\,\text{kPa}$, $T_9 = 20°\text{C}$, $p_{11} = 1200\,\text{kPa}$, $T_{11} = 1300°\text{C}$, $p_{13} = 800\,\text{kPa}$, $T_{13} = 1300°\text{C}$, $p_{15} = 500\,\text{kPa}$, $T_{15} = 1300°\text{C}$, $p_{17} = 300\,\text{kPa}$, $T_{17} = 1300°\text{C}$, $p_{19} = 200\,\text{kPa}$, $T_{19} = 1300°\text{C}$, $p_{20} = 100\,\text{kPa}$, $T_{21} = 550°\text{C}$, $mdot_1 = 1\,\text{kg/sec}$, $\eta_{tur1} = \eta_{tur2} = \eta_{tur3} = \eta_{tur4} = \eta_{tur5} = 85\%$, and $\eta_{cmp1} = \eta_{cmp2} = \eta_{cmp3} = \eta_{cmp4} = \eta_{cmp5} = 85\%$.

Rankine cycle

$p_{22} = 7\,\text{kPa}$, $x_{22} = 0$, $p_{24} = 2\,\text{MPa}$, $x_{24} = 0$, $p_{26} = 4\,\text{MPa}$, $x_{26} = 0$,
$p_{28} = 8\,\text{MPa}$, $x_{28} = 0$, $p_{30} = 12\,\text{MPa}$, $T_{30} = 500°\text{C}$, $\eta_{\text{tur}6} = \eta_{\text{tur}7} = \eta_{\text{tur}8} =$
$\eta_{\text{tur}9} = 85\%$, and $\eta_{\text{pmp}1} = \eta_{\text{pmp}2} = \eta_{\text{pmp}3} = \eta_{\text{pmp}4} = 85\%$.
The following output results are obtained from his design:

Combined cycle

$\eta_{\text{cycle}} = 43.26\%$, $W\text{dot}_{\text{input}} = -283.2\,\text{kW}$, $W\text{dot}_{\text{output}} = 1099\,\text{kW}$, $W\text{dot}_{\text{net}}$
$_{\text{output}} = 815.9\,\text{kW}$, $Q\text{dot}_{\text{add}} = 1886\,\text{kW}$, $Q\text{dot}_{\text{remove}} = -538.5\,\text{kW}$.

Try to improve his design (use p_3, p_5, p_7, p_9, p_{24}, p_{26}, and p_{28} as design parameters only) and get a better cycle thermal efficiency than his $\eta_{\text{cycle}} = 43.26\%$.

2. A three-stage regenerative steam Rankine cycle and a four-stage inter-cool and four-stage reheat air Brayton cycle combined with a heat exchanger, as shown in Fig. 5.28a, has been designed by a junior engineer with the following design input information:

Brayton cycle

$p_1 = 100\,\text{kPa}$, $T_1 = 20°\text{C}$, $p_3 = 200\,\text{kPa}$, $T_3 = 20°\text{C}$, $p_5 = 300\,\text{kPa}$, $T_5 = 20°\text{C}$,
$p_7 = 500\,\text{kPa}$, $T_7 = 20°\text{C}$, $p_9 = 800\,\text{kPa}$, $T_9 = 20°\text{C}$, $p_{11} = 1200\,\text{kPa}$,
$T_{11} = 1300°\text{C}$, $p_{13} = 800\,\text{kPa}$, $T_{13} = 1300°\text{C}$, $p_{15} = 500\,\text{kPa}$, $T_{15} = 1300°\text{C}$,
$p_{17} = 300\,\text{kPa}$, $T_{17} = 1300°\text{C}$, $p_{19} = 200\,\text{kPa}$, $T_{19} = 1300°\text{C}$, $p_{20} = 100\,\text{kPa}$,
$T_{21} = 550°\text{C}$, $m\text{dot}_1 = 1\,\text{kg/sec}$, $\eta_{\text{tur}1} = \eta_{\text{tur}2} = \eta_{\text{tur}3} = \eta_{\text{tur}4} = \eta_{\text{tur}5} = 85\%$, and
$\eta_{\text{cmp}1} = \eta_{\text{cmp}2} = \eta_{\text{cmp}3} = \eta_{\text{cmp}4} = \eta_{\text{cmp}5} = 88\%$.

Rankine cycle

$p_{22} = 7\,\text{kPa}$, $x_{22} = 0$, $p_{24} = 2\,\text{MPa}$, $x_{24} = 0$, $p_{26} = 4\,\text{MPa}$, $x_{26} = 0$,
$p_{28} = 8\,\text{MPa}$, $x_{28} = 0$, $p_{30} = 12\,\text{MPa}$, $T_{30} = 500°\text{C}$, $\eta_{\text{tur}6} = \eta_{\text{tur}7} = \eta_{\text{tur}8} = \eta_{\text{tur}9} =$
85%, and $\eta_{\text{pmp}1} = \eta_{\text{pmp}2} = \eta_{\text{pmp}3} = \eta_{\text{pmp}4} = 85\%$.
The following output results are obtained from his design:

Combined cycle

$\eta_{\text{cycle}} = 43.70\%$, $W\text{dot}_{\text{input}} = -274.1\,\text{kW}$, $W\text{dot}_{\text{output}} = 1099\,\text{kW}$,
$W\text{dot}_{\text{net output}} = 824.9\,\text{kW}$, $Q\text{dot}_{\text{add}} = 1888\,\text{kW}$, $Q\text{dot}_{\text{remove}} = -530.9\,\text{kW}$.

Try to improve his design (use p_3, p_5, p_7, p_9, p_{24}, p_{26}, and p_{28} as design parameters only) and get a better cycle thermal efficiency than his $\eta_{\text{cycle}} = 43.70\%$.

5.11 SUMMARY

Combined- and cascaded-cycle heat engines of several types discussed in this chapter can improve cycle efficiency and reduce the fuel required for producing work or electrical energy. In a similar fashion, cogeneration, which produces thermal energy and electrical energy, can also result in significant energy savings.

6
Refrigeration and Heat Pump Cycles

6.1 CARNOT REFRIGERATOR AND HEAT PUMP

A system is called a refrigerator or a heat pump depending on the purpose of the system. If the purpose of the system is to remove heat from a low-temperature thermal reservoir, it is a refrigerator. If the purpose of the system is to deliver heat to a high-temperature thermal reservoir, it is a heat pump.

The Carnot cycle is a reversible cycle. Reversing the cycle will also reverse the directions of heat and work interactions. The reversed Carnot heat engine cycles are Carnot refrigeration and heat pump cycles. Therefore, a reversed Carnot vapor heat engine is either a Carnot vapor refrigerator or a Carnot vapor heat pump, depending on the function of the cycle.

A schematic diagram of the *Carnot refrigerator* or *Carnot heat pump* is illustrated in Fig. 6.1.

1-2 Isentropic compression
2-3 Isothermal cooling
3-4 Isentropic expansion
4-1 Isothermal heating

Applying the first and second laws of thermodynamics of the open system to each of the four processes of the basic vapor refrigeration cycle under steady-flow and steady-state conditions yields:

$$Q_{12} = 0 \tag{6.1}$$

$$0 - W_{12} = m(h_2 - h_1) \tag{6.2}$$

$$W_{23} = 0 \tag{6.3}$$

$$Q_{23} - 0 = m(h_3 - h_2) \tag{6.4}$$

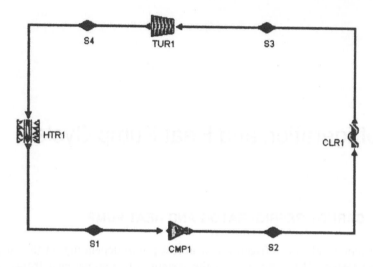

Figure 6.1 Carnot refrigerator or Carnot heat pump.

$$Q_{34} = 0 \tag{6.5}$$

$$0 - W_{34} = m(h_3 - h_4) \tag{6.6}$$

$$W_{41} = 0 \tag{6.7}$$

and

$$Q_{41} - 0 = m(h_1 - h_4) \tag{6.8}$$

The desirable energy output of the refrigeration cycle is the heat added to the evaporator (or heat removed from the inner space or the low-temperature reservoir of the refrigerator). The energy input to the cycle is the compressor work required. The energy produced is the turbine work. The net work (W_{net}) required to operate the cycle is ($W_{12} + W_{34}$). Thus, the coefficient of performance (COP) of the cycle is

$$\beta_R = Q_{41} / W_{net} \tag{6.9}$$

The rate of heat removed from the inner space of the refrigerator is called the cooling load or cooling capacity. The cooling load of a refrigeration system is sometimes given a unit in tons of refrigeration. A *ton of refrigeration* is the removal of heat from the cold space at a rate of 200 Btu/min (12,000 Btu/hr) or 211 kJ/min (3.52 kW). A ton of refrigeration is the rate of cooling required to make a ton of ice per day.

Example 6.1

Determine the COP, horsepower required, and cooling load of a Carnot vapor refrigeration cycle using R-12 as the working fluid and in which the condenser temperature is 100°F and the evaporation temperature is 20°F. The circulation rate of fluid is 0.1 lbm/sec. Determine the compressor power required, turbine power produced, net power required, cooling load, quality at the inlet of the evaporator, quality at the inlet of the compressor, and COP of the refrigerator.

To solve this problem by CyclePad, we take the following steps:

1. Build:
   a. Take a compressor, a cooler (condenser), a turbine, and a heater (evaporator) from the open-system inventory shop and connect the four devices to form the basic vapor refrigeration cycle.
   b. Switch to analysis mode.
2. Analysis:
   a. Assume a process for each of the four devices: (1) compressor as isentropic, (2) condenser as isobaric, (3) turbine as isentropic, and (4) evaporator as isobaric.
   b. Input the given information: (1) working fluid is R-12, (2) inlet temperature and quality of the compressor are 20°F and 1, (3) inlet temperature and quality of the turbine are 100°F and 0, (4) phase of the exit refrigerant from the turbine is a saturated mixture, and (5) mass flow rate is 0.1 lbm/sec.
3. Display the cycle properties' results. The cycle is a refrigerator.

The answers are: compressor power required = −1.36 hp, turbine power produced = 0.2234 hp, net power required = −1.13 hp, cooling load = 4.79 Btu/sec, quality at the inlet of the evaporator = 0.2504, and COP = 5.99. (See Fig. 6.2.)

COMMENTS: (1) The turbine work produced is very small. It does not pay to install an expansion device to produce a small amount of work. The expansion process can be achieved by a simple throttling valve. (2) The compressor handles the refrigerant as a mixture of saturated liquid and saturated vapor. It is not practical. Therefore, the compression process should be moved out of the mixture region to the superheated region.

Review Problems 6.1   Carnot Refrigerator and Heat Pump

1. Does the area enclosed by the Carnot heat pump cycle on a *T–s* diagram represent the network input for the heat pump?

280

Chapter 6

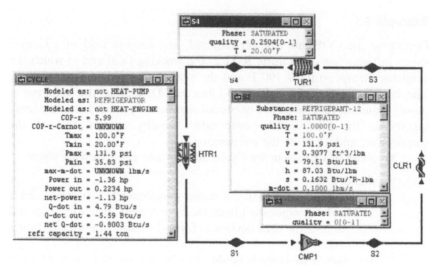

Figure 6.2 Carnot vapor refrigeration cycle.

2. Does the Carnot heat pump cycle involve any internal irreversibilities?

6.2 BASIC VAPOR REFRIGERATION CYCLE

The Carnot refrigerator is not a practical cycle, because the compressor is designed to handle superheated vapor or gas. The turbine in a small temperature and pressure range produces a very small amount of work. It is not worth having an expensive turbine in the cycle to produce a very small amount of work. Therefore, the Carnot refrigeration cycle is modified to have the compression process completely in the superheated region and the turbine is replaced by an inexpensive throttling valve termed a thermal expansion valve to form a basic vapor refrigeration cycle. The schematic diagram of the basic vapor refrigeration system is shown in Fig. 6.3. The components of the *basic vapor refrigeration cycle* include a compressor, a condenser, an expansion valve, and an evaporator. The *T–s* diagram of the cycle is shown in Fig. 6.4. Notice that the throttling process 3-4 is an irreversible process and is indicated by a broken line on the *T–s* diagram.

The basic vapor refrigeration cycle consists of the following four processes:

1-2 Isentropic compression
2-3 Isobaric cooling

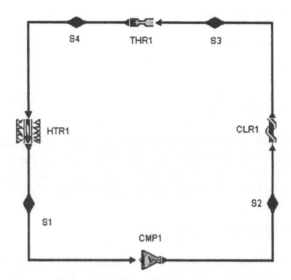

Figure 6.3 Basic vapor refrigeration system.

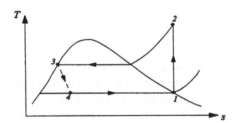

Figure 6.4 *T–s* diagram of the basic vapor refrigeration cycle.

3-4 Throttling
4-1 Isobaric heating

Applying the first and second laws of thermodynamics of the open system to each of the four processes of the basic vapor refrigeration cycle yields:

$$Q_{12} = 0 \qquad\qquad\qquad (6.10)$$

$$0 - W_{12} = m(h_2 - h_1) \qquad\qquad\qquad (6.11)$$

$$W_{23} = 0 \qquad\qquad\qquad (6.12)$$

$$Q_{23} - 0 = m(h_3 - h_2) \qquad\qquad\qquad (6.13)$$

$$Q_{34} = 0 \text{ and } W_{34} = 0 \tag{6.14}$$

$$0 - 0 = m(h_3 - h_4) \tag{6.15}$$

$$W_{41} = 0 \tag{6.16}$$

and

$$Q_{41} - 0 = m(h_1 - h_4) \tag{6.17}$$

The desirable energy output of the basic vapor refrigeration cycle is the heat added to the evaporator (or heat removed from the inner space of the refrigerator, Q_{41}). The energy input to the cycle is the compressor work required (W_{12}). Thus, the coefficient of performance (COP) of the basic refrigeration cycle is

$$\beta_R = Q_{41}/W_{12} = (h_4 - h_1)/(h_2 - h_1) \tag{6.18}$$

The rate of heat removed from the inner space of the refrigerator is called the cooling load or cooling capacity. The cooling load of a refrigeration system is sometimes given a unit in tons of refrigeration. A ton of refrigeration is the removal of heat from the cold space at a rate of 12,000 Btu/hr.

Example 6.2

Determine the COP, horsepower required, and cooling load of a basic vapor refrigeration cycle using R-12 as the working fluid and in which the condenser pressure is 130 psia and the evaporation pressure is 35 psia. The circulation rate of fluid is 0.1 lbm/sec. Determine the compressor power required, cooling load, quality at the inlet of the evaporator, and COP of the refrigerator.

To solve this problem by CyclePad, we take the following steps:

1. Build:
    a. Take a compressor, a condenser, a valve, and a heater (evaporator) from the open-system inventory shop and connect the four devices to form the basic vapor refrigeration cycle.
    b. Switch to analysis mode.
2. Analysis:
    a. Assume a process for each of the four devices: (1) compressor as isentropic, (2) condenser as isobaric, (3) valve as constant enthalpy, and (4) evaporator as isobaric.
    b. Input the given information: (1) working fluid is R-12, (2) inlet pressure and quality of the compressor are 35 psia 20°F and 1,

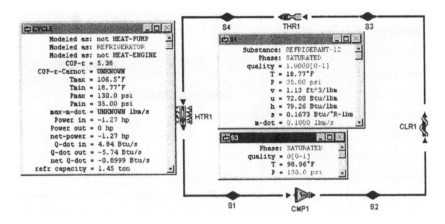

```
┌─ CYCLE ────────────── _□×┐
│  Modeled as: not HEAT-PUMP  ▲│
│  Modeled as: REFRIGERATOR    │
│  Modeled as: not HEAT-ENGINE │
│        COP-r = 5.38          │
│  COP-r-Carnot = UNKNOWN      │
│         Tmax = 106.5°F       │
│         Tmin = 18.77°F       │
│         Pmax = 130.0 psi     │
│         Pmin = 35.00 psi     │
│      max-m-dot = UNKNOWN lbm/s│
│       Power in = -1.27 hp    │
│      Power out = 0 hp        │
│      net-power = -1.27 hp    │
│       Q-dot in = 4.84 Btu/s  │
│      Q-dot out = -5.74 Btu/s │
│      net Q-dot = -0.8999 Btu/s│
│  refr capacity = 1.45 ton   ▼│
└──────────────────────────────┘
```

```
┌─ S4 ──────────── _□×┐
│  Substance: REFRIGERANT-12  ▲│
│      Phase: SATURATED        │
│   quality = 1.0000[0-1]      │
│         T = 18.77°F          │
│         P = 35.00 psi        │
│         v = 1.13 ft^3/lbm    │
│         u = 72.00 Btu/lbm    │
│         h = 79.26 Btu/lbm    │
│         s = 0.1673 Btu/°R-lbm │
│     m-dot = 0.1000 lbm/s    ▼│
└──────────────────────────────┘
```

```
┌─ S3 ──────────── _□×┐
│      Phase: SATURATED   ▲│
│   quality = 0[0-1]       │
│         T = 98.96°F      │
│         P = 130.0 psi   ▼│
└──────────────────────────┘
```

Figure 6.5 Basic vapor refrigeration cycle.

(3) inlet pressure and quality of the valve are 130 psia and 0, and (4) mass flow rate is 0.1 lbm/sec.

3. Display the cycle properties' results. The cycle is a refrigerator.

The answers are: compressor power required $= -1.27$ hp, cooling load $= 4.84$ Btu/sec $= 1.45$ ton, quality at the inlet of the evaporator $= 0.2738$, and COP $= 5.38$. (See Fig. 6.5.)

Example 6.3

Determine the COP, horsepower required, and cooling load of a basic vapor refrigeration cycle using R-12 as the working fluid and in which the condenser pressure is 130 psia and the evaporation pressure is 35 psia. The circulation rate of fluid is 0.1 lbm/sec. The temperature of the refrigerant at the exit of the compressor is 117°F. Determine the compressor power required, cooling load, quality at the inlet of the evaporator, and COP of the refrigerator.

To solve this problem by CyclePad, we take the following steps:

1. Build:
    a. Take a compressor, a condenser, a valve, and a heater (evaporator) from the open-system inventory shop and connect the four devices to form the basic vapor refrigeration cycle.
    b. Switch to analysis mode.
2. Analysis:
    a. Assume a process for each of the four devices: (1) compressor as adiabatic, (2) condenser as isobaric, (3) valve as constant enthalpy, and (4) evaporator as isobaric.

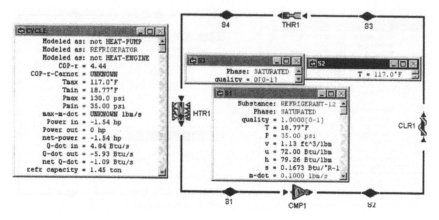

Figure 6.6   Actual vapor refrigeration cycle.

b.   Input the given information: (1) working fluid is R-12, (2) inlet pressure and quality of the compressor are 35 psia and 1, (3) temperature of the refrigerant at the exit of the compressor is 117°F, (4) inlet pressure and quality of the valve are 130 psia and 0, and (5) mass flow rate is 0.1 lbm/sec.

3.   Display the cycle properties' results. The cycle is a refrigerator.

The answers are: compressor power required = −1.54 hp, cooling load = 4.84 Btu/sec = 1.45 ton, quality at the inlet of the evaporator = 0.2738, and COP = 4.44. (See Fig. 6.6.)

Review Problems 6.2   Basic Vapor Refrigeration Cycle

1.   Why is the Carnot refrigeration cycle executed within the saturation dome not a realistic model for refrigeration cycles?

2.   What is the difference between a refrigerator and a heat pump?

3.   Why is the throttling valve not replaced by an isentropic turbine in the ideal refrigeration cycle?

4.   What is the area enclosed by the refrigeration cycle on a *T–s* diagram?

5.   Does the ideal vapor compression refrigeration cycle involve any internal irreversibility?

6.   A steady-flow ideal 0.4 ton refrigerator uses refrigerant R-134a as the working fluid. The evaporator pressure is 120 kPa. The condenser pressure is 600 kPa. Determine (a) the mass rate flow, (b) the compressor power required, (c) the rate of heat absorbed

from the refrigerated space, (d) the rate of heat removed from the condenser, and (e) the COP.

7. An actual vapor compression refrigeration cycle operates at steady state with R-134a as the working fluid. Saturated vapor enters the compressor at 263 K. Superheated vapor enters the condenser at 311 K. Saturated liquid leaves the condenser at 301 K. The mass flow rate of refrigerant is 0.1 kg/sec. Determine (a) the cooling load, (b) the compressor work required, (c) the condenser pressure, (d) the rate of heat removed from the condenser, (e) the compressor efficiency, and (f) the COP.

8. An actual vapor compression refrigeration cycle operates at steady state with R-134a as the working fluid. The evaporator pressure is 120 kPa and the condenser pressure is 600 kPa. The mass flow rate of refrigerant is 0.1 kg/sec. The efficiency of the compressor is 85%. Determine (a) the compressor power, (b) the refrigerating capacity, and (c) the COP.

9. R-134a enters the compressor of a steady-flow vapor compression refrigeration cycle as superheated vapor at 0.14 MPa and −10°C at a rate of 0.04 kg/sec, and it leaves at 0.7 MPa and 50°C. The refrigerant is cooled in the condenser to 24°C and saturated liquid. Determine (a) the compressor power required, (b) the rate of heat absorbed from the refrigerated space, (c) the compressor efficiency, and (d) the COP.

10. Find the compressor power required, quality of the refrigerant at the end of the throttling process, cooling load, and COP for a refrigerator that uses R-12 as the working fluid and is designed to operate at an evaporator temperature of 5°C and a condenser temperature of 30°C. The compressor efficiency is 68%. The mass rate flow of R-12 is 0.22 kg/sec.

11. Consider an ideal refrigerator that uses R-12 as the working fluid. The temperature of the refrigerant in the evaporator is −10°C and in the condenser it is 38°C. The refrigerant is circulated at a rate of 0.031 kg/sec. Determine the compressor power required, quality of the refrigerant at the end of the throttling process, cooling load, and COP of the refrigerator.

12. An ideal refrigerator uses ammonia as the working fluid. The temperature of the refrigerant in the evaporator is 20°F and the pressure in the condenser is 140 psia. The refrigerant is circulated at the rate of 0.051 lbm/sec. Determine the compressor power required, cooling load, and COP of the refrigerator.

13. An ice-making machine operates on an ideal refrigeration cycle using R-134a. The refrigerant enters the compressor as saturated vapor at 20 psia and leaves the condenser as saturated liquid at 80 psia. For 1 ton (12,000 Btu/hr) of refrigeration, determine the compressor power required, mass rate flow of R-134a, and COP of the refrigerator.

14. An ice-making machine operates on an ideal refrigeration cycle using R-134a. The refrigerant enters the compressor at 18 psia and 0°F, and leaves the condenser at 125 psia and 90°F. For 1 ton (12,000 Btu/hr) of refrigeration, determine the compressor power required, mass rate flow of R-134a, and COP of the refrigerator.

15. Find the compressor power required, cooling load, and COP for a refrigerator that uses R-12 as the working fluid and is designed to operate at an evaporator temperature of 5°C and a condenser temperature of 30°C. The mass flow rate of R-12 is 0.22 kg/sec.

16. Consider a 2-ton (24,000 Btu/hr) air-conditioning unit that operates on an ideal refrigeration cycle with R-134a as the working fluid. The refrigerant enters the compressor as saturated vapor at 140 kPa and is compressed to 800 kPa. Determine the compressor power required, mass rate flow of R-134a, quality of the refrigerant at the end of the throttling process, and COP of the refrigerator.

6.3 ACTUAL VAPOR REFRIGERATION CYCLE

The actual vapor refrigeration cycle deviates from the ideal cycle primarily because of the inefficiency of the compressor, as shown in Fig. 6.7.

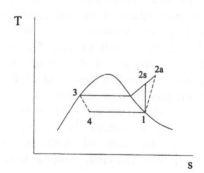

Figure 6.7 *T–s* diagram of actual vapor refrigeration cycle.

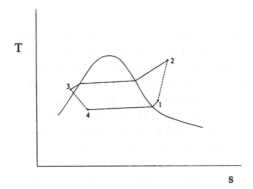

Figure 6.8 *T–s* diagram of actual vapor refrigeration cycle.

In industry, pressure drops associated with fluid flow and heat transfer to or from the surroundings are also considered. The vapor that enters the compressor is usually superheated rather than at saturated vapor state. The degree of superheat of the refrigerant at the inlet of the compressor determines the extent of opening of the expansion valve. This is a principal way to control the refrigeration cycle. The refrigerant that enters the throttling valve is usually compressed rather than at saturated liquid state. The *T–s* diagram of the actual vapor refrigeration cycle is shown in Fig. 6.8.

Example 6.4

Determine the COP, horsepower required, and cooling load of an actual air conditioning unit using R-12 as the working fluid. The refrigerant enters the compressor at 100 kPa and 5°C. The compressor efficiency is 87%. The refrigerant enters the throttling valve at 1.2 MPa and 45°C. The circulation rate of fluid is 0.05 kg/sec. Show the cycle on a *T–s* diagram. Determine the COP and cooling load of the air-conditioning unit, and the power required for the compressor. Plot the sensitivity diagram of COP versus condenser pressure.

To solve this problem by CyclePad, we take the following steps:

1. Build:
   a. Take a compressor, a condenser, a valve, and a heater (evaporator) from the open-system inventory shop and connect the four devices to form the actual vapor refrigeration cycle.
   b. Switch to analysis mode.

2. Analysis:
    a. Assume a process for each of the four devices: (1) compressor as adiabatic, (2) condenser as isobaric, (3) valve as constant enthalpy, and (4) evaporator as isobaric.
    b. Input the given information: (1) working fluid is R-12, (2) inlet temperature and pressure of the compressor are 5°C and 100 kPa, (3) inlet pressure and temperature of the valve are 1.2 MPa and 45°C, (4) mass flow rate is 0.05 kg/sec, and (5) compressor efficiency is 87%.
3. Display results:
    a. Display the *T–s* diagram and cycle properties' results. The cycle is a refrigerator. The answers are: COP = 2.09, cooling load = 5.88 kW = 1.67 tons, and net power input = −2.81 kW.
    b. Display the sensitivity diagram of cycle COP versus condenser pressure. (See Figs. 6.9a and 6.9b.)

Review Problems 6.3   Actual Vapor Refrigeration Cycle

1. Does the ideal vapor refrigeration cycle involve any internal irreversibility?
2. Why is the inlet state of the compressor of the actual vapor refrigeration cycle in the superheated vapor region?
3. Why is the inlet state of the throttling process of the actual vapor refrigeration cycle in the compressed liquid region?
4. R-134a enters the compressor of an actual refrigerator at 140 kPa and −10°C at a rate of 0.05 kg/sec and leaves at 800 kPa.

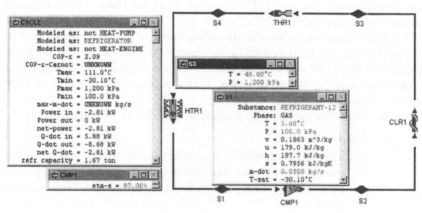

Figure 6.9a   Actual refrigeration cycle.

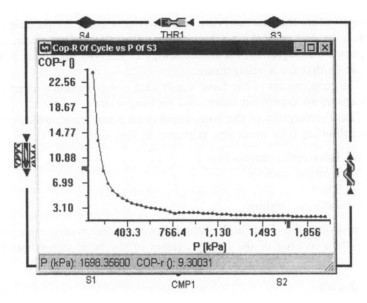

Figure 6.9b Actual refrigeration cycle sensitivity analysis.

The compressor efficiency is 80%. The refrigerant is cooled in the condenser to 26°C. Determine the rate of heat added, rate of heat removed, power input, cooling load, and COP of the actual refrigerator.

5. R-22 enters the compressor of an actual refrigerator at 140 kPa and −10°C at a rate of 0.05 kg/sec and leaves at 800 kPa. The compressor efficiency is 80%. The refrigerant is cooled in the condenser to 26°C. Determine the rate of heat added, rate of heat removed, power input, cooling load, and COP of the actual refrigerator.

6. Ammonia enters the compressor of an actual refrigerator at 140 kPa and −10°C at a rate of 0.05 kg/sec and leaves at 800 kPa. The compressor efficiency is 80%. The refrigerant is cooled in the condenser to 26°C. Determine the rate of heat added, rate of heat removed, power input, cooling load, and COP of the actual refrigerator.

6.4 BASIC VAPOR HEAT PUMP CYCLE

A system is called a refrigerator or a heat pump depending on the purpose of the system. If the purpose of the system is to remove heat from a low-temperature thermal reservoir, it is a refrigerator. If the purpose of the system is to deliver heat to a high-temperature thermal reservoir, it is a heat pump. Consequently, the methodology of analysis for a heat pump is identical to that for a refrigerator.

The components of the basic vapor heat pump include a compressor, a condenser, an expansion valve, and an evaporator.

The T–s diagram of the basic vapor heat pump cycle, which consists of the following four processes, is shown in Fig. 6.10:

1-2 Isentropic compression
2-3 Isobaric cooling
3-4 Throttling
4-1 Isobaric heating

Applying the first and second laws of thermodynamics of the open system to each of the four processes of the basic vapor heat pump yields:

$$Q_{12} = 0 \qquad\qquad\qquad (6.19)$$

$$0 - W_{12} = m(h_2 - h_1) \qquad\qquad\qquad (6.20)$$

$$W_{23} = 0 \qquad\qquad\qquad (6.21)$$

$$Q_{23} - 0 = m(h_3 - h_2) \qquad\qquad\qquad (6.22)$$

$$Q_{34} = 0 \text{ and } W_{34} = 0 \qquad\qquad\qquad (6.23)$$

$$0 - 0 = m(h_3 - h_4) \qquad\qquad\qquad (6.24)$$

$$W_{41} = 0 \qquad\qquad\qquad (6.25)$$

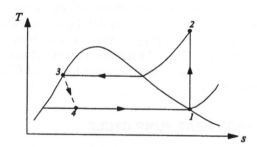

Figure 6.10 Basic vapor heat pump T–S diagram.

and

$$Q_{41} - 0 = m(h_1 - h_4) \tag{6.26}$$

The desirable energy output of the basic vapor heat pump is the heat removed from the condenser (or heat added to the high-temperature thermal reservoir). The energy input to the cycle is the compressor work required. Thus, the COP of the cycle is

$$\beta_{HP} = Q_{23}/W_{12} = (h_3 - h_2)/(h_2 - h_1) \tag{6.27}$$

The rate of heat removed from the inner space of the heat pump is called the heating load or heating capacity.

Example 6.5

Determine the COP, horsepower required, and heating load of a basic vapor heat pump cycle using R-134a as the working fluid and in which the condenser pressure is 900 kPa and the evaporation pressure is 240 kPa. The circulation rate of fluid is 0.1 kg/sec. Show the cycle on a T–s diagram. Plot the sensitivity diagram of COP versus condenser pressure.

To solve this problem by CyclePad, we take the following steps:

1. Build:
   a. Take a compressor, a condenser, a valve, and a heater (evaporator) from the open-system inventory shop and connect the four devices to form the basic heat pump cycle.
   b. Switch to analysis mode.
2. Analysis:
   a. Assume a process for each of the four devices: (1) compressor as isentropic, (2) condenser as isobaric, (3) valve as constant enthalpy, and (4) evaporator as isobaric.
   b. Input the given information: (1) working fluid is R-134a, (2) inlet pressure and quality of the compressor are 240 kPa and 1, (3) inlet pressure and quality of the valve are 900 kPa and 0, and (4) mass flow rate is 0.1 kg/sec.
3. Display results:
   a. Display the T–s diagram and cycle properties' results. The cycle is a heat pump. The answers are COP = 6.28, heating load = −17.27 kW, and net power input = −2.75 kW.

b. Display the sensitivity diagram of cycle COP versus evaporation pressure. (See Figs. 6.11a and 6.11b.)

Review Problems 6.4   Heat Pump

1. Find the compressor power required, quality of the refrigerant at the end of the throttling process, heating load, and COP for a heat pump that uses R-12 as the working fluid and is designed to operate at an evaporator saturation temperature of 10°C and a

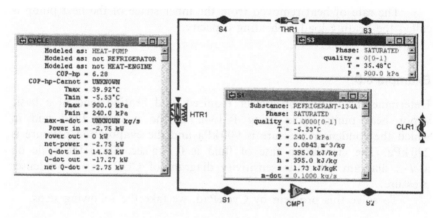

Figure 6.11a   Basic vapor heat pump cycle.

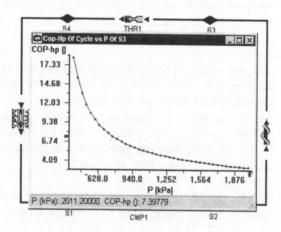

Figure 6.11b   Basic vapor heat pump cycle sensitivity analysis.

condenser saturation temperature of 40°C. The mass rate flow of R-12 is 0.22 kg/sec.

2. Consider an ideal heat pump which uses R-12 as the working fluid. The saturation temperature of the refrigerant in the evaporator is 6°C and in the condenser it is 58°C. The refrigerant is circulated at the rate of 0.021 kg/sec. Determine the compressor power required, quality of the refrigerant at the end of the throttling process, heating load, and COP of the heat pump.

3. An ideal heat pump uses ammonia as the working fluid. The saturation temperature of the refrigerant in the evaporator is 22°F and in the condenser it is 98°F. The refrigerant is circulated at the rate of 0.051 lbm/sec. Determine the compressor power required, heating load, and COP of the heat pump.

4. Find the compressor power required, heating load, and COP for a heat pump that uses R-12 as the working fluid and is designed to operate at an evaporator saturation temperature of 2°C and a condenser temperature of 39°C. The compressor efficiency is 78%. The mass rate flow of R-12 is 0.32 kg/sec.

5. Find the compressor power required, turbine power produced, heating load, and COP for a heat pump that uses R-12 as the working fluid and is designed to operate at an evaporator saturation temperature of 2°C and a condenser saturation temperature of 39°C. The mass rate flow of R-12 is 0.32 kg/sec. The throttling valve is replaced by an adiabatic turbine with 74% efficiency.

6.5 ACTUAL VAPOR HEAT PUMP CYCLE

The actual vapor heat pump cycle deviates from the ideal cycle primarily because of inefficiency of the compressor, pressure drops associated with fluid flow and heat transfer to or from the surroundings. The vapor entering the compressor must be superheated slightly rather than a saturated vapor. The refrigerant entering the throttling valve is usually compressed liquid rather than a saturated liquid.

Example 6.6

Determine the COP, horsepower required, and heating load of a basic vapor heat pump cycle using R-134a as the working fluid and in which the condenser pressure is 900 kPa and the evaporation pressure is 240 kPa. The circulation rate of fluid is 0.1 kg/sec. The compressor efficiency is 88%. Show the cycle on a *T–s* diagram.

To solve this problem by CyclePad, we take the following steps:

1. Build:
   a. Take a compressor, a cooler (condenser), a valve, and a heater (evaporator) from the open-system inventory shop and connect the four devices to form the basic heat pump cycle.
   b. Switch to analysis mode.
2. Analysis:
   a. Assume a process for each of the four devices: (1) compressor as isentropic, (2) condenser as isobaric, (3) valve as constant enthalpy, and (4) evaporator as isobaric.
   b. Input the given information: (1) working fluid is R-134a, (2) inlet pressure and quality of the compressor are 240 kPa and 1, (3) inlet pressure and quality of the valve are 900 kPa and 0, (4) compressor efficiency is 88%, and (5) mass flow rate is 0.1 kg/sec.
3. Display the *T–s* diagram and cycle properties' results. The cycle is a heat pump.

The answers are COP = 5.65, heating load = 14.52 kW, and net power input = −3.12 kW. (See Fig. 6.12.)

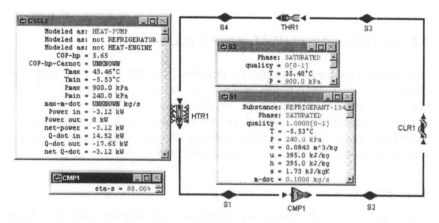

Figure 6.12 Basic vapor heat pump cycle.

Review Problems 6.5   Actual Vapor Heat Pump

1. Find the compressor power required, quality of the refrigerant at the end of the throttling process, heating load, and COP for a heat pump that uses R-12 as the working fluid and is designed to operate at an evaporator saturation temperature of 10°C and a condenser saturation temperature of 40°C. The compressor efficiency is 68%. The mass rate flow of R-12 is 0.22 kg/sec.

2. Consider a heat pump that uses R-12 as the working fluid. The compressor efficiency is 80%. The saturation temperature of the refrigerant in the evaporator is 6°C and in the condenser it is 58°C. The refrigerant is circulated at the rate of 0.021 kg/sec. Determine the compressor power required, quality of the refrigerant at the end of the throttling process, heating load, and COP of the heat pump.

3. A heat pump uses ammonia as the working fluid. The compressor efficiency is 80%. The saturation temperature of the refrigerant in the evaporator is 22°F and in the condenser it is 98°F. The refrigerant is circulated at the rate of 0.051 lbm/sec. Determine the compressor power required, heating load, and COP of the heat pump.

4. Find the compressor power required, heating load, and COP for a heat pump that uses R-12 as the working fluid and is designed to operate at an evaporator saturation temperature of 2°C and a condenser temperature of 39°C. The compressor efficiency is 78%. The mass rate flow of R-12 is 0.32 kg/sec.

5. Find the compressor power required, turbine power produced, heating load, and COP for a heat pump that uses R-12 as the working fluid and is designed to operate at an evaporator temperature of 2°C and a condenser temperature of 39°C. The compressor efficiency is 78%. The mass rate flow of R-12 is 0.32 kg/sec. The throttling valve is replaced by an adiabatic turbine with 74% efficiency.

6. Determine the COP, horsepower required, and heating load of a basic vapor heat pump cycle using R-134a as the working fluid in which the condenser pressure is 900 kPa and the evaporator pressure is 240 kPa. The circulation rate of fluid is 0.1 kg/sec. The compressor efficiency is 78%.

7. R-134a enters the compressor of a refrigerator at 0.14 MPa and −10°C at a rate of 0.1 kg/sec and leaves at 0.7 MPa and 50°C. The refrigerant is cooled in the condenser to 24°C and 0.65 MPa. The refrigerant is throttled to 0.15 MPa. Determine the

compressor efficiency, power input to the compressor, cooling effect, and COP of the refrigerator.

8. R-134a enters the compressor of a refrigerator at 0.14 MPa and $-10°C$ at a rate of 0.1 kg/sec and leaves at 0.7 MPa. The compressor efficiency is 85%. The refrigerant is cooled in the condenser to 24°C and 0.65 MPa. The refrigerant is throttled to 0.15 MPa. Determine the power input to the compressor, cooling effect, and COP of the refrigerator.

9. R-134a enters the compressor of a refrigerator at 0.14 MPa and $-10°C$ at a rate of 0.1 kg/sec and leaves at 0.8 MPa. The compressor efficiency is 85%. The refrigerant is cooled in the condenser to 26°C and 0.75 MPa. The refrigerant is throttled to 0.15 MPa. Determine the power input to the compressor, cooling effect, and COP of the refrigerator.

6.6 WORKING FLUIDS FOR VAPOR REFRIGERATION AND HEAT PUMP SYSTEMS

Ammonia, carbon dioxide, and sulphur dioxide were used widely in the early years of refrigeration in industrial refrigeration applications. For domestic and industrial applications now, the principal refrigerants have been synthetic freons. This family of substances are known by an R number of the general form RN, R signifying refrigerant and the number N specifically identifying the chemical compound. The number allocated to the halogenated hydrocarbons (freons) are derived as follows: for refrigerants derived from methane (CH_4), N is a two-digit integer. The first digit indicates the number of hydrogen atoms +1 and the second digit indicates the number of fluorine atoms, e.g., CCl_2F_2 is R-12. For refrigerants derived from ethane (C_2H_6), N is a three-digit integer. The first digit is always 1, the second digit is the number of hydrogen atoms +1, and the third digit is the number of fluorine atoms, e.g., $C_2Cl_2F_4$ is R-114.

There are three R number refrigerants on the substance menu of CyclePad. The three refrigerants are R-12, R-22, and R-134a.

The desirable properties of working fluids for vapor refrigeration and heat pump systems include high critical temperature and low pressure, low specific volume, inexpensive, nonflammable, nonexplosive, nontoxic, noncorrosive, inert and stable, etc.

In recent years, the effects of freons on the ozone layer have been critically evaluated. Some freons such as R-12, having leaked from refrigeration systems into the atmosphere, spend many years slowly diffusing upward into the stratosphere. There it is broken down, releasing chlorine, which depletes the protective ozone layer surrounding the Earth's stratosphere. Ozone is a critical component of the atmospheric system both for climate control and for reducing solar radiation. It is, therefore, important to human beings to ban these freons such as the widely used, but life threatening, R-12. New desirable refrigerants which contain no chlorine atoms are found to be suitable and acceptable replacements.

Review Problems 6.6   Working Fluids for Vapor Refrigeration and Heat Pump Systems

1. Why are ammonia, carbon dioxide, and sulphur dioxide no longer used in domestic refrigerators and heat pumps?
2. What devastating consequence results in our environment by refrigerants leaking out from refrigeration systems into the atmosphere?
3. List five desirable properties of working fluids for vapor refrigeration and heat pump systems.
4. What does the refrigerant number mean?

6.7   CASCADE AND MULTISTAGED VAPOR REFRIGERATORS

There are several variations of the basic vapor refrigeration cycle. A cascade cycle is used when the temperature difference between the evaporator and the condenser is quite large. The multistaged cycle is used to reduce the required compressor power input.

6.7.1   Cascade Vapor Refrigerators

There are applications when the temperature difference between the evaporator and the condenser is quite large. A single vapor refrigeration cycle usually cannot be used to achieve the large difference. To solve this problem and still using vapor refrigeration cycles, a cascade vapor refrigeration cycle must be used. A cascade cycle is several vapor refrigeration cycles connected in series. A cascade cycle made of three cycles in series is illustrated in Fig. 6.13. The condenser of the lowest-temperature cycle (cycle A, 1-2-3-4-1) provides the heat input to the evaporator of the midtemperature cycle (cycle B, 5-6-7-8-5) and the condenser of the midtemperature cycle (cycle B) provides the heat input to

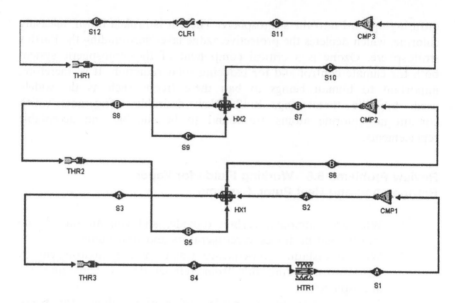

Figure 6.13 Cascade vapor refrigerator.

the evaporator of the highest-temperature cycle (cycle C, 9-10-11-12-9). Different working fluids may be used in each of the individual cycles.

Neglecting kinetic and potential energy changes, a steady-state and steady-flow mass and energy balance on the components of the cascade vapor refrigeration cycle have the general forms:

$$\sum mdot_e = \sum mdot_i \tag{6.28}$$

and

$$Qdot - Wdot = \sum mdot_e h_e - \sum mdot_i h_i \tag{6.29}$$

The cooling load of the cascade vapor refrigeration cycle is the rate of heat added in the evaporator of the lowest temperature cycle. The power added to the cycle is the sum of the power added to the individual compressors:

$$Wdot = \sum Wdot_{compressor} \tag{6.30}$$

and the COP of the cycle is

$$\beta = Qdot_{\text{lowest T evaporator}} / Wdot \tag{6.31}$$

The following example illustrates the analysis of the cascade vapor refrigeration cycle.

Example 6.7

A cascade vapor refrigeration cycle made of two separate vapor refrigeration cycles has the following information:

Top cycle: working fluid = R-134a, $p_5 = 200$ kPa, $x_5 = 1$, $p_7 = 500$ kPa, and $x_7 = 0$.

Bottom cycle: working fluid = R-134a, $p_1 = 85$ kPa, $mdot_1 = 1$ kg/sec, $x_1 = 1$, $p_3 = 250$ kPa, and $x_3 = 0$

Determine the mass rate flow of the top cycle, power required by compressors #1 and #2, total power required by the compressors, rate of heat added to the evaporator, cooling load, and COP of the cascade vapor refrigeration cycle.

To solve the problem by CyclePad, we take the following steps:

1. Build the cycle as shown in Fig. 6.14a.
2. Assume compressors are adiabatic with 100% efficiency, heater and cooler are isobaric, and both hot and cold sides of the heat exchanger are isobaric.
3. Input working fluid is R-134a at state 1, $p_1 = 85$ kPa, $mdot_1 = 1$ kg/sec, $x_1 = 1$, $p_3 = 250$ kPa, and $x_3 = 0$; working fluid is R-134a at state 5, $p_5 = 200$ kPa, $x_5 = 1$, $p_7 = 500$ kPa, and $x_7 = 0$.

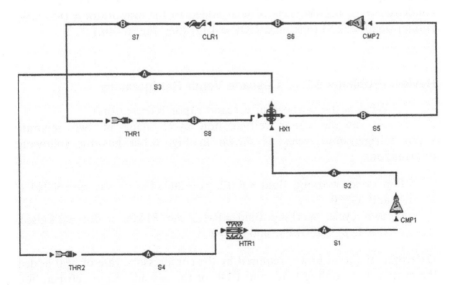

Figure 6.14a Cascade refrigerator.

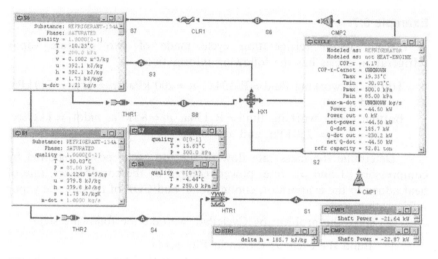

Figure 6.14b Cascade refrigerator.

4. Display the cycle properties' results:

The answers are: $mdot_5 = 1.21$ kg/sec, COP = 4.17, power input by compressor 1 = -21.64 kW, power input by compressor 2 = -22.87 kW, power input by compressors = -44.5 kW, rate of heat removed by the condenser = -230.2 kW, rate of heat added to the evaporator = 185.7 kW, cooling load = 52.81 tons, and COP = 4.17. (See Fig. 6.14b.)

Review Problems 6.7.1   Cascade Vapor Refrigerators

1. What is the purpose of cascade vapor refrigerators?
2. A cascade vapor refrigeration cycle made of two separate vapor refrigeration cycles as shown in Fig. 6.14a has the following information:

Top cycle: working fluid = R-12, $p_5 = 200$ kPa, $x_5 = 1$, $p_7 = 500$ kPa, and $x_7 = 0$

Bottom cycle: working fluid = R-12, $p_1 = 85$ kPa, $mdot_1 = 1$ kg/sec, $x_1 = 1$, $p_3 = 250$ kPa, and $x_3 = 0$

Determine the total power required by the compressors, rate of heat added the evaporator, cooling load, and COP of the cascade vapor refrigeration cycle.

ANSWERS: power input by compressors $= -44.5$ kW, rate of heat removed by the condenser $= -179.9$ kW, rate of heat added to the evaporator $= 142.2$ kW, cooling load $= 40.42$ tons, and COP $= 3.77$.

3. A cascade vapor refrigeration cycle made of two separate vapor refrigeration cycles as shown in Fig. 6.14a has the following information:

Top cycle: working fluid $=$ ammonia, $p_5 = 200$ kPa, $x_5 = 1$, $p_7 = 500$ kPa, and $x_7 = 0$

Bottom cycle: working fluid $=$ ammonia, $p_1 = 85$ kPa, $m$dot$_1 = 1$ kg/sec, $x_1 = 1$, $p_3 = 250$ kPa, and $x_3 = 0$

Determine total power required by the compressors, rate of heat added to the evaporator, cooling load, and COP of the cascade vapor refrigeration cycle.

ANSWERS: power input by compressors $= -279.7$ kW, rate of heat added to the evaporator $= 1276$ kW, cooling load $= 362.9$ tons, and COP $= 4.56$.

6.7.2 Multistaged Vapor Refrigerators

A flash chamber may have better heat-transfer characteristics than those of the heat exchanger employed between the upstream cycle and downstream cycle of the cascaded cycle. It is, therefore, used to replace the heat exchanger in *multistaged vapor refrigerators*. In this arrangement, the working fluid flowing throughout the whole system must be the same.

A schematic diagram of a two-stage vapor refrigerator is shown in Fig. 6.15. The liquid leaving the condenser is throttled into a flashing chamber (separator used to separate mixture into vapor and liquid) maintained at a pressure between the evaporator pressure and condenser pressure. Saturated vapor separated from the liquid in the flashing chamber enters a mixing chamber, where it mixes with the vapor leaving the low-pressure compressor at state 2. The saturated liquid is throttled to the evaporator pressure at state 9. By adjusting the mass flow rate in the separator, the cooling load of the refrigeration cycle can be controlled. The analysis of the cycle is illustrated in Example 6.8.

Neglect kinetic and potential energy changes. A steady-state and steady-flow mass and energy balance on the components of the cascade vapor refrigeration cycle have the general forms:

$$\sum m\text{dot}_e = \sum m\text{dot}_i \tag{6.32}$$

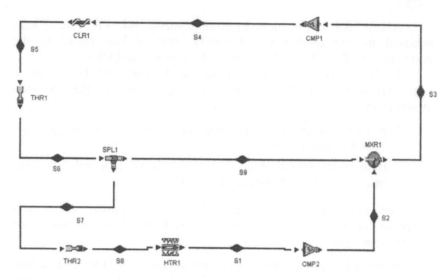

Figure 6.15   Multistage vapor refrigerator.

and

$$Q\text{dot} - W\text{dot} = \sum m\text{dot}_e h_e - \sum m\text{dot}_i h_i \tag{6.33}$$

The cooling load of the cascade vapor refrigeration cycle is the rate of heat added to the evaporator of the lowest temperature cycle. The power added to the cycle is the sum of the power added to the individual compressors:

$$W\text{dot} = \sum W\text{dot}_{\text{compressor}} \tag{6.34}$$

and the COP of the cycle is

$$\beta = Q\text{dot}_{\text{lowest T evaporator}}/W\text{dot} \tag{6.35}$$

The following examples illustrate the analysis of the cascade vapor refrigeration cycle.

Example 6.8

A two-stage vapor refrigeration cycle as shown in Fig. 6.15 has the following information: working fluid = R-134a, $p_1 = 85\,\text{kPa}$, $x_1 = 0$, $p_2 = 200\,\text{kPa}$, $p_4 = 500\,\text{kPa}$, $m_4 = 1\,\text{kg/sec}$, $m_7 = 0.8\,\text{kg/sec}$, and $x_5 = 0$.

Determine the power required by compressors #1 and #2, total power required by the compressors, rate of heat added to the evaporator, cooling load, and COP of the cycle. Plot the cooling load versus m_7 sensitivity diagram.

To solve the problem by CyclePad, we take the following steps:

1. Build the cycle as shown in Fig. 6.16.
2. Assume compressors are adiabatic with 100% efficiency, heater and cooler are isobaric, and both hot and cold sides of the heat exchanger are isobaric.
3. Input working fluid is R-134a at state 1, $p_1 = 85$ kPa, $x_1 = 0$, $p_2 = 200$ kPa, $p_4 = 500$ kPa, $m_4 = 1$ kg/sec, $m_7 = 0.8$ kg/sec, and $x_5 = 0$.
4. Display the cycle properties' results:

The answers are: COP = 4.84, power input by compressor 1 = −18.35 kW, power input by compressor 2 = −13.58 kW, power input by compressors = −31.93 kW, rate of heat removed by the condenser = −186.6 kW, rate of heat added to the evaporator = 154.7 kW, and cooling load = 43.99 tons. (See Fig. 6.17.)

An arrangement of either a cascaded or multistaged refrigerator can be made as illustrated in Fig. 6.18. In this arrangement, the system can be either a cascaded refrigerator or a multistaged refrigerator.

Suppose that $mdot_3 = 0$ and $mdot_8 = 0$, then the working fluids of the top and that of the bottom cycle do not mix. Therefore, it is a cascaded refrigerator.

Suppose that $mdot_3 = 1$, then the working fluids of the top and that of the bottom cycle do mix. It becomes a multistaged refrigerator.

Example 6.9 illustrates the system as a multistaged refrigerator if $mdot_3 = 1$.

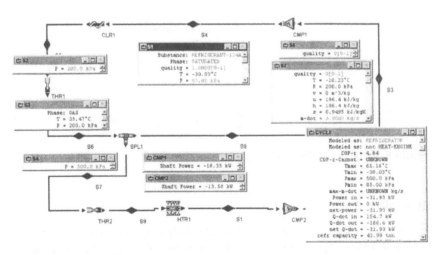

Figure 6.16 Two-stage vapor refrigerator.

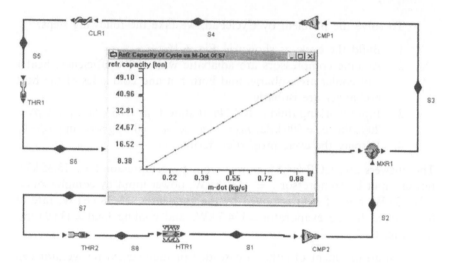

Figure 6.17 Two-stage vapor refrigerator cooling load sensitivity analysis.

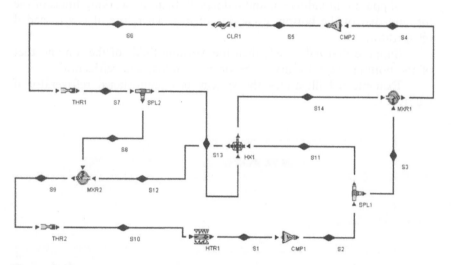

Figure 6.18 Cascaded or multistaged refrigerator.

Example 6.9

A cycle as shown in Fig. 6.18 has the following information: working fluid = R-134a, $p_1 = 85\,\text{kPa}$, $x_1 = 0$, $p_2 = 200\,\text{kPa}$, $p_5 = 500\,\text{kPa}$, $m_4 = 1\,\text{kg/sec}$, $m_{11} = 0\,\text{kg/sec}$, $x_6 = 0$, $x_8 = 0$, and $x_{13} = 1$.

Determine the power required by compressors #1 and #2, total power required by the compressors, rate of heat added to the evaporator, cooling load, and COP of the cycle.

To solve the problem by CyclePad, we take the following steps:

1. Build the cycle as shown in Fig. 6.18.
2. Assume compressors are adiabatic with 100% efficiency, heater and cooler are isobaric, and both hot and cold sides of the heat exchanger are isobaric.
3. Input working fluid is R-134a at state 1, $p_1 = 85\,\text{kPa}$, $x_1 = 0$, $p_2 = 200\,\text{kPa}$, $p_5 = 500\,\text{kPa}$, $m_4 = 1\,\text{kg/sec}$, $m_{11} = 0\,\text{kg/sec}$, $x_6 = 0$, $x_8 = 0$, and $x_{13} = 1$.
4. Display the cycle properties' results.

The answers are: COP = 4.81, power input by compressor 1 = −33.37 kW, rate of heat removed by the condenser = −193.9 kW, rate of heat added to the evaporator = 160.5 kW, and cooling load = 45.64 tons. (See Fig. 6.19.)

Review Problems 6.7.2   Multistage Vapor Refrigerators

1. What is the purpose of multistaged vapor refrigerators?
2. What is the difference between cascaded and multistaged vapor refrigerators?

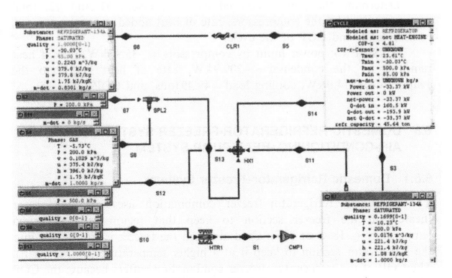

Figure 6.19 Cascaded or multistaged refrigerator.

3. A two-stage vapor refrigeration cycle as shown in Fig. 6.19 has the following information: working fluid $= $ R-22, $p_1 = 85\,$kPa, $x_1 = 0$, $p_2 = 200\,$kPa, $p_4 = 500\,$kPa, $m_4 = 1\,$kg/sec, $m_7 = 0.8\,$kg/sec, and $x_5 = 0$.

Determine the power required by compressors #1 and #2, total power required by the compressors, rate of heat added to the evaporator, cooling load, and COP of the cycle.

ANSWERS: power input by compressors $= -38.12\,$kW, rate of heat removed by the condenser $= -210.2\,$kW, rate of heat added to the evaporator $= 172.1\,$kW, cooling load $= 48.92\,$tons, and COP $= 4.51$.

4. A two-stage vapor refrigeration cycle as shown in Fig. 6.19 has the following information: working fluid $= $ ammonia, $p_1 = 85\,$kPa, $x_1 = 0$, $p_2 = 200\,$kPa, $p_4 = 500\,$kPa, $m_4 = 1\,$kg/sec, $m_7 = 0.8\,$kg/sec, and $x_5 = 0$.

Determine the power required by compressors #1 and #2, total power required by the compressors, rate of heat added in the evaporator, cooling load, and COP of the cycle.

ANSWERS: power input by compressors $= -217.5\,$kW, rate of heat removed by the condenser $= -1257\,$kW, rate of heat added to the evaporator $= 1040\,$kW, cooling load $= 295.6\,$tons, and COP $= 4.78$.

5. A two-stage vapor refrigeration cycle as shown in Fig. 6.19 has the following information: working fluid $= $ R-134a, $p_1 = 85\,$kPa, $x_1 = 0$, $p_2 = 200\,$kPa, $p_4 = 500\,$kPa, $m_4 = 1\,$kg/sec, $m_7 = 0.9\,$kg/sec, and $x_5 = 0$.

Determine the power required by compressors #1 and #2, total power required by the compressors, rate of heat added in the evaporator, cooling load, and COP of the cycle.

ANSWERS: power input by compressors $= -35.92\,$kW, rate of heat removed by the condenser $= -209.9\,$kW, rate of heat added to the evaporator $= 174.0\,$kW, cooling load $= 49.49\,$tons, and COP $= 4.84$.

6.8 DOMESTIC REFRIGERATOR-FREEZER SYSTEM AND AIR-CONDITIONING–HEAT PUMP SYSTEM

6.8.1 Domestic Refrigerator–Freezer System

The household refrigerator–freezer combination uses one evaporator (heater) in the freezer section to keep that region at the desired temperature ($-18°$C or $0°$F). Cold air from the freezer is transferred into the refrigerator section to keep it at a higher temperature ($2°$C or $35°$F). The COP of the refrigerator–freezer combination suffers because the COP of the combination is equal to the COP of the freezer. It is known that the

COP of a refrigeration cycle is inversely proportional to $(T_H - T_L)$. The lower the T_L, the lower the COP.

One method of improving the COP of the refrigerator–freezer combination is to employ an evaporator for both the refrigerator region and the freezer region with a single compressor, as illustrated in Fig. 6.20. A numerical example of this arrangement is shown in Example 6.10.

Example 6.10

A two-region section refrigerator requires refrigeration at $-37°C$ and $-19°C$. Using ammonia as the refrigerant, design a dual evaporator refrigerator and find the COP, compressor input power, and cooling load of the refrigerator based on one unit mass flow rate of refrigerant.

To design the refrigerator by CyclePad, we take the following steps:

1. Build the two-region section refrigerator as shown in Fig. 6.20.
2. Assume that the compressor is adiabatic and 100% efficient, and cooler and heaters are isobaric.
3. Let working fluid be ammonia, $T_1 = -37°C$, $x_1 = 1$, $mdot = 1\,kg/sec$, $p_2 = 800\,kPa$, $x_3 = 0$, $T_4 = -19°C$, and $x_6 = 0.4$.
4. Display cycle properties' results.

The answers are: COP $= 3.39$, $Qdot_{htr\#1} = 301.3\,kW$, $Qdot_{htr\#2} = 828.2\,kW$, compressor input power $= -333.4\,kW$, and cooling load $= 321.2\,tons$. (See Fig. 6.21.)

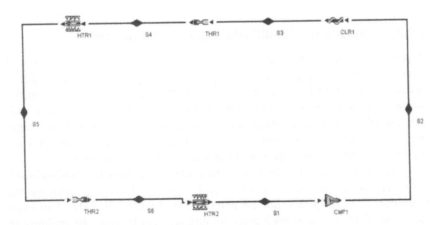

Figure 6.20 Refrigerator and freezer with dual evaporator.

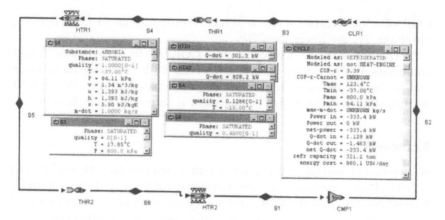

Figure 6.21   Refrigerator and freezer with dual evaporator.

Review Problems 6.8.1   Refrigerator and Freezer with Dual Evaporator

1. What is the purpose of the refrigerator and freezer with dual evaporator?

2. A two-region section refrigerator requires refrigeration at $-17°C$ and $2°C$. Using ammonia as the refrigerant, design a dual evaporator refrigerator and find the COP, compressor input power, and cooling load of the refrigerator based on one unit mass flow rate of refrigerant.

6.8.2   Domestic Air-Conditioning–Heat Pump System

Refrigerators and heat pumps have the same energy flow diagram and have the same components. A domestic air-conditioning and heat pump system as shown in Fig. 6.22 can, therefore, be used as a heat pump in the winter as well as an air-conditioning unit in the summer. Notice that both the domestic air-conditioning and heat pump system share the same equipment. Thus, the investment in the heat pump can also be used for air-conditioning to provide year-round house comfort control.

In the air-conditioning mode, the cycle (cycle A) is 1-2-3-4-5-6-7-8-9-10-1. The heat exchanger removes heat from the building, replacing the evaporator. Atmospheric hot air entering the dwelling at state 11 of cycle B is cooled by the heat exchanger by removing heat from the building to vaporize the refrigerant and leaving the dwelling at state 12 of cycle B.

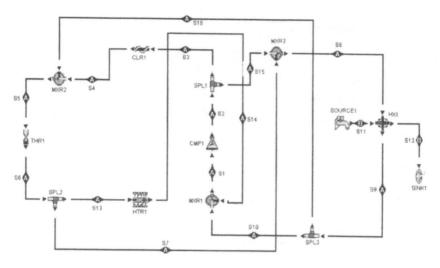

Figure 6.22 Domestic air-conditioning and heat pump system.

In the heat-pump mode, the cycle (cycle A) is 1-2-15-8-9-16-5-6-13-14-1. The heat exchanger adds heat to the building, replacing the condenser. Atmospheric cold air entering the dwelling at state 11 of cycle B is heated by the heat exchanger by discharging heat to the building to condense the refrigerant and leaving the dwelling at state 12 of cycle B.

Notice that the heat exchanger is an evaporator in the air-conditioning mode, and a condenser in the heat-pump mode. Therefore, the refrigerant is on the hot side in the heat-pump mode and on the cold side in the air-conditioning mode when the system is built using CyclePad.

Example 6.11

In the air-conditioning mode, a domestic air-conditioning and heat-pump system as shown in Fig. 6.22 uses R-134a as the refrigerant. The refrigerant-saturated vapor is compressed from 140 to 700 kPa. Summer ambient air at 30°C is to be cooled down to 17°C. Find the compressor power required, heat removed from the ambient air in the heat exchanger, COP of the system, and mass rate of air flow per unit of mass rate of refrigerant flow.

To solve this problem by CyclePad, we take the following steps:

1. Build the system as shown in Fig. 6.22.
2. Assume that the compressor is adiabatic and 100% efficient, and cooler and heaters are isobaric.

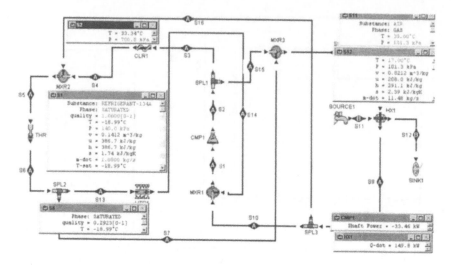

Figure 6.23 Domestic air-conditioning system.

3. Let working fluid be R-134a, $p_1 = 140\,\text{kPa}$, $x_1 = 1$, $mdot = 1\,\text{kg/sec}$, $p_2 = 700\,\text{kPa}$, $x_3 = 0$, $T_{11} = 30°\text{C}$, $T_{12} = 17°\text{C}$, $mdot_{13} = 0$, $mdot_{15} = 0$, and $mdot_{16} = 0$.

4. Display results. The answers are: compressor input power $= -33.46\,\text{kW}$, $Qdot_{HX1} = -149.8\,\text{kW}$, and $mdot_{11} = 11.49\,\text{kg/sec}$.

5. The COP of the system is $Qdot_{HX1}/\text{compressor}$ input power $= 149.8/33.46 = 4.477$. (See Fig. 6.23.)

6.9 ABSORPTION AIR-CONDITIONING

The *absorption air-conditioning or refrigeration system* shown in Fig. 6.24 is a system in which heat instead of work is employed to produce a refrigeration effect. In a conventional refrigeration system, high quality and expensive electrical work is consumed by the compressor, which compresses vapor from a low to a high pressure. Since pumping involves only a liquid, the pump consumes very little electrical work. Therefore, the compressor of the basic vapor refrigeration cycle is replaced by a refrigerant generator (heater), an absorber (mixing chamber), a separator (splitter), a valve, and a liquid pump in the absorption air-conditioning system. The major energy input to the absorption system is heat added to the refrigerant generator, which generates refrigerant.

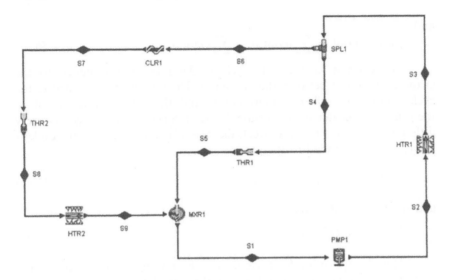

Figure 6.24   Absorption air-conditioning.

To illustrate the operation of the absorption air-conditioning or refrigeration system, consider the working fluids employed to be ammonia–water. Ammonia is the refrigerant and water is the absorber in the absorption air-conditioning or refrigeration system shown in Fig. 6.24. The vaporized ammonia refrigerant leaving the evaporator at state 9 is absorbed by the weak absorber solution at state 5 and is accompanied by a release of heat. The absorbent solution with a high concentration of refrigerant at state 1 is pumped to an upper pressure at state 2 corresponding to that of the condenser system. The heat input to the refrigerant generator from state 2 to state 3 boils off the refrigerant to state 6, leaving a weak absorber solution at state 4. The vaporized refrigerant that enters the condenser at state 6, the expansion valve at state 7, and then the evaporator at state 9 completes the refrigeration cycle as in the basic vapor refrigeration cycle. The most commonly used working fluid combinations in the absorption air-conditioning or refrigeration systems is water and lithium bromide, where water is the refrigerant and lithium bromide is the absorber.

Neglect kinetic and potential energy changes. A steady-state and steady-flow mass and energy balance on the components of the cycle have the general forms:

$$\sum mdot_e = \sum mdot_i \qquad (6.36)$$

and

$$Q\text{dot} - W\text{dot} = \sum m\text{dot}_e h_e - \sum m\text{dot}_i h_i \qquad (6.37)$$

The cooling load of the cycle is the rate of heat added to the evaporator (heater between state 8 and 9). The rate of energy added to the cycle is the sum of the pump power and the rate of heat added to the generator (heater between state 2 and 3). Since the pump requires very little power (neglect pump power), the rate of energy added to the cycle is the rate of heat added to the generator:

$$Q\text{dot}_\text{in} = Q\text{dot}_{23} \qquad (6.38)$$

and the COP of the cycle is

$$\beta = Q\text{dot}_{89}/Q\text{dot}_{23} \qquad (6.39)$$

The absorption cycle efficiency can be improved by placing a heat exchanger between the generator and the absorber as a regenerator, as shown in Fig. 6.25.

Review Problems 6.9   Absorption Air-Conditioning

1. What is absorption refrigeration?
2. How does an absorption refrigeration system differ from a vapor compression refrigeration system?
3. What are the advantages of absorption refrigeration?

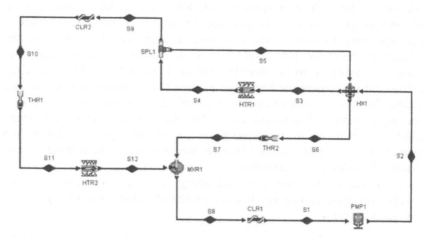

Figure 6.25   Absorption air-conditioning with regenerator.

6.10 BRAYTON GAS REFRIGERATION CYCLE

The *basic gas Brayton refrigeration cycle* is the reversed Brayton gas power cycle. The components of the basic gas refrigeration cycle include a compressor, a cooler, a turbine, and a heater, as shown in Fig. 6.26.

The T–s diagram of the basic gas Brayton refrigeration cycle, which consists of the following four processes, is shown in Fig. 6.27:

1-2 Isentropic compression
2-3 Isobaric cooling
3-4 Isentropic expansion
4-1 Isobaric heating

Applying the first and second laws of thermodynamics of the open system to each of the four processes of the basic gas refrigeration yields:

$$Q_{12} = 0 \tag{6.40}$$

$$0 - W_{12} = m(h_2 - h_1) \tag{6.41}$$

$$W_{23} = 0 \tag{6.42}$$

$$Q_{23} - 0 = m(h_3 - h_2) \tag{6.43}$$

$$Q_{34} = 0 \tag{6.44}$$

$$0 - W_{34} = m(h_3 - h_4) \tag{6.45}$$

$$W_{41} = 0 \tag{6.46}$$

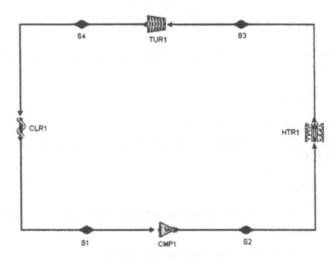

Figure 6.26 Basic Brayton refrigeration cycle.

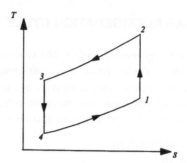

Figure 6.27 T–s diagram of the basic Brayton refrigeration cycle.

and

$$Q_{41} - 0 = m(h_1 - h_4) \tag{6.47}$$

The net work (W_{net}) added to the cycle is the sum of the compressor work (W_{12}) and the turbine work (W_{34}):

$$W_{net} = W_{12} + W_{34} \tag{6.48}$$

The desirable energy output of the basic gas refrigeration cycle is the heat added to the heater (or heat removed from the inner space of the refrigerator). The energy input to the cycle is the net work required. Thus, the coefficient of performance (COP) of the cycle is

$$\beta_R = Q_{41}/W_{net} = (h_4 - h_1)/[(h_1 - h_2) + (h_4 - h_3)] \tag{6.49}$$

Assuming that the gas has constant specific heat, Eq. (6.49) can be simplified to

$$\beta_R = 1/\{[r_p]^{(k-1)/k} - 1\} \tag{6.50}$$

where $r_p = p_2/p_1 = p_3/p_4$ is the pressure ratio.

The basic gas Brayton refrigeration cycle analysis is given by Example 6.12.

Example 6.12

Consider the design of an ideal air refrigeration cycle according to the following specifications:

Pressure of air at compressor inlet $= 15\,\mathrm{psia}$
Pressure of air at turbine inlet $= 60\,\mathrm{psia}$

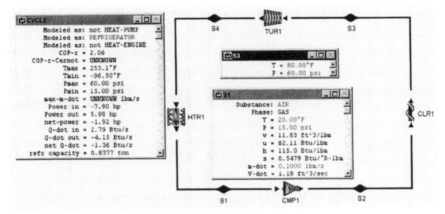

Figure 6.28 Ideal gas refrigeration cycle.

Temperature of air at compressor inlet $= 20°F$
Temperature of air at turbine inlet $= 80°F$
Mass rate of air flow $= 0.1\,lbm/sec$

Determine the COP, the compressor horsepower required, turbine power produced, net power required, and cooling load for the cycle.
To solve this problem by CyclePad, we take the following steps:

1. Build:
   a. Take a compressor, a cooler, a turbine, and a heater from the open-system inventory shop and connect the four devices to form the gas refrigeration cycle.
   b. Switch to analysis mode.
2. Analysis:
   a. Assume a process for each of the four devices: (1) compressor as isentropic, (2) cooler as isobaric, (3) turbine as isentropic, and (4) heater as isobaric.
   b. Input the given information: (1) working fluid is air, (2) inlet temperature and pressure of the compressor are 20°F and 15 psia, (3) inlet temperature and pressure of the turbine are 80°F and 60 psia, and (4) mass flow rate is 0.1 lbm/sec.
3. Display results:
   a. Display the *T–s* diagram and cycle properties' results. The cycle is a refrigerator. The answers as shown in Fig. 6.29a are COP = 2.06, compressor horsepower required $= -7.90\,hp$, turbine power produced $= 5.98\,hp$, net power required $= -1.92\,hp$, cooling load $= 2.79\,Btu/sec = 0.8377\,ton$, and net power input $= -7.90\,hp$.

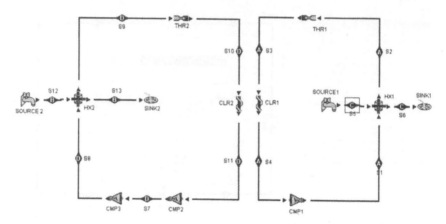

Figure 6.29a   Gas refrigeration systems comparison.

b.   Display the sensitivity diagram of cycle COP versus compression pressure ratio. (See Fig. 6.29b.)

COMMENT: The sensitivity diagram of cycle COP versus compression ratio indicates that the COP is increased as the compression pressure ratio is decreased. Unfortunately, the volume of the gas also increases when the compression pressure ratio decreases. Thus, this type of air refrigeration cycle is very bulky when the compression pressure ratio is too low.

Example 6.13

The refrigeration cycle shown at the left in Fig. 6.29a is proposed to replace the more conventional cycle at the right. The compressor of the conventional system is isentropic. The low-pressure compressor of the nonconventional system is isentropic, but the high-pressure compressor is isothermal. The circulating fluid is CO_2 and the temperature entering (100°F) and leaving (10°F) the cooler are to be the same for both systems (same refrigeration cooling load), and the cooling water temperature entering (50°F) and leaving (70°F) the heat exchanger are also to be the same for both systems. The heater pressure (1100 psia) and cooler pressure (150 psia) are also to be the same for both systems. The CO_2 pressure between the two compressors is 420 psia. With the same refrigeration cooling load, find the compressor power required for both systems.

To solve this problem by CyclePad, we take the following steps:

1.   Build:

a.   Take three compressors, two coolers, two throttling valves, two heat exchangers, two sources, and two sinks from the

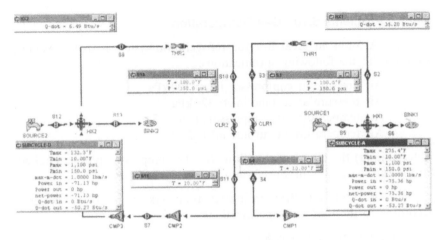

Figure 6.29b Gas refrigeration systems comparison.

open-system inventory shop and connect the devices to form the two gas refrigeration cycles (cycles A and D).

b. Switch to analysis mode.

2. Analysis:

a. Assume a process for each of the following devices: (1) compressors #1 and #2 as isentropic, (2) coolers and heat exchangers as isobaric, and (3) compressor #3 as isothermal.

b. Input the given information: (1) working fluid is CO_2 and the mass flow rate is 1 lbm/sec ($mdot_3 = mdot_{10}$), (2) inlet temperatures and pressure of the coolers are 100°F ($T_3 = T_{10}$) and 10°F ($T_4 = T_{11}$) and 150 psia ($p_3 = p_{10}$), (3) water inlet and outlet temperature and pressure of the heat exchangers are 50°F ($T_5 = T_{12}$) and 70°F ($T_6 = T_{13}$) at atmospheric pressure 14.7 psia ($p_5 = p_{12}$), (4) working fluid high pressure of the cycles are 1100 psia ($p_9 = p_2$), (5) working fluid low pressure of the cycles are 150 psia ($p_4 = p_{11}$), and (6) working fluid pressure between the two compressors of the cycle at the left is 420 psia (p_7).

3. Display results.

The answers are: compressor power required for the system (cycle D) at the left $= -71.13$ hp, and compressor power required for the system (cycle A) at the right $= -75.36$ hp. (See Fig. 6.29b.)

Review Problems 6.10 Gas Refrigeration

1. A 5-ton air ideal Brayton refrigeration system is to be designed according to the following specifications:

> Air pressure at compressor inlet: 100 kPa
> Air pressure at turbine inlet: 420 kPa
> Air temperature at compressor inlet: −5°C
> Air temperature at turbine inlet: 25°C

Determine (a) the air mass rate flow, (b) the compressor power required, (c) the turbine power produced, and (d) the cycle COP.

2. A 5-ton air Brayton refrigeration system is to be designed according to the following specifications:

> Compressor efficiency: 82%
> Turbine efficiency: 84%
> Air pressure at compressor inlet: 100 kPa
> Air pressure at turbine inlet: 420 kPa
> Air temperature at compressor inlet: −5°C
> Air temperature at turbine inlet: 25°C

Determine (a) the air mass rate flow, (b) the compressor power required, (c) the turbine power produced, and (d) the cycle COP.

3. An ideal Brayton refrigeration system uses air as a refrigerant. The pressure and temperature of air at compressor inlet are 14.7 psia and 100°F. The pressure and temperature of air at the turbine inlet are 60 psia and 260°F. The mass rate of air flow is 0.03 lbm/sec. Determine (a) the cooling load, (b) the compressor power required, (c) the turbine power produced, and (d) the cycle COP.

4. An ideal air Brayton refrigeration system is operated between −10°F and 120°F. The air pressure at the compressor inlet is 14.5 psia, and the air pressure at the turbine inlet is 75.5 psia. The mass rate of air flow is 0.031 lbm/sec. Determine (a) the cooling load, (b) the compressor power required, (c) the turbine power produced, and (d) the cycle COP.

6.11 STIRLING REFRIGERATION CYCLE

An ideal reciprocating Stirling refrigeration cycle is shown in Fig. 6.30. It is the reversible Stirling heat engine cycle, which is composed of two isothermal processes and two isochoric processes. Working fluid is compressed in an isothermal process 1-2 at T_H. Heat is then removed at a constant volume process 2-3. Working fluid is expanded in an isothermal process 3-4 at T_L. The cycle is completed by a constant-volume heat-addition process 4-1. The T–s diagram of the cycle is illustrated in Fig. 6.31. Work has been done recently on developing a practical refrigeration device based on the Stirling refrigeration cycle at extremely low temperatures (less than 200 K or $-100°F$). In the presence of an ideal regenerator in the cycle, the heat quantity Q_{23} and Q_{41}, which are equal in magnitude but opposite in sign, are exchanged between fluid streams within the device. Hence, the only external heat transfer occurs in

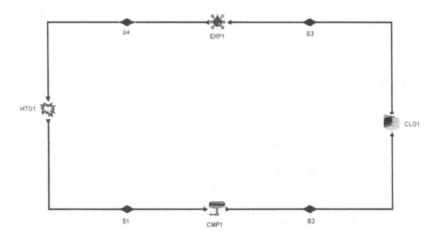

Figure 6.30 Stirling refrigeration cycle.

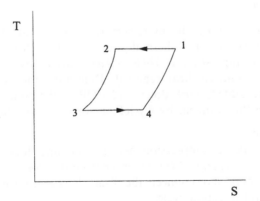

Figure 6.31 Stirling refrigeration cycle T–s diagram.

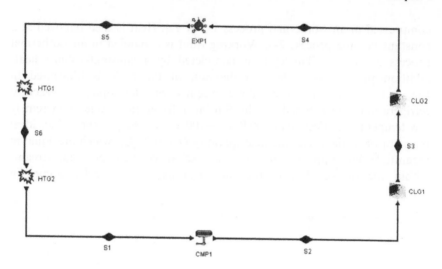

Figure 6.32 Stirling refrigeration cycle with regeneration.

processes 1-2 at constant temperature T_H and 3-4 at constant temperature T_L. Consequently, the coefficient of performance (COP) of the Stirling refrigeration cycle theoretically equals that of the Carnot refrigeration cycle, $T_L/(T_H - T_L)$. The Stirling refrigeration cycle with regenerator is shown in Fig. 6.32. In this figure, the heater #1 and cooler #1 comprise the regenerator. Heat removed from cooler #1 is added to heater #1.

Example 6.14

In an ideal reciprocating Stirling refrigeration cycle, 0.01 kg of air at 235 K and 10 bars is expanded isothermally to 1 bar. It is then heated to 320 K isometrically. Compression at 320 K isothermally follows, and the cycle is completed by isometric heat removal. Determine the heat added, heat removed, work added, work done, and COP of the cycle.

To solve this problem by CyclePad, we take the following steps:

1. Build:
   a. Take a compression device, a cooling device, an expansion device, and a heating device from the closed-system inventory shop and connect the four devices to form the Stirling refrigeration cycle.
   b. Switch to analysis mode.

  2. Analysis:

     a. Assume a process for each of the four devices: (1) compression and expansion devices as isothermal, and (2) cooling and heating devices as isochoric.

     b. Input the given information: (1) working fluid is air, (2) inlet temperature and pressure of the expansion are 235 K and 10 bars, $m = 0.01$ kg, (3) inlet temperature of the compression is 320 K, and exit pressure of the expansion is 1 bar.

  3. Display the cycle properties' results. The cycle is a refrigerator.

The answers are: COP $= 3.05$, $Q_{in} = 2.16$ kJ, $Q_{out} = -2.72$ kJ, $W_{in} = -2.11$ kJ, and $W_{out} = 1.55$ kJ. (See Fig. 6.33.)

Review Problems 6.11   Stirling Refrigeration Cycle

  1. What are the four basic processes of the Stirling refrigeration cycle?

  2. What temperature range is the main application of the Stirling refrigeration cycle?

  3. In an ideal reciprocating Stirling refrigeration cycle, 0.01 kg of CO_2 at 235 K and 10 bars is expanded isothermally to 1 bar. It is then heated to 320 K isometrically. Compression at 320 K isothermally follows, and the cycle is completed by isometric heat

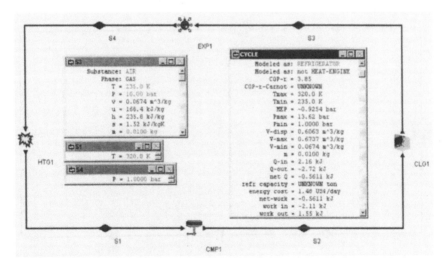

Figure 6.33   Stirling refrigeration cycle.

removal. Determine the heat added, heat removed, work added, work done, and COP of the cycle.

ANSWER: COP = 4.26

4. In an ideal reciprocating Stirling refrigeration cycle, 0.01 kg of helium at 235 K and 10 bars is expanded isothermally to 1 bar. It is then heated to 320 K isometrically. Compression at 320 K isothermally follows, and the cycle is completed by isometric heat removal. Determine the heat added, heat removed, work added, work done, and COP of the cycle.

ANSWER: COP = 3.41

6.12 ERICSSON CYCLE

The Ericsson refrigeration cycle is a reversible Ericsson power cycle. The schematic Ericsson cycle is shown in Fig. 6.34. The working fluid usually used in the Ericsson refrigerator is helium or hydrogen; the refrigerator may be used for very low-temperature application. The cycle consists of two isothermal processes and two isobaric processes. The four processes of the Ericsson refrigeration cycle are isothermal compression process 1-2 (compressor), isobaric cooling process 2-3 (cooler), isothermal expansion process 3-4 (turbine), and isobaric heating process 4-1 (heater).

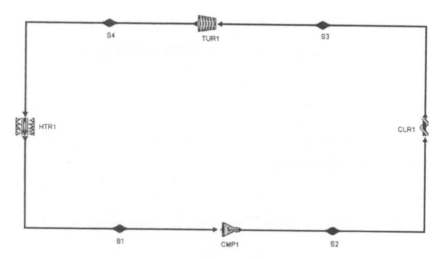

Figure 6.34 Ericsson refrigeration cycle.

Applying the basic laws of thermodynamics, we have

$$q_{12} - w_{12} = h_2 - h_1 \tag{6.51}$$

$$q_{23} - w_{23} = h_3 - h_2, \ w_{23} = 0 \tag{6.52}$$

$$q_{34} - w_{34} = h_4 - h_3 \tag{6.53}$$

$$q_{41} - w_{41} = h_1 - h_4, \ w_{41} = 0 \tag{6.54}$$

The net work produced by the cycle is

$$w_{net} = w_{12} + w_{34} \tag{6.55}$$

The heat added to the cycle in the heater is q_{41}, and the cycle COP is

$$\beta_R = q_{41}/w_{net} \tag{6.56}$$

Example 6.15

The mass flow rate (1 g/sec) of helium is compressed and heated from 100 kPa and 300 K in an Ericsson refrigeration cycle to a turbine inlet at 800 kPa and 100 K. Determine power required by the compressor, power produced by the turbine, rate of heat removed from the cooler, rate of heat added in the heater, and cycle COP. Draw the T–s diagram of the cycle.

To solve this problem by CyclePad, we take the following steps:

1. Build the cycle as shown in Fig. 6.34. Assume the compressor is isothermal, the heater is isobaric, the turbine is isothermal, and the cooler is isobaric.
2. Input working fluid = helium, mass flow rate = 1 g/sec, compressor inlet pressure = 100 kPa, compressor inlet temperature = 300 K, turbine inlet pressure = 800 kPa, and turbine inlet temperature = 100 K.
3. Display results.

The answers are: $Q\text{dot}_{htr} = 1.47$ kW, $Q\text{dot}_{clr} = -2.33$ kW, $W\text{dot}_{cmp} = -1.3$ kW, $W\text{dot}_{tur} = 0.4319$ kW, $W\text{dot}_{net} = -0.8638$ kW, and $\beta_R = 1.7$. (See Fig. 6.35.)

Review Problems 6.12   Ericsson Refrigeration Cycle

A mass flow rate (1 g/sec) of helium is compressed and heated from 100 kPa and 300 K in an Ericsson refrigeration cycle to a turbine inlet at 500 kPa and 100 K. Determine power required by the compressor, power

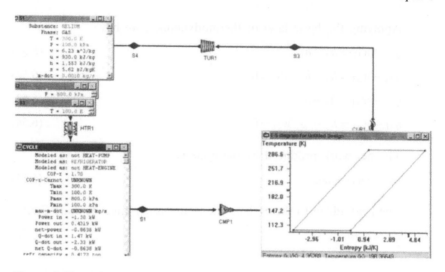

Figure 6.35   Ericsson refrigeration cycle.

produced by the turbine, rate of heat removed from the cooler, rate of heat added in the heater, and cycle COP.

ANSWERS: $Q\text{dot}_{htr} = 1.37\,\text{kW}$, $Q\text{dot}_{clr} = -2.04\,\text{kW}$, $W\text{dot}_{cmp} = -1.0\,\text{kW}$, $W\text{dot}_{tur} = 0.3343\,\text{kW}$, $W\text{dot}_{net} = -0.6685\,\text{kW}$, and $\beta_R = 2.05$.

6.13   LIQUEFACTION OF GASES

The liquefaction of gases is a very important area in refrigeration at very low temperature. Methods of producing very low temperatures refrigeration, liquefying gases, or solidification solids are based on the adiabatic expansion of a high-pressure gas either through a throttling valve or in an expansion turbine. The schematic and T–s diagrams for an ideal Hampson–Linde gas liquefaction system are shown in Figs. 6.36 and 6.37 (Linde, C., The refrigerating machine of today, *ASME Trans.*, 1893). Makeup gas at state 1 is mixed with the uncondensed portion of the gas at state 15 from the previous cycle, and the mixture at state 2 is compressed by a four-stage compressor with intercoolers (compressor 1, intercooler 1, compressor 2, intercooler 2, compressor 3, intercooler 3, and compressor 4) to state 9. After the multistage compression, the gas is cooled from state 9 to state 10 at constant pressure in a cooler. The gas is further cooled to state 11 in a regenerative heat exchanger. After expansion through a throttle valve, the fluid at state 12 is in the liquid–vapor mixture state and is separated into liquid (state 13) and vapor (state 14) states. The liquid at

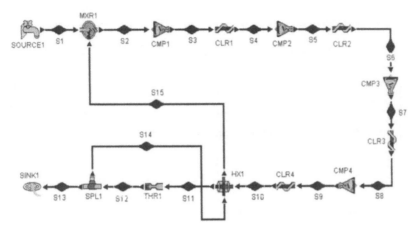

Figure 6.36 Hampson–Linde gas liquefaction system schematic diagram.

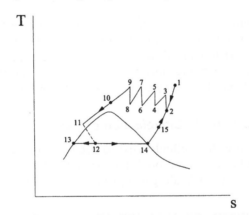

Figure 6.37 Hampson–Linde gas liquefaction system *T–s* diagram.

state 13 is drawn off as the desired product, and the vapor at state 14 flows through the regenerative heat exchanger to cool high-pressure gas flowing toward the throttle valve. The gas at state 15 is finally mixed with fresh makeup gas, and the cycle is repeated. This cycle can also be used for the solidification of gases at even lower temperature.

Figure 6.38 is the Claude gas liquefaction system, a modification of the Hampson–Linde gas liquefaction system. The Claude system has a turbine in the expansion process to replace a part of the highly irreversible throttling process of the Hampson–Linde system. From state 1 to state 10, the Claude system processes are the same as those of the Hampson–Linde system. After the gas is cooled to state by the regenerative cooler (heat

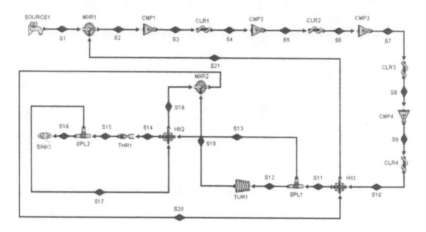

Figure 6.38   Claude gas liquefaction system.

exchanger 1), most of it is expanded through a turbine from state 12 to state 19 and then is mixed with vapor at state 11 from the separator (splitter 2) and the mixture flows back toward the compressor through a heat exchanger, which precools the small fraction of the flow that is directed toward the throttle valve instead of the turbine.

Review Problems 6.13   Liquefaction of Gases

1. What are the methods used to produce liquefying gases?
2. Describe the Hampson–Linde gas liquefaction system.
3. Describe the Claude gas liquefaction system.

6.14   NONAZEOTROPIC MIXTURE REFRIGERATION CYCLE

The thermodynamic performance of a single working fluid vapor refrigeration cycle may be improved potentially by using a nonazeotropic mixture working fluid such as ammonia–water (Wu, C., Non-azeotropic mixture energy conversion. *Energy Conversion and Management*, vol. 25, no. 2, pp. 199–206, 1985). A mixture of two or more different fluids is classified as an azeotrope when such a mixture possesses its own thermodynamic properties, quite unlike the thermal and chemical characteristics of its components. A distinguishing feature of this type of fluid is its ability to maintain a permanent composition and uniform boiling point during evaporation, much the same as a pure simple fluid in that its transition from liquid to vapor phase (or vice versa) occurs at a constant pressure and temperature without any change in the composition.

Otherwise, the mixture is called a nonazeotrope. A nonazeotropic mixture has a temperature distribution parallel to that of the thermal reservoir. Note that one of the requirements for the nonazeotropic mixture energy conversion improvement is to have a nonconstant temperature heat source and heat sink. The proper choice of best combination of the nonazeotropic mixture is still not entirely understood. Uncertainties in modeling the thermodynamic and heat-transfer aspects of the nonazeotropic mixture refrigeration cycle are such that the probability of realizing significant net benefits in actual application is also not fully known.

An ideal nonazeotropic mixture refrigeration cycle and an ideal Carnot refrigeration cycle operating between a nonconstant temperature heat source and a nonconstant temperature heat sink are shown in Fig. 6.39. The ideal Carnot refrigeration cycle consists of an isentropic compression process from state 1 to state 2, an isobaric heat-removing process from state 2 to state 3, an isentropic expansion process from state 3 to state 4, and an isobaric heat-addition process from state 4 to state 1. The ideal nonazeotropic mixture refrigeration cycle consists of an isentropic compression process from state 6 to state 2, an isobaric heat-removing process from state 2 to state 5, an isentropic expansion process from state 5 to state 4, and an isobaric heat-addition process from state 4 to state 6, respectively. The inlet and exit temperature of the cooling fluid (finite-heat-capacity heat sink) in the hot-side heat exchanger are T_b and T_a, and the inlet and exit temperature of the heating fluid (finite-heat-capacity heat source) in the cold-side heat exchanger are T_d and T_c, respectively. It is clearly demonstrated that the temperature distribution curves of the ideal nonazeotropic mixture refrigeration cycle (curves 4-6

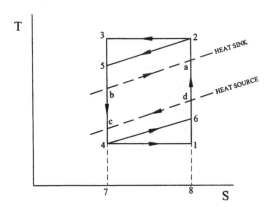

Figure 6.39 *T–s* diagram of ideal nonazeotropic refrigeration cycle and Carnot cycle.

and 2-5) are more closely matched to the temperature distribution curves of the heat source and heat sink (curves d-c and b-a) than the temperature distribution curves of the Carnot refrigeration cycle (curves 4-1 and 2-3).

Referring to Fig. 6.39, the net work added and heat removed to the ideal nonazeotropic mixture refrigeration cycle are $W_{net,nonaze}$ = area 46254 and Q_{nonaze} = area 46871, and the net work added and heat removed to the Carnot refrigeration cycle are $W_{net\,Carnot}$ = area 12341 and Q_{Carnot} = area 4178, respectively. The cycle COP of the ideal nonazeotropic mixture refrigeration cycle is $\beta_{nonaze} = Q_{nonaze}/W_{net,nonaze}$ = area 46871/area 46254. Similarly, the cycle COP of the Carnot refrigeration cycle is $\beta_{Carnot} = Q_{Carnot}/W_{net,Carnot}$ = area 4178/area 12341.

Since Q_{nonaze} = area 46871 is larger than Q_{Carnot} = area 4178 and $W_{net,nonaze}$ = area 46254 is smaller than $W_{net\,Carnot}$ = area 12341, it is apparent that β_{nonaze} is larger than β_{Carnot}.

A numerical example (Wu, C., Non-azeotropic mixture energy conversion. *Energy Conversion and Management*, vol. 25, no. 2, pp. 199–206, 1985), using an R-114 and R-12 nonazeotropic mixture to predict the COP of the ideal nonazeotropic mixture refrigeration cycle, is carried out. The COP performance results of the cycle are displayed in Fig. 6.40. At an R-114 mass concentration of 25%, a mixed R-12/R-114 nonazeotrope refrigeration cycle experiences COP improvements of 4.4 and 9.2% over single R-12 and single R-114 refrigeration cycles, respectively.

A nonazeotropic mixture of two refrigerants does not always make a better refrigerant. The R-12/R-114 pair may not be the best combination for a nonazeotropic mixture. The proper choice of the best combination is still not entirely understood. Uncertainties in modeling the thermodynamics and heat-transfer aspects of the nonazeotropic mixture

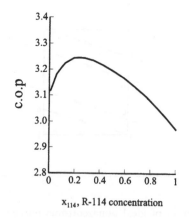

x_{114}, R-114 concentration

Figure 6.40 COP of ideal nonazeotropic R-114 and R-12 refrigeration cycle.

refrigeration cycle are such that the probability of realizing significant net benefits in actual applications is also not fully known.

Review Problems 6.14 Nonazeotropic Mixture Refrigeration Cycle

1. What is a nonazeotropic mixture refrigerant?
2. Draw an isobaric heating process on a T–s diagram for a nonazeotropic mixture from a superheated vapor state to a compressed liquid state. Does temperature remain the same during the condensation?
3. Why may the COP of a vapor refrigeration cycle be improved potentially by using a nonazeotropic mixture working fluid?

6.15 DESIGN EXAMPLES

The design of refrigeration and heat pump cycles would typically be tedious. With the use of the intelligent computer-aided software CyclePad, however, the design can be completed rapidly and modifications are simple to make. CyclePad makes designing a refrigeration or heat pump plant as simple as point and click. To demonstrate the power CyclePad offers designers in the design of refrigeration and heat-pump cycles, the following design examples are considered.

Example 6.16

A combined split-shaft gas turbine power plant and gas refrigeration system to be used in an airplane, as illustrated in Fig. 6.41a, has been designed by a junior engineer.

The following design information is provided: $m\mathrm{dot}_1 = 1\,\mathrm{kg/sec}$, $p_1 = 101\,\mathrm{kPa}$, $T_1 = 15^\circ\mathrm{C}$, $p_2 = 1\mathrm{M\,Pa}$, $T_5 = 1200^\circ\mathrm{C}$, $p_7 = 102\,\mathrm{kPa}$, $T_{11} = 400^\circ\mathrm{C}$, $p_{12} = 102\,\mathrm{kPa}$, $T_{13} = 15^\circ\mathrm{C}$, $\eta_{\mathrm{turbine}} = 85\%$, and $\eta_{\mathrm{compressor}} = 85\%$.

1. During the cruise condition, the split-shaft gas turbine power plant required to produce 240 kW and 10% of the compressed air, is used in the gas refrigeration system, which is required to remove 7 kW from the cabin.
2. During the take-off condition, the split-shaft gas turbine power plant is required to produce 300 kW.
3. During the high-wind condition, the split-shaft gas turbine power plant is required to produce 270 kW while at least 3.5 kW of cabin refrigeration is to be provided.

Check if all these conditions can be met by the design.

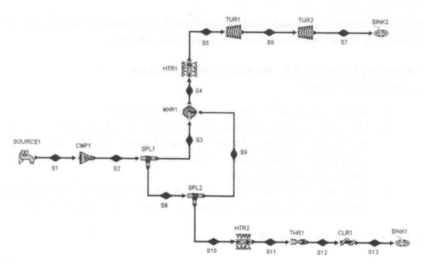

Figure 6.41a Combined gas turbine power plant and gas refrigeration system design.

To check this design by CyclePad, we take the following steps:

1. Build the cycle as shown in Fig. 6.41a. Assume that the compressor and turbines are adiabatic with 85% efficiency, the heaters, mixing chambers, and cooler are isobaric, and the splitters are isoparametric.

2. Input working fluid = air, $mdot_1 = 1$ kg/sec, $p_1 = 101$ kPa, $T_1 = 15°C$, $p_2 = 1$ MPa, $T_5 = 1200°C$, $p_7 = 102$ kPa, $T_{11} = 400°C$, $p_{12} = 102$ kPa, $T_{13} = 15°C$, $\eta_{turbine} = 85\%$, and $\eta_{compressor} = 85\%$. Display the compressor power (-314.7 kW) and input $Wdot_{turbine\ \#1} = 314.7$ kW, as shown in Fig. 6.41b.

3. Cruise condition: input $mdot_8 = 0.1$ kg/sec and $mdot_9 = 0$ kg/sec. Display results. The answers are: $Wdot_{turbine\ \#2} = 240.2$ kW and $Qdot_{heater\ \#2} = 7.16$ kW as shown in Fig. 6.41c. The design requirement is met.

4. Take-off condition: input $mdot_9 = 0.01$ kg/sec and $mdot_{10} = 0$ kg/sec. Display results. The answers are: $Wdot_{turbine\ \#2} = 301.7$ kW and $Qdot_{heater\#2} = 0$ kW as shown in Fig. 6.41d. The design requirement is met.

5. High-wind condition: input $mdot_9 = 0.05$ kg/sec and $mdot_{10} = 0.05$ kg/sec. Display results. The answers are: $Wdot_{turbine\#2} = 271.0$ kW and $Qdot_{heater\#2} = 3.58$ kW as shown in Fig. 6.41e. The design requirement is met.

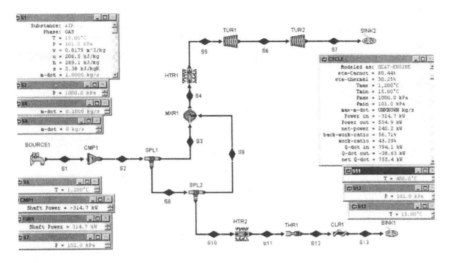

Figure 6.41b Combined gas turbine power plant and gas refrigeration system design input.

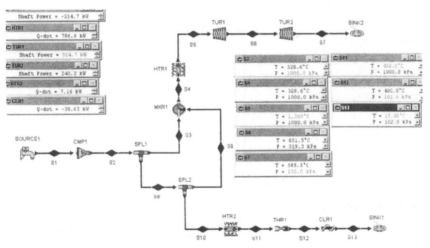

Figure 6.41c Combined gas turbine power plant and gas refrigeration system design at cruise condition.

Example 6.17

An engineer claims that the performance of a simple refrigeration cycle can be improved by using his three-stage compression process as shown in Fig. 6.42a. Refrigerant enters all compressors as a saturated vapor, and

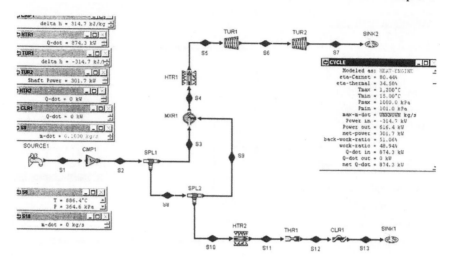

Figure 6.41d Combined gas turbine power plant and gas refrigeration system design at take-off condition.

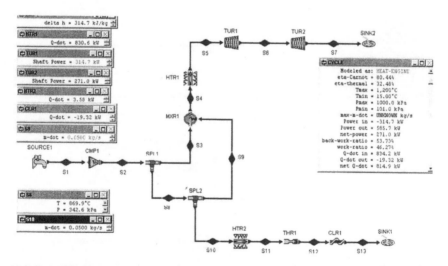

Figure 6.41e Combined gas turbine power plant and gas refrigeration system design at high-wind condition.

enters the throttling valve as a saturated liquid. The highest and lowest pressure of the cycle are 1200 and 150 kPa, respectively. His design information is: Refrigerant: R-12, $mdot_1 = 1$ kg/sec, $p_1 = 150$ kPa, $x_1 = 1$, $p_3 = 300$ kPa, $x_3 = 1$, $p_5 = 600$ kPa, $x_5 = 1$, $p_7 = 1200$ kPa, $x_7 = 0$, and $\eta_{compressor} = 85\%$.

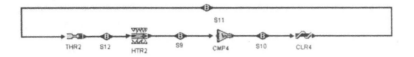

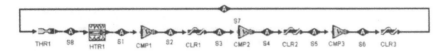

Figure 6.42a Single-stage compressor and three-stage compressor refrigeration systems.

His design results are: β(COP) = 2.16, Wdot$_{in}$ = −43.64 kW, Qdot$_{in}$ = 94.48 kW, Qdot$_{out}$ = −138.1 kW, and cooling load = 26.87 tons.

• Check on his claim.
• What is the performance of the cycle if ammonia, R-134a, or R-22 is used instead of R-12.
• Try to improve the COP by varying p_3 and p_5.

To check this design by CyclePad, we take the following steps:

1. Build the three-stage-compressor refrigeration cycle as shown in Fig. 6.42a. Assume the compressors are adiabatic with 85% efficiency, and the heater and cooler are isobaric.
2. Input working fluid = R-12, $m$dot$_1$ = 1 kg/sec, p_1 = 150 kPa, x_1 = 1, p_3 = 300 kPa, x_3 = 1, p_5 = 600 kPa, x_5 = 1, p_7 = 1200 kPa, x_7 = 0, and $\eta_{compressor}$ = 85%.
3. Display results: The answers are: COP = 2.22, compressor power (−42.56 kW), Qdot$_{in}$ = 94.48 kW, Qdot$_{out}$ = −137.0 kW and cooling capacity = 26.87 tons as shown in Fig. 6.42b.

For the one-compressor refrigeration system using R-12:

1. Retract p_3 = 300 kPa, x_3 = 1, p_5 = 600 kPa, and x_5 = 1.
2. Let Wdot$_{compressor\,\#1}$ = 0, Wdot$_{compressor\,\#2}$ = 0, Qdot$_{cooler\,\#1}$ = 0, and Qdot$_{cooler\,\#2}$ = 0.

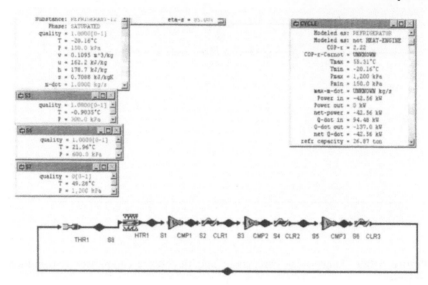

Figure 6.42b   Three-stage compressor refrigeration system using R-12.

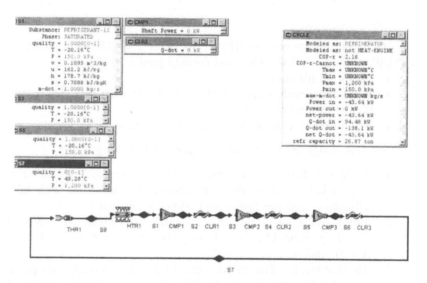

Figure 6.42c   One-stage compressor refrigeration system using R-12.

3. Display results:

The answers are: $COP = 2.16$, compressor power $(-43.64\,\text{kW})$, $\dot{Q}\text{dot}_{in} = 94.48\,\text{kW}$, $\dot{Q}\text{dot}_{out} = -138.1\,\text{kW}$ and cooling capacity $= 26.87\,\text{tons}$ tons as shown in Fig. 6.42c. The COP is indeed improved.

For the three-compressor refrigeration system using ammonia:

1. Retract working fluid and let the working fluid be ammonia.
2. Display results:

The answers are: COP = 3.36, compressor power (−323.0 kW), Qdot$_{in}$ = 1084 kW, Qdot$_{out}$ = −1407 kW, and cooling capacity = 308.2 tons as shown in Fig. 6.42d.

For the three-compressor refrigeration system using R-134a:

1. Retract working fluid and let the working fluid be R-134a.
2. Display results:

The answers are: COP = 2.41, compressor power (−50.41 kW), Qdot$_{in}$ = 121.6 kW, Qdot$_{out}$ = −172.1 kW, and cooling capacity = 34.59 tons as shown in Fig. 6.42e.

For the three-compressor refrigeration system using R-22:

1. Retract working fluid and let the working fluid be R-22.
2. Display results:

The answers are: COP = 2.64, compressor power (−58.82 kW), Qdot$_{in}$ = 155.2 kW, Qdot$_{out}$ = −214.0 kW, and cooling capacity = 44.13 tons as shown in Fig. 6.42f.

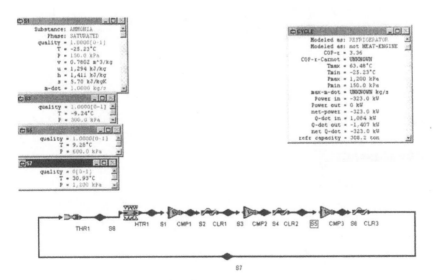

Figure 6.42d Three-stage compressor refrigeration system using ammonia.

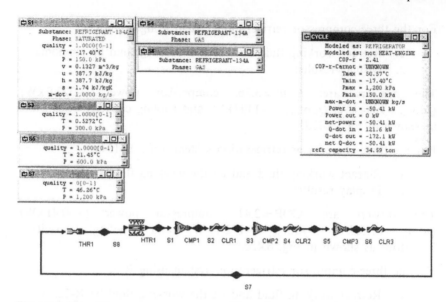

Figure 6.42e   Three-stage compressor refrigeration system using R-134a.

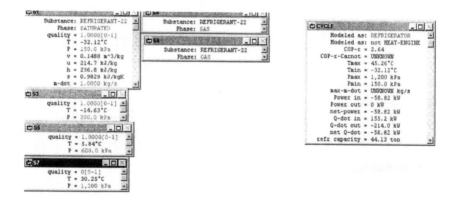

Figure 6.42f   Three-stage-compressor refrigeration system using R-22.

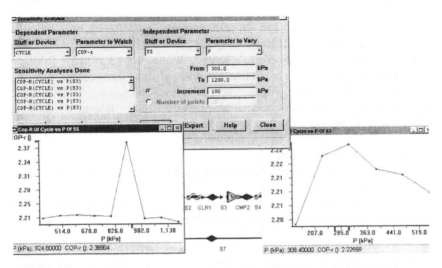

Figure 6.42g Three-stage compressor refrigeration system sensitivity diagrams.

To improve the COP by varying p_3 and p_5, draw the COP versus p_3 and COP versus p_5 sensitivity diagrams as shown in Fig. 6.42g. The maximum COP is about 2.225 when p_3 is about 309.7 kPa, and the maximum COP is about 2.384 when p_5 is about 898.8 kPa.

Review Problems 6.15 Design Examples

1. The performance of a simple refrigeration cycle can be improved by using a cascaded refrigeration cycle. An engineer claims that he has developed a separate three-loop cascaded refrigeration cycle. The cascaded cycle consists of three separate loops—one at high pressure, one at low pressure, and one at midpressure using R-12 as working fluid in all three loops. The three loops are connected by two heat exchangers. His design information is:

Low-pressure loop—$\dot{m} = 1$ lbm/sec, high pressure = 60 psia, low pressure = 20 psia, $\eta_{compressor} = 85\%$, R-12 quality at inlet of compressor = 1, and R-12 quality at inlet of throttling valve = 0. Midpressure loop—high pressure = 110 psia, low pressure = 60 psia, $\eta_{compressor} = 85\%$, R-12 quality at inlet of compressor = 1, and R-12 quality at inlet of throttling valve = 0. High-pressure loop—high pressure = 160 psia, low pressure = 110 psia, $\eta_{compressor} = 85\%$, R-12 quality at inlet of compressor = 1, and R-12 quality at inlet of throttling valve = 0.

His design results are—whole system: $\beta(\text{COP}) = 2.74$, $W\text{dot}_{in} = -29.59\,\text{hp}$, $Q\text{dot}_{in} = 57.21\,\text{Btu/sec}$, $Q\text{dot}_{out} = -78.12\,\text{Btu/sec}$, and cooling load $= 17.16\,\text{tons}$; low-pressure loop: $W\text{dot}_{in} = -13.53\,\text{hp}$, $Q\text{dot}_{in} = 57.21\,\text{Btu/sec}$, $Q\text{dot}_{out} = -66.78\,\text{Btu/sec}$, and $m\text{dot} = 1\,\text{lbm/sec}$; midpressure loop: $W\text{dot}_{in} = -9.41\,\text{hp}$, $Q\text{dot}_{in} = 66.78\,\text{Btu/sec}$, $Q\text{dot}_{out} = -73.43\,\text{Btu/sec}$, and $m\text{dot} = 1.23\,\text{lbm/sec}$; and high-pressure loop: $W\text{dot}_{in} = -6.65\,\text{hp}$, $Q\text{dot}_{in} = 73.43\,\text{Btu/sec}$, $Q\text{dot}_{out} = -78.12\,\text{Btu/sec}$, and $m\text{dot} = 1.43\,\text{lbm/sec}$.

- Check on his claim.
- What is the performance of the cycle if ammonia, or R-22 is used instead of R-12?
- Try to improve the COP by varying the two pressures of the midpressure loop as design variables.

ANSWERS: Ammonia—whole system: $\beta(\text{COP}) = 3.42$, $W\text{dot}_{in} = -219.0\,\text{hp}$, $Q\text{dot}_{in} = 529.9\,\text{Btu/sec}$, $Q\text{dot}_{out} = -684.7\,\text{Btu/sec}$, and cooling load $= 159.0\,\text{tons}$; low-pressure loop: $W\text{dot}_{in} = -104.2\,\text{hp}$, $Q\text{dot}_{in} = 529.9\,\text{Btu/sec}$, $Q\text{dot}_{out} = -603.6\,\text{Btu/sec}$, and $m\text{dot} = 1\,\text{lbm/sec}$; midpressure loop: $W\text{dot}_{in} = -68.13\,\text{hp}$, $Q\text{dot}_{in} = 603.6\,\text{Btu/sec}$, $Q\text{dot}_{out} = -651.7\,\text{Btu/sec}$, and $m\text{dot} = 1.18\,\text{lbm/sec}$; and high-pressure loop: $W\text{dot}_{in} = -46.65\,\text{hp}$, $Q\text{dot}_{in} = 651.7\,\text{Btu/sec}$, $Q\text{dot}_{out} = -684.7\,\text{Btu/sec}$, and $m\text{dot} = 1.33\,\text{lbm/sec}$.

R-22—whole system: $\beta(\text{COP}) = 2.95$, $W\text{dot}_{in} = -40.72\,\text{hp}$, $Q\text{dot}_{in} = 85.05\,\text{Btu/sec}$, $Q\text{dot}_{out} = -113.8\,\text{Btu/sec}$, and cooling load $= 25.51\,\text{tons}$; low-pressure loop: $W\text{dot}_{in} = -18.78\,\text{hp}$, $Q\text{dot}_{in} = 85.05\,\text{Btu/sec}$, $Q\text{dot}_{out} = -98.32\,\text{Btu/sec}$, and $m\text{dot} = 1\,\text{lbm/sec}$; midpressure loop: $W\text{dot}_{in} = -12.84\,\text{hp}$, $Q\text{dot}_{in} = 98.32\,\text{Btu/sec}$, $Q\text{dot}_{out} = -107.4\,\text{Btu/sec}$, and $m\text{dot} = 1.22\,\text{lbm/sec}$; and high-pressure loop: $W\text{dot}_{in} = -9.10\,\text{hp}$, $Q\text{dot}_{in} = 107.4\,\text{Btu/sec}$, $Q\text{dot}_{out} = -113.8\,\text{Btu/sec}$, and $m\text{dot} = 1.41\,\text{lbm/sec}$.

2. The performance of a simple refrigeration cycle can be improved by using a cascaded refrigeration cycle. An engineer claims that he has developed a separate four-loop cascaded refrigeration cycle. The cascaded cycle consists of four separate loops—one between 20 and 40 psia, one between 40 and 80 psia, one between 80 and 120 psia, and one between 120 and 160 psia. The four loops are connected by three heat exchangers. His design information is:

Loop A—$m\text{dot} = 1\,\text{lbm/sec}$, high pressure $= 40$ psia, low pressure $= 20$ psia, $\eta_{compressor} = 85\%$, R-12 quality at inlet of compressor $= 1$, and R-12 quality at inlet of throttling valve $= 0$.

Loop B—high pressure $= 80$ psia, low pressure $= 40$ psia, $\eta_{compressor} = 85\%$, R-12 quality at inlet of compressor $= 1$, and R-12 quality at inlet of throttling valve $= 0$.

Loop C—high pressure $= 120$ psia, low pressure $= 80$ psia, $\eta_{\text{compressor}} = 85\%$, R-12 quality at inlet of compressor $= 1$, and R-12 quality at inlet of throttling valve $= 0$.

Loop D— high pressure $= 160$ psia, low pressure $= 120$ psia, $\eta_{\text{compressor}} = 85\%$, R-12 quality at inlet of compressor $= 1$, and R-12 quality at inlet of throttling valve $= 0$.

His design results are—whole system: $\beta(\text{COP}) = 2.84$, $W\text{dot}_{\text{in}} = -31.06$ hp, $Q\text{dot}_{\text{in}} = 62.26$ Btu/sec, $Q\text{dot}_{\text{out}} = -84.21$ Btu/sec, and cooling load $= 18.68$ tons; loop A: $W\text{dot}_{\text{in}} = -8.33$ hp, $Q\text{dot}_{\text{in}} = 62.26$ Btu/sec, $Q\text{dot}_{\text{out}} = -68.15$ Btu/sec, and $m\text{dot} = 1$ lbm/sec; loop B: $W\text{dot}_{\text{in}} = -10.37$ hp, $Q\text{dot}_{\text{in}} = 68.15$ Btu/sec, $Q\text{dot}_{\text{out}} = -75.48$ Btu/sec, and $m\text{dot} = 1.20$ lbm/sec; loop C: $W\text{dot}_{\text{in}} = -6.86$ hp, $Q\text{dot}_{\text{in}} = 75.48$ Btu/sec, $Q\text{dot}_{\text{out}} = -80.32$ Btu/sec, and $m\text{dot} = 1.38$ lbm/sec; and loop D: $W\text{dot}_{\text{in}} = -5.50$ hp, $Q\text{dot}_{\text{in}} = 80.32$ Btu/sec, $Q\text{dot}_{\text{out}} = -84.21$ Btu/sec, and $m\text{dot} = 1.55$ lbm/sec.

- Check on his claim.
- What are the performance of the cycle if ammonia, R-134a, or R-22 is used instead of R-12.
- Try to improve the COP by varying the three pressures (40, 80, and 120 psia) of the loops as design variables.

6.16 SUMMARY

The reversed Carnot cycle is modified for the most widely used vapor heat pump and refrigerator. The basic vapor heat pump and refrigerator cycle consists of an isentropic compression process, an isobaric cooling process, an irreversible throttling process, and an isobaric heating process. The coefficient of performance (COP) of refrigerators is defined as Q_L (desirable heat output or cooling effect)$/W_{\text{net}}$. The coefficient of performance (COP) of heat pumps is defined as Q_H (desirable heat output or heating effect)$/W_{\text{net}}$.

Large temperature differences can be achieved by cascaded refrigerators and heat pumps.

Multistaged refrigerators and heat pumps reduce the compressor power.

Stirling and Ericsson refrigerators have practical applications at very low temperatures.

Domestic refrigerator–freezer and air-conditioning–heat pump systems share equipments to reduce cost.

An absorption refrigerator or heat pump is economically attractive because it uses inexpensive heat input rather than the expensive electrical work input to produce the refrigeration or heat pump effect.

The Brayton gas refrigeration cycle is a reversed Brayton gas power cycle.

Liquefaction and solidification of gases are obtained by compression of gas followed by cooling and throttling, leading to a change of phase for part of the fluid.

7

Finite-Time Thermodynamics

7.1 INTRODUCTION

Among the important topics in thermodynamics is the formulation of criteria for comparing the performance of real and ideal processes. Carnot showed that any heat engine absorbing heat from a high-temperature heat source reservoir to produce work must transfer some heat to a heat sink reservoir of lower temperature. He also showed that no heat engine could be better than the Carnot heat engine. The early tradition was carried on by Clausius, Kelvin, and others, using thermodynamics as a tool to find limits on work, heat transfer, efficiency, coefficient of performance, energy effectiveness, and energy figure of merit of energy conversion devices. The basic laws of thermodynamics were all conceived on the basis of irreversible processes. However, since Gibbs, the subsequent development of thermodynamics has turned from the process variable of heat and work toward state variables. The Carnot–Clausius–Kelvin view emphasizes the interaction of a thermodynamic system with its surroundings, while the Gibbs view makes the properties of the system dominant and focuses on equilibrium states. Contemporary classical thermodynamics gives a fairly complete description of equilibrium states and reversible processes. The only fact that it tells about real processes is that these irreversible processes always produce less work and more entropy than the corresponding reversible processes. Reversible processes are defined only in the limit of infinitely slow execution.

In the real engineering world, actual changes in enthalpy and free energy in an irreversible process rarely approach the corresponding ideal enthalpy and free energy changes. No practicing engineer wants to design a heat engine that runs infinitely slowly without producing power. The need to produce power in real energy conversion devices is one reason why the high efficiency of ideal, reversible performance is seldom approached.

Classical equilibrium thermodynamics can be extended to quasi-static processes. Conventional irreversible thermodynamics has become increasingly powerful, but its microscopic view does not lend itself to the macroscopic view preferred by practicing engineers. This is a significant extension, since quasistatic processes happen in finite time, produce entropy, and provide a better approximation of real processes than provided by equilibrium thermodynamics. System parameters in equilibrium thermodynamics are the measurable quantities: volume, temperature, pressure, and heat capacity. To model real-time dependent processes rigorously, the set of parameters must also include transport properties, relaxation time, etc. In general, irreversible thermodynamic problems are too difficult for practicing engineers to solve exactly.

The literature of finite-time thermodynamics started with Curzon and Ahlborn (Curzon, F.L. and Ahlborn, B., Efficiency of a Carnot engine at maximum power output. *American Journal of Physics*, 1975, vol. 41, no.1, pp. 22–24) in 1975. They treated an endo-reversible Carnot engine power output being limited by the rates of heat transfer to and from the working substance. They remarked that the Carnot efficiency $[\eta_{Carnot} = 1 - (T_L/T_H)$, where T_L and T_H are the temperatures of the heat sink and heat source for the engine] is realized only by a completely reversible heat engine operating at zero speed and hence at zero power. They showed that the efficiency of an engine operating at maximum power is given by a remarkably simple formula $[\eta_{Curzon-Ahlborn} = 1 - (T_L/T_H)^{1/2}]$, which of course always gives a lower value than the Carnot formula. They also verified that their formula agrees much better with the measured efficiencies of operating installations.

Finite-time thermodynamics is an extension to traditional thermodynamics in order to obtain more realistic limits to the performance of real processes, and to deal with processes or devices with finite-time characteristics. Finite-time thermodynamics is a method for the modeling and optimization of real devices that owe their thermodynamic imperfection to heat transfer, mass transfer, and fluid flow irreversibility.

A literature survey of finite-time thermodynamics is given by Wu, Chen, and Chen (Wu, C., Chen, L., and Chen, J., Recent Advances in Finite-Time Thermodynamics, Nova Science Publ. Inc., New York, 1999).

Engineering thermodynamic cycle analysis is based on the concept of equilibrium and does not deal with time. Heat transfer does deal with time but not cycle analysis. Finite-time thermodynamics fills in a gap that has long existed between equilibrium thermodynamics and heat transfer.

Review Problems 7.1   Introduction

1. What is the basic concept of classical engineering equilibrium thermodynamics? Does engineering thermodynamic cycle analysis deal with time?
2. Are the heat transfers between the Carnot heat engine and its surrounding heat source and heat sink reversible?
3. Why cannot the Carnot cycle efficiency be approached in the real world?
4. What is finite-time thermodynamics?

7.2 RATE OF HEAT TRANSFER

Heat is an amount of microscopic energy transfer across the boundary of a system in an energy interaction with its surroundings. The symbol Q is used to denote heat.

The *rate of heat transfer* is an amount of microscopic energy transfer per unit time across the boundary of a system in an energy interaction with its surroundings. The symbol Qdot is used to denote rate of heat transfer.

The relationships between Q and Qdot are

$$Q\text{dot} = \delta Q / dt \tag{7.1}$$

and

$$Q = \int (Q\text{dot}) dt \tag{7.2}$$

There are three modes of rate of heat transfer: conduction ($Q\text{dot}_k$), convection ($Q\text{dot}_c$), and radiation ($Q\text{dot}_r$).

Conduction is the rate of heat transfer through a medium without mass transfer. The basic rate of conduction heat-transfer equation is Fourier's law:

$$Q\text{dot}_k = -kA(dT/dx) \tag{7.3}$$

where k is a heat-transfer property called *thermal conductivity*, A is the cross-sectional area normal to the heat transfer, and dT/dx is the temperature gradient in the direction of the heat transfer.

For example, in the case of a linear temperature gradient, the rate of conduction heat transfer from a high temperature T_H on one side to a low temperature T_L on the other side through a solid wall with thickness L is

$$Q\text{dot}_k = kA(T_H - T_L)/L \tag{7.4}$$

Convection is the rate of heat transfer leaving a surface to a fluid. The basic rate of convection heat-transfer equation is Newton's law:

$$Q\text{dot}_c = hA(T_{\text{surface}} - T_{\text{fluid}}) \tag{7.5}$$

where h is a heat-transfer transport property called the *convection coefficient*, A is the surface area, and $(T_{\text{surface}} - T_{\text{fluid}})$ is the temperature difference between the surface and the fluid.

Radiation is the rate of heat transfer by electromagnetic waves emitted by matter. Unlike conduction and convection, radiation does not require an intervening medium to propagate. The basic rate of radiation heat-transfer equation between a high temperature (T_H) black body and a low temperature (T_L) black body is Stefan–Boltzmann's law:

$$Q\text{dot}_r = \sigma A[(T_H)^4 - (T_L)^4] \tag{7.6}$$

where σ is the Stefan–Boltzmann constant and A is the emitting surface area.

Radiation heat transfer is usually not important in ordinary heat exchanger design and analysis, unless significant temperature differences are present.

Equations (7.4) and (7.5) can be rewritten in the form of Ohm's law:

$$Q\text{dot}_k = (T_H - T_L)/(L/kA) = (T_H - T_L)/R_k \tag{7.7}$$

and

$$Q\text{dot}_c = (T_{\text{surface}} - T_{\text{fluid}})/(1/hA) = (T_{\text{surface}} - T_{\text{fluid}})/R_c \tag{7.8}$$

where R_k is the conduction heat-transfer resistance and R_c is the convection heat-transfer resistance.

Usually, the rate of heat transfer is a combination of conduction and convection in a heat exchanger system as illustrated in Fig. 7.1 and only the fluid temperature on either side of the solid surface is known. For steady state, the rate of conduction heat transfer and the rate of convection heat transfer are equal. The total resistance (R) of the combined rate of heat transfer is

$$R = \sum R_k + \sum R_c \tag{7.9}$$

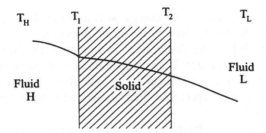

Figure 7.1 Rate of heat transfer in a heat exchanger.

The rate of the combined heat transfer (Qdot) is

$$Q\text{dot} = (T_H - T_L)/(1/UA) = (T_H - T_L)/R \qquad (7.10)$$

where U is the overall heat transfer coefficient.

Review Problems 7.2   Rate of Heat Transfer

1. What are the three modes of heat transfer?
2. How does conduction differ from convection?
3. What is the mechanism of radiation heat transfer?
4. What is the overall heat-transfer coefficient, U?

7.3   HEAT EXCHANGER

One of the most important thermodynamic devices is the heat exchanger. Heat exchangers can be classified as mixed flow, recuperative, and regenerative types.

In the *mixed flow-type heat exchanger*, one fluid being cooled and another fluid being heated are mixed together.

In the *recuperative flow-type heat exchanger*, the fluid being cooled is physically separated from the fluid being heated by some solid boundary.

In the *regenerative flow-type heat exchanger*, there is only one set of flow channels. The hot fluid enters through the channels and heats the material in the heat exchanger surrounding the channels. The hot fluid then exits the heat exchanger. Next, the cold fluid enters through the channels and is heated by the material in the heat exchanger surrounding the channels. The cold fluid then exits the heat exchanger.

The three commonly used recuperative flow-type heat exchangers in the power and refrigeration industry are *parallel-flow*, *counter-flow*, and *cross-flow* heat exchangers. Consider the case where a fluid is flowing through a pipe and exchanging energy with another fluid flowing around the pipe. When the fluids flow in the same direction, it is a parallel-flow heat exchanger. When the fluids flow in the opposite directions, it is a counter-flow heat exchanger. When the fluids flow in the normal direction, it is a cross-flow heat exchanger. The operation of parallel-flow and counter-flow heat exchangers and their associated temperature profiles are shown in Fig. 7.2.

As can be seen from the nonlinear temperature profiles, the temperature difference between the fluids varies from one end of the heat exchanger to the other. To find an effective temperature difference between the two fluids, a logarithmic mean temperature difference (LMTD) is defined as

$$\text{LMTD} = [(T_{B1} - T_{A1}) - (T_{B2} - T_{A2})]/\ln[(T_{B1} - T_{A1})$$
$$/(T_{B2} - T_{A2})] \tag{7.11}$$

The rate of heat transfer between the two fluids is

$$Q\text{dot} = UA(\text{LMTD}) \tag{7.12}$$

where U is the overall heat-transfer coefficient and A is the surface area of the tube(s), respectively.

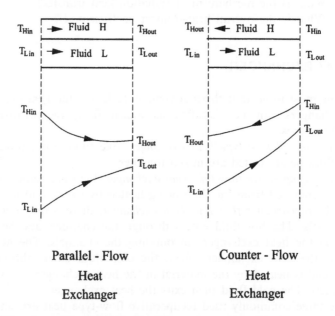

Figure 7.2 Operation of parallel-flow and counter-flow heat exchangers and their associated temperature profiles.

Usually, the counter-flow heat exchanger is smaller than the parallel-flow heat exchanger because the LMTD of the former exchanger is larger than that of the latter exchanger when the inlet and outlet temperatures of the hot fluid and the inlet and outlet temperatures of the cold fluid are identical. The following example illustrates this comparison.

Example 7.1

A counter-flow lubricating oil cooler with a net heat-transfer area of $258\,\text{ft}^2$ cools $60,000\,\text{lbm}$ of oil per hour from a temperature of $145°F$ to $120°F$. The temperature of the cooling water entering and leaving are $75°F$ and $90°F$. The specific heat of the oil is $0.5\,\text{Btu/[lbm (°F)]}$. Find the LMTD and overall heat-transfer coefficient under these operating conditions. Also, find the required area for a parallel-flow heat exchanger under these identical operating conditions. The temperature profiles of the counter-flow and parallel-flow heat exchangers are shown in Fig. 7.2.

Solution:

$$Q\text{dot} = [m\text{dot}(c)(T_{in} - T_{exit})]_{oil} = 60,000(0.5)(145 - 120)$$
$$= 750,000\,\text{Btu/hr}$$
$$\text{LMTD} = [(145 - 90) - (120 - 75)]/\ln[(145 - 90)/(120 - 75)] = 49.8°F$$
$$U = Q\text{dot}/[A\ (\text{LMTD})] = 750,000/[258(49.8)] = 58.4\,\text{Btu/[hr (ft}^2)°F]}$$

For the parallel-flow heat exchanger:

$$\text{LMTD} = [(145 - 75) - (120 - 90)]/\ln[(145 - 75)/(120 - 90)] = 47.2°F$$

and

$A = Q\text{dot}/[U(\text{LMTD})] = 750,000/[58.4(47.2)] = 272\,\text{ft}^2$. This area is larger than $258\,\text{ft}^2$.

Example 7.2

A counter-flow heater as shown in Fig. 7.3a heats helium at $101\,\text{kPa}$ from a temperature of $20°C$ to $800°C$. The temperature of the heating flue gas (air) entering and leaving are $1800°C$ and $1200°C$ at $101\,\text{kPa}$. Find (A) the LMTD, rate of helium flow, and heat transfer based on a unit of heating flue gas, and (B) the LMTD, rate of helium flow, and heat transfer for a parallel-flow heat exchanger under these identical operating conditions.

Problem (A)

To solve this problem by CyclePad, we take the following steps:

1. Build the heat exchanger as shown in Fig. 7.3a.

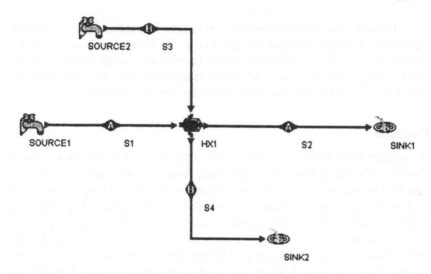

Figure 7.3a   Heat exchanger.

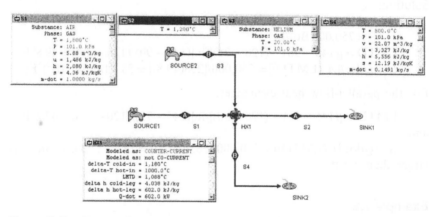

Figure 7.3b   Heat exchanger input and output results.

2. Analysis:
    a. Assume that both hot and cold sides of the heat exchanger are isobaric, and the type is counter flow.
    b. Input hot-side fluid = air, $T_1 = 1800°C$, $p_1 = 101$ kPa, $mdot_1 = 1$ kg/sec, and $T_2 = 1200°C$; cold-side fluid = helium, $T_3 = 20°C$, $p_3 = 101$ kPa, and $T_4 = 800°C$.
3. Display results:

The answers are: LMTD = 1088°C, Qdot = 602 kW, and $mdot_3 =$ 0.1491 kg/sec, as shown in Fig. 7.3b.

Problem (B)

   1. Analysis:
      a. Retract the heat exchanger type to counter flow.
      b. Input the heat exchanger type as co-current (parallel) flow.
   2. Display results:

The answers are: $\text{LMTD} = 924.4°\text{C}$, $\dot{Q} = 602\,\text{kW}$, and $\dot{m}\text{dot}_3 = 0.1491\,\text{kg/sec}$, as shown in Fig. 7.3c.

Example 7.3

A counter-flow heat exchanger heats water at 101 kPa from saturated liquid state to saturated vapor state. The temperature of the heating flue gas (air) entering and leaving are 1800°C and 1200°C at 101 kPa. Find (A) the LMTD, rate of water flow, and heat transfer based on a unit mass of heating flue gas, and (B) the LMTD, rate of helium flow, and heat transfer for a parallel-flow heat exchanger under these identical operating conditions.

Problem (A)

To solve this problem by CyclePad, we take the following steps:

   1. Build the heat exchanger as shown in Fig. 7.3a.
   2. Analysis:
      a. Assume that both hot and cold sides of the heat exchanger are isobaric, and the type is counter flow.

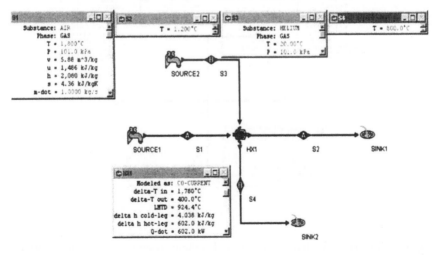

Figure 7.3c Heat exchanger input and output results.

b. Input   hot-side   fluid = air, $T_1 = 1800°C$, $p_1 = 101$ kPa,
    $mdot_1 = 1$ kg/sec, and $T_2 = 1200°C$; cold-side fluid = water,
    $p_3 = 101$ kPa, $x_3 = 0$, and $x_4 = 1$.
3. Display results:

The answers are: LMTD = 1378°C,   Qdot = 602 kW,   and $mdot_3 =$
0.2668 kg/sec, as shown in Fig. 7.4a.

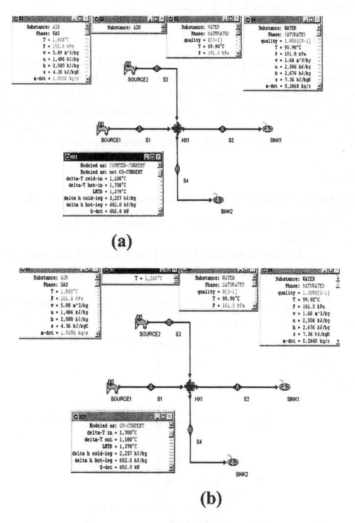

(a)

(b)

Figure 7.4 Heat exchanger input and output results. (a) Counter current;
(b) co-current.

Problem (B)

1. Analysis:
    a. Retract the heat exchanger of counter-flow.
    b. Input the heat exchanger type as co-current (parallel) flow.
2. Display results:

The answers are: LMTD $= 1378°C$, $Q\mathrm{dot} = 602\,\mathrm{kW}$, and $m\mathrm{dot}_3 = 0.2668\,\mathrm{kg/sec}$, as shown in Fig. 7.4b.

Review Problems 7.3   Heat Exchanger

A counter-flow heater heats ammonia at 101 kPa from saturated liquid state to saturated vapor state. The temperature of the heating air entering and leaving are 33°C and 17°C at 101 kPa. Find (a) the LMTD, rate of water flow, and heat transfer based on a unit of heating flue gas, and (b) the LMTD, rate of helium flow, and heat transfer for a parallel-flow heat exchanger under these identical operating conditions.

ANSWERS: (a) LMTD $= 58.05°C$, $Q\mathrm{dot} = 16.05\,\mathrm{kW}$, and $m\mathrm{dot}_3 = 0.0117\,\mathrm{kg/sec}$; (b) LMTD $= 58.05°C$, $Q\mathrm{dot} = 16.05\,\mathrm{kW}$, and $m\mathrm{dot}_3 = 0.0117\,\mathrm{kg/sec}$.

7.4 CURZON AND AHLBORN (ENDOREVERSIBLE CARNOT) CYCLE

The *T–s* diagram and schematic diagram of the Curzon and Ahlborn (endoreversible Carnot) cycle are shown in Figs. 7.5 and 7.6, respectively (Cuzon, F.L. and Ahlborn, B., Efficiency of a Carnot engine at maximum

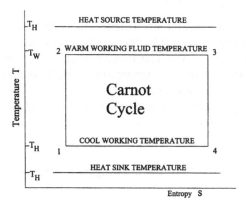

Figure 7.5 Curzon and Ahlborn cycle *T–s* diagram.

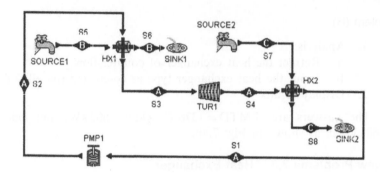

Figure 7.6 Curzon and Ahlborn cycle schematic diagram.

output, *Am. Jrnl. of Phys.*, 1975). The cycle operates between a heat source at temperature T_H and a heat sink at temperature T_L. The temperatures of the working fluid in the isothermal heat-addition and heat-rejection processes are T_W and T_C. The finite temperature difference $(T_H - T_W)$ allows a finite heat transfer from the heat source to the working fluid in the heat-addition process. Similarly, the finite temperature difference $(T_C - T_L)$ allows a heat transfer from the working fluid to the heat sink in the heat-rejection process. The cycle is a modified Carnot cycle. Other than the external irreversibility due to the two heat-transfer processes, the modified cycle is an internal reversible heat engine.

Assume that the working fluid flows through the heat engine in a steady-state fashion. The rates of heat rejection and addition of the heat engine are

$$Q\text{dot}_H = U_H A_H (T_H - T_W) \tag{7.13}$$

$$Q\text{dot}_L = U_L A_L (T_C - T_L) \tag{7.14}$$

where U_H is the heat-transfer coefficient, A_H is the heat-transfer surface area of the high-temperature side heat exchanger between the heat engine and the heat source, U_L is the heat transfer coefficient, and A_L is the heat-transfer surface area of the low-temperature side heat exchanger between the heat engine and the heat sink.

The total heat-transfer surface area (A) of the two heat exchangers is assumed to be a constant:

$$A = A_H + A_L \tag{7.15}$$

The power output (P) of the heat engine according to the first law of thermodynamics is

$$P = Q\text{dot}_H - Q\text{dot}_L \tag{7.16}$$

The second law of thermodynamics requires that

$$Q\mathrm{dot}_H/T_W = Q\mathrm{dot}_L/T_C \qquad (7.17)$$

The efficiency (η) of the heat engine is

$$\eta = P/Q\mathrm{dot}_H \qquad (7.18)$$

We define a heat-transfer surface area ratio (f) as

$$f = A_H/A_L \qquad (7.19)$$

Combining Eqs. (7.13)–(7.19) gives the optimum value of f at maximum power output (f_a) and the optimal power output for given values of T_H, T_L, U_H, U_L, A, and η.

$$f = f_a = (U_L/U_H)^{1/2} \qquad (7.20)$$

$$P = [T_H - T_L/(1 - \eta)]\eta B_1 \qquad (7.21)$$

where

$$B_1 = (U_H A)/[1 + (U_L/U_H)^{-1/2}]^2 \qquad (7.22)$$

Equation (7.21) is the optimal performance characteristics of the endoreversible Carnot heat engine. It indicates that $P = 0$ when $\eta = 0$ and $\eta = \eta_c = 1 - T_L/T_H$. Taking the derivative of P with respect to η and setting it equal to zero ($dP/d\eta = 0$) gives

$$\eta = \eta_{C-A} = 1 - (T_L/T_H)^{1/2} \qquad (7.23)$$

and the optimal power output delivered by the cycle is

$$P_{\max} = B_1[(T_H)^{1/2} - (T_L)^{1/2}]^2 \qquad (7.24)$$

The power versus efficiency characteristics of the endoreversible Carnot heat engine is a parabolic curve. The endoreversible heat engine is a simple model, which considers the external heat-transfer irreversibility between the heat engine and its surrounding heat reservoirs only.

The required optimum intermediate temperatures at maximum power condition are

$$T_W = \{(T_H)^{0.5} + [U_L A_L/(U_H A_H)](T_L)^{0.5}\}/\{1 + [U_L A_L/(U_H A_H)]\}(T_H)^{0.5} \qquad (7.25)$$

and

$$T_C = \{(T_H)^{0.5} + [U_L A_L/(U_H A_H)](T_L)^{0.5}\}/\{1 + [U_L A_L/(U_H A_H)]\}(T_L)^{0.5} \qquad (7.26)$$

By taking into account the rate of heat transfer associated with the endoreversible cycle, the upper bound of the power output of the cycle can be found. This bound provides a practical basis for a real power plant design. The industrial view is that the heat engine efficiency is secondary to the power output in power plants whose worth is constrained by economic considerations.

Another important industrial design objective function for power cycle design is net power per unit conductance of heat exchangers. The conductance of the high-temperature side heat exchanger is $U_H A_H$ in Eq. (7.13), and the conductance of the low-temperature side heat exchanger is $U_L A_L$ in Eq. (7.14). The net power per unit conductance of the heat exchanger $[W\mathrm{dot}_{\mathrm{net}}/(U_H A_H + U_L A_L)]$ represents the initial and operational costs of the heat exchanger, which is a very important part of the power plant.

For many power plants, such as waste-heat plants, geothermal, plants, and OTEC (ocean thermal energy conversion) plants, which have relatively small fuel costs, the industrial design objective function for these power cycle designs is specific net power. The specific net power is defined as $W\mathrm{dot}_{\mathrm{net}}/(A_H + A_L)$. The total cost of these plants is determined mainly by the construction cost. From the viewpoints of cost and size, the most important component of these plants is the heat exchanger. The volume of supporting structure, the weight of buoyance-adjusting ballast, the length of pipe, head loss, pump power, and all other major components increase as the size of the heat exchanger increases. High performance of these plants is essential for producing power. For this reason, the performance evaluation objective function of these plants is taken to be the specific power.

Example 7.4

An endoreversible (Curzon and Ahlborn) cycle operates between a heat source at temperature $T_H = 1600\,\mathrm{K}$ and a heat sink at temperature $T_L = 400\,\mathrm{K}$. Suppose that $U_H = 100\,\mathrm{kW/m^2}$ (overall heat-transfer coefficient of the high-temperature side heat exchanger between the heat engine and the heat source), $A_H = 1\,\mathrm{m^2}$ (heat-transfer surface area of the high-temperature side heat exchanger between the heat engine and the heat source), and $U_L = 100\,\mathrm{kW/m^2}$ (overall heat-transfer coefficient of the low-temperature side heat exchanger between the heat engine and the heat sink).

Determine the maximum power output of the cycle. Find the heat-transfer added, heat transfer removed, heat transfer surface area of the low-temperature side heat exchanger between the heat engine and the heat sink, and efficiency of the cycle at the maximum power output condition.

Solution:

$$Q\text{dot}_H = U_H A_H (T_H - T_W) = 100(1)(1600 - 1200) = 40,000 \text{ kW} = 40 \text{ MW}.$$

$$Q\text{dot}_H / T_W = 40,000/1200 = Q\text{dot}_L / T_C = Q\text{dot}_L / 600 \quad \text{gives} \quad Q\text{dot}_L = 20,000 \text{ kW} = 2 \text{ MW}$$

$$Q\text{dot}_L = 20,000 = U_L A_L (T_C - T_L) = 100 A_L (600 - 400) \text{ gives } A_L = 1 \text{ m}^2$$

$$P = Q\text{dot}_H - Q\text{dot}_L = 40,000 - 20,000 = 20,000 \text{ kW} = 20 \text{ MW}$$

and

$$\eta = 1 - (T_C/T_W) = 1 - (400/1600)^{1/2} = 50\%.$$

Example 7.5

An endoreversible (Curzon and Ahlborn) steam cycle operates between a heat source at temperature $T_H = 640$ K and a heat sink at temperature $T_L = 300$ K. The following information is given:

Heat source: fluid = water, $T_5 = 640$ K, $x_5 = 1$, $T_6 = 640$ K, and $x_6 = 0$
Heat sink: fluid = water, $T_7 = 300$ K, $x_7 = 0$, $T_8 = 300$ K, and $x_8 = 1$
Steam cycle: fluid = water, $x_2 = 0$, $T_3 = 500$ K, $x_3 = 1$, $T_4 = 400$ K, and $m\text{dot} = 1$ kg/sec.

Determine the rate of heat added from the heat source, rate of heat removed to the heat sink, power required by the isentropic pump, power produced by the isentropic turbine, net power produced, and efficiency of the cycle.

To solve this problem by CyclePad, we take the following steps:

1. Build the cycle and its surroundings as shown in Fig. 7.6.
2. Analysis:
   a. Assume that the heat exchangers are isobaric (notice that isobaric is also isothermal in the saturated mixture region), and the turbine and pump are isentropic.
   b. Input heat source fluid = water, $T_5 = 640$ K, $x_5 = 1$, $T_6 = 640$ K, and $x_6 = 0$; heat sink fluid = water, $T_7 = 300$ K, $x_7 = 0$, $T_8 = 300$ K, and $x_8 = 1$; steam cycle fluid = water, $x_2 = 0$, $T_3 = 500$ K, $x_3 = 1$, $T_4 = 400$ K, and $m\text{dot} = 1$ kg/sec.
3. Display results:

The answers are: rate of heat added from the heat source = 1827 kW, rate of heat removed to the heat sink = −1462 kW, power required by the isentropic pump = −50.53 kW, power produced by the isentropic turbine = 416.0 kW, net power produced = 365.4 kW, and efficiency of the cycle = 20%. (See Fig. 7.7.)

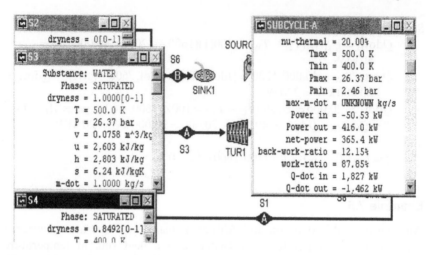

Figure 7.7 Endoreversible (Curzon and Ahlborn) steam cycle.

Example 7.6

An endoreversible (Curzon and Ahlborn) steam cycle operates between a heat source at temperature $T_H = 640$ K and a heat sink at temperature $T_L = 300$ K. The following information is given:

> Heat source: fluid = water, $T_5 = 640$ K, $x_5 = 1$, $T_6 = 640$ K, and $x_6 = 0$
> Heat sink: fluid = water, $T_7 = 300$ K, $x_7 = 0$, $T_8 = 300$ K, and $x_8 = 1$
> Steam cycle: fluid = water, $x_2 = 0$, $x_3 = 1$, $T_4 = 400$ K, and $m$dot = 1 kg/sec.

Determine the maximum net power produced and working fluid temperature at the inlet of the turbine.

To solve this problem by CyclePad, we take the following steps:

1. Build the cycle and its surroundings as shown in Fig. 7.6.
2. Analysis:
   a. Assume that the heat exchangers are isobaric (notice that isobaric is also isothermal in the saturated mixture region), and the turbine and pump are isentropic.
   b. Input heat source fluid = water, $T_5 = 640$ K, $x_5 = 1$, $T_6 = 640$ K, and $x_6 = 0$; heat sink fluid = water, $T_7 = 300$ K, $x_7 = 0$, $T_8 = 300$ K, and $x_8 = 1$; steam cycle fluid = water, $x_2 = 0$, $T_3 = 500$ K, $x_3 = 1$, $T_4 = 400$ K, and $m$dot = 1 kg/sec.
3. Sensitivity analysis: plot net power versus T_3 diagram as shown in Fig. 7.8.

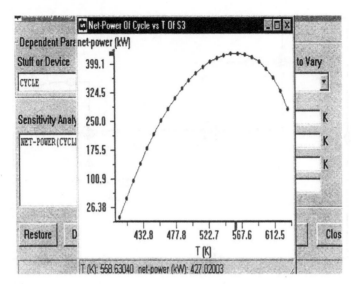

Figure 7.8 Endoreversible cycle sensitivity analysis.

The answers are: maximum net power is about 427.0 kW and T_3 is about 558.6 K.

COMMENT: The partial optimization is only for $\partial(net\ power)/\partial(T_3) = 0$. To have the full optimization, we must let $\partial(net\ power)/\partial(T_4) = 0$ also.

Example 7.7

An endoreversible (Curzon and Ahlborn) steam cycle as shown in Fig. 7.9a operates between a heat source at temperature $T_H = 300°C$ and a heat sink at temperature $T_L = 20°C$. The following information is given:

Heat source: fluid = water, $T_1 = 300°C$, $x_1 = 1$, $T_2 = 300°C$, and $x_2 = 0$

Heat sink: fluid = water, $T_3 = 20°C$, $x_3 = 0$, $T_4 = 20°C$, and $x_4 = 1$

Carnot cycle: fluid = water, $T_5 = 100°C$, $s_5 = 3.25\ kJ/kg(K)$, $T_7 = 200°C$, $s_7 = 5.70\ kJ/kg(K)$, and $m$dot $= 1\ kg/sec$.

The heat exchangers are counter-flow type: $U_H = 0.4\ kJ/(m^2)K$ and $U_L = 0.4\ kJ/(m^2)K$

Determine the rate of heat added from the heat source, rate of heat removed to the heat sink, power required by the isentropic pump, power produced by the isentropic turbine, net power produced, and efficiency of the cycle.

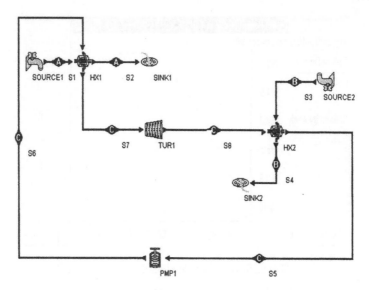

Figure 7.9a   Finite-time Carnot cycle.

Taking the specific net power output (net output power per unit total heat exchanger surface area) as the design objective function, optimize the warm-side (heater or high-temperature side heat exchanger) and cold-side (cooler or low-temperature side heat exchanger) working fluid temperatures.

To solve this problem by CyclePad, we take the following steps:

1. Build the cycle and its surroundings as shown in Fig. 7.6.

2. Analysis: (a) Assume that the heat exchangers are isobaric and counter flow, and turbine and pump are isentropic. (b) Input heat source: fluid = water, $T_1 = 300°C$, $x_1 = 1$, $T_2 = 300°C$, and $x_2 = 0$; heat sink: fluid = water, $T_3 = 20°C$, $x_3 = 0$, $T_4 = 20°C$, and $x_4 = 1$; Carnot cycle: fluid = water, $T_5 = 100°C$, $s_5 = 3.25$ kJ/kg(K), $T_7 = 200°C$, $s_7 = 5.70$ kJ/ kg(K), and $m\text{dot} = 1$ kg/sec.

3. Display results: The answers are: $\text{LMTD}_H = 100$ K, $\text{LMTD}_H = 80$ K, rate of heat added from the heat source = 1159 kW, rate of heat removed to the heat sink = −914.2 kW, power required by the isentropic pump = −143.2 kW, power produced by the isentropic turbine = 388.2 kW, net power produced = 245.0 kW, and efficiency of the cycle = 21.14%, as shown in Fig. 7.9b.

4. Calculate $U_H A_H = Q_H / \text{LMTD}_H = 1159/100 = 11.59$ kW/K, $U_L A_L = Q_L / \text{LMTD}_L = 914.2/80 = 11.43$ kW/K, $U_H A_H + U_L A_L = 11.59 + 11.43 = 23.02$ kW/K, and specific net power output = $W\text{dot}_{net}/(A_H + A_L) = 245.0/(11.59/0.4 + 11.43/0.4) = 13.31$ kW/m^2.

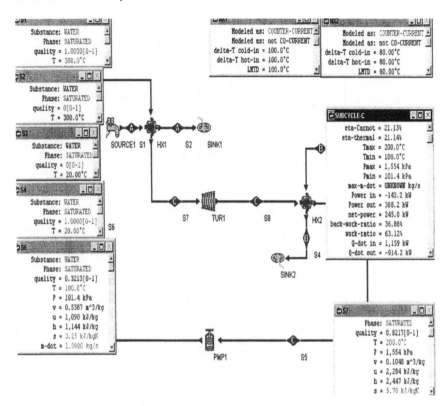

Figure 7.9b   Finite-time Carnot cycle input and output.

5. To optimize the specific power output of the cycle, we let ∂(specific power output)$/\partial T_7 = 0$ first and then ∂(specific power output)$/\partial T_7 = 0$ as shown in Tables 7.1a and 7.1b.

It is seen that ∂ (specific power output)$/\partial T_7 = 0$ occurs at $T_7 = 220°C$ as shown in Table 7.1a.

6. Let ∂ (specific power output)$/\partial T_5 = 0$. It is seen that ∂ (specific power output)$/\partial T_5 = 0$ occurs at $T_5 = 80°C$ as shown in Table 7.1b.

7. Let $T_5 = 80°C$ and $T_7 = 220°C$; the optimized specific power output of the cycle is $14.53\,\text{kW/m}^2$. At the maximum optimized specific power output condition, $\text{LMTD}_H = 80\,\text{K}$, $\text{LMTD}_H = 60\,\text{K}$, rate of heat added from the heat source $= 1208\,\text{kW}$, rate of heat removed to the heat sink $= -865.2\,\text{kW}$, power required by the isentropic pump $= -201.8\,\text{kW}$, power produced by the isentropic turbine $= 544.9\,\text{kW}$, net power produced $= 343.0\,\text{kW}$, and efficiency of the cycle $= 28.39\%$, as shown in Fig. 7.9c.

Table 7.1a Specific Power Optimization with Respect to T_7

		$T_1 = T_2 = 300$		$T_3 = T_4 = 20$									
T_5	T_7	p_{net}/U_A	p_{net}/A	efficiency	power in	power out	powernet	heat in	heat out	LMTD$_H$	LMTD$_L$	$U_H A_H$	$U_L A_L$
100	100.1	0.015276	0.019096	0.0267	−0.195	0.4395	0.2444	914.2	−914.2	200	80	4.571	11.4275
100	120	2.920981	3.651226	5.09	−36.72	85.74	49.01	963.2	−914.2	180	80	5.351111	11.4275
100	140	5.522602	6.903253	9.69	−69.22	167.3	98.04	1012	−914.2	160	80	6.325	11.4275
100	160	7.734371	9.667963	13.85	−97.68	244.7	147	1061	−914.2	140	80	7.578571	11.4275
100	180	9.478902	11.84863	17.66	−122.3	318.3	196	1110	−914.2	120	80	9.25	11.4275
100	200	10.64408	13.30509	21.14	−143.2	388.2	245	1159	−914.2	100	80	11.59	11.4275
100	**220**	**11.08284**	**13.85355**	**24.34**	**−160.6**	**454.6**	**294**	**1208**	**−914.2**	**80**	**80**	**15.1**	**11.4275**
100	240	10.59378	13.24222	27.28	−174.6	517.6	343	1257	−914.2	60	80	20.95	11.4275
100	260	8.893426	11.11678	30.01	−185.5	577.5	392	1306	−914.2	40	80	32.65	11.4275
100	280	5.569764	6.962205	32.54	−193.2	634.2	441	1355	−914.2	20	80	67.75	11.4275
100	299.9	0.034822	0.043528	34.86	−198.4	687.7	489.3	1404	−914.2	0.1	80	14040	11.4275

assume $U_H = U_L = 0.4 \, \text{kJ/m}^2(\text{K})$.

Table 7.1b Specific Power Optimization with Respect to T_5

T_5	T_7	p_{net}/U_A	p_{net}/A	efficiency	power in	power out	powernet	heat in	heat out	$LMTD_H$	$LMTD_L$	$U_H A_H$	$U_L A_L$
20.1	220	0.068036	0.085045	40.54	−354.9	844.5	489.8	1208	−718.4	80	0.1	15.1	71.84
40	220	8.249158	10.31145	36.5	−298.7	739.8	441	1208	−767.2	80	20	15.1	38.36
60	220	11.0407	13.80087	32.45	−247.8	639.8	392	1208	−816.2	80	40	15.1	20.405
80	**220**	**11.61924**	**14.52405**	**28.39**	**−201.8**	**544.9**	**343**	**1208**	**−865.2**	**80**	**60**	**15.1**	**14.42**
100	220	11.08284	13.85355	24.34	−160.6	454.6	294	1208	−914.2	80	80	15.1	11.4275
120	220	9.906194	12.38274	20.28	−123.9	368.9	245	1208	−963.2	80	100	15.1	9.632
140	220	8.328612	10.41076	16.22	−91.38	287.4	196	1208	−1012	80	120	15.1	8.433333
160	220	6.48189	8.102362	12.17	−62.92	209.9	147	1208	−1061	80	140	15.1	7.578571
180	220	4.446965	5.558707	8.11	−38.35	136.3	98	1208	−1110	80	160	15.1	6.9375
200	220	2.274955	2.843694	4.06	−17.4	66.41	49	1208	−1159	80	180	15.1	6.438889
219.9	220	0.011569	0.014461	0.0202	−0.0787	0.3233	0.2446	1208	−1208	80	199.9	15.1	6.043022

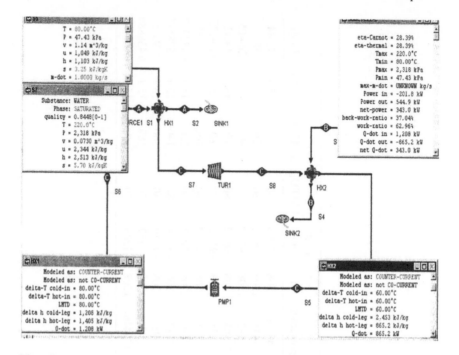

Figure 7.9c   Finite-time Carnot cycle optimization result.

Review Problems 7.4   Curzon and Ahlborn Cycle with Infinite Heat Capacity Heat Source and Sink

1. An endoreversible (Curzon and Ahlborn) cycle operates between a heat source at temperature $T_H = 3600$ K and a heat sink at temperature $T_L = 300$ K. The temperatures of the working fluid in the isothermal heat-addition and heat-rejection processes are $T_W = 2000$ K and $T_C = 400$ K. Suppose that $U_H = 100$ kW/m^2 (overall heat-transfer coefficient of the high-temperature side heat exchanger between the heat engine and the heat source), $A_H = 1$ m^2 (heat-transfer surface area of the high-temperature side heat exchanger between the heat engine and the heat source), and $U_L = 100$ kW/m^2 (overall heat-transfer coefficient of the low-temperature side heat exchanger between the heat engine and the heat sink). Determine the heat transfer added, heat transfer removed, heat-transfer surface area of the low-temperature side heat exchanger between the heat engine and the heat sink, power output, and efficiency of the cycle.

2. Referring to Problem 1 and with fixed heat source and heat sink temperatures, determine the maximum power output of the cycle. Find the working fluid temperatures in the isothermal heat addition and heat

rejection processes, heat transfer added, heat transfer removed, heat transfer surface area of the low-temperature side heat exchanger between the heat engine and the heat sink, and efficiency of the cycle at the maximum power output condition.

3. An endoreversible (Curzon and Ahlborn) steam cycle operates between a heat source at temperature $T_H = 640\,K$ and a heat sink at temperature $T_L = 300\,K$. The following information is given:

Heat source: fluid = water, $T_5 = 600\,K$, $x_5 = 1$, $T_6 = 600\,K$, and $x_6 = 0$
Heat sink: fluid = water, $T_7 = 290\,K$, $x_7 = 0$, $T_8 = 290\,K$, and $x_8 = 1$
Steam cycle: fluid = water, $x_2 = 0$, $T_3 = 550\,K$, $x_3 = 1$, $T_4 = 300\,K$, and $m$dot $= 1\,kg/sec$.

Determine the rate of heat added from the heat source, rate of heat removed to the heat sink, power required by the isentropic pump, power produced, by the isentropic turbine, net power produced, and efficiency of the cycle.

4. Referring to Problem 1 and with fixed heat source and heat sink temperatures as well as working fluid temperatures in the isothermal heat-rejection processes, determine the maximum power output of the cycle. Find the working fluid temperatures in the isothermal heat addition, heat transfer added, heat transfer removed, and efficiency of the cycle at the maximum power output condition. Draw the sensitivity diagram.

5. Referring to Problem 3 and with fixed heat source and heat sink temperatures as well as working fluid temperatures in the isothermal heat-addition processes, determine the maximum power output of the cycle. Find the working fluid temperatures in the isothermal heat removal, heat transfer added, heat transfer removed, and efficiency of the cycle at the maximum power output condition. Draw the sensitivity diagram.

6. An endoreversible (Curzon and Ahlborn) steam cycle as shown in Fig. 7.10 operates between a heat source at temperature $T_H = 300°C$ and a heat sink at temperature $T_L = 40°C$. The following information is given:

Heat source: fluid = water, $T_1 = 300°C$, $x_1 = 1$, $T_2 = 300°C$, and $x_2 = 0$
Heat sink: fluid = water, $T_3 = 40°C$, $x_3 = 0$, $T_4 = 40°C$, and $x_4 = 1$
Carnot cycle: fluid = water, $T_5 = 100°C$, $s_5 = 3.25\,kJ/kg(K)$, $T_7 = 200°C$, $s_7 = 5.70\,kJ/kg(K)$, and $m$dot $= 1\,kg/sec$.
The heat exchangers are counter-flow type, $U_H = 0.4\,kJ/(m^2)K$ and $U_L = 0.4\,kJ/(m^2)K$

Determine the rate of heat added from the heat source, rate of heat removed to the heat sink, power required by the isentropic pump, power produced by the isentropic turbine, net power produced, and efficiency of the cycle.

Taking the specific net power output (net output power per unit total heat exchanger surface area) as the design objective function, optimize the warm-side (heater or high-temperature side heat exchanger) and cold-side (cooler or low-temperature side heat exchanger) working fluid temperatures.

7.5 CURZON AND AHLBORN CYCLE WITH FINITE HEAT CAPACITY HEAT SOURCE AND SINK

The T-s diagram and schematic diagram of the *Curzon and Ahlborn (endo-reversible Carnot) cycle* are shown in Fig. 7.10 and Fig. 7.11, respectively. The cycle operates between a heat source and a heat sink with finite heat capacity. The fluid of the heat source enters the hot-side heat exchanger at T_5 and exits at T_6. The fluid of the heat sink enters the cold-side heat exchanger at T_7 and exits at T_8. The temperatures of the working fluid in the isothermal heat-addition and heat-rejection processes are T_W and T_C. The finite mean temperature difference LMTD_H $\{\text{LMTD}_H = (T_5 - T_6)/[\ln \ (T_5 - T_W)/(T_6 - T_W)]\}$ allows a finite heat transfer from the heat source to the working fluid in the heat-addition process. Similarly, the finite mean temperature difference LMTD_L $\{\text{LMTD}_L = (T_8 - T_7)/[\ln(T_C - T_7)/(T_C - T_8)]\}$ allows a heat transfer from the working fluid to the heat sink in the heat-rejection process. The cycle is a modified Carnot cycle. Other than the external

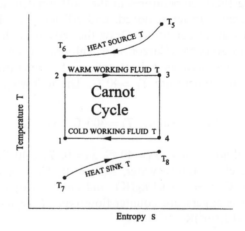

Figure 7.10 Curzon and Ahlborn cycle with finite heat capacity source and sink T-s diagram.

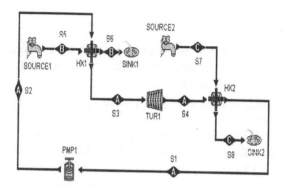

Figure 7.11 Curzon and Ahlborn cycle with finite heat capacity source and sink schematic diagram.

irreversibility due to the two heat-transfer processes, the modified cycle is an internal reversible heat engine.

Assume that the working fluid flows through the heat engine in a steady-state fashion. The rates of heat rejection and addition of the heat engine are

$$Q\text{dot}_H = U_H A_H \text{LMTD}_H \tag{7.27}$$

$$Q\text{dot}_L = U_L A_L \text{LMTD}_L \tag{7.28}$$

where U_H is the heat-transfer coefficient, A_H is the heat-transfer surface area of the high-temperature side heat exchanger between the heat engine and the heat source, U_L is the heat-transfer coefficient, and A_L is the heat-transfer surface area of the low-temperature side heat exchanger between the heat engine and the heat sink.

The total heat transfer surface area (A) of the two heat exchangers is assumed to be a constant:

$$A = A_H + A_L \tag{7.29}$$

The power output (P) of the heat engine according to the first law of thermodynamics is

$$P = Q\text{dot}_H - Q\text{dot}_L \tag{7.30}$$

The second law of thermodynamics requires that

$$Q\text{dot}_H / T_W = Q\text{dot}_L / T_C \tag{7.31}$$

The efficiency (η) of the heat engine is

$$\eta = P / Q\text{dot}_H \tag{7.32}$$

Example 7.8

An endoreversible (Curzon and Ahlborn) steam cycle operates between a finite heat capacity heat source and a finite heat capacity heat sink. The following information is given:

> Heat source: fluid = air, $T_5 = 1500$ K, $p_5 = 1$ bar, $T_6 = 650$ K, and $p_6 = 1$ bar
>
> Heat sink: fluid = air, $T_7 = 290$ K, $p_7 = 1$ bar, $T_8 = 400$ K, and $p_8 = 1$ bar
>
> Steam cycle: fluid = water, $x_2 = 0$, $T_3 = 500$ K, $x_3 = 1$, $T_4 = 420$ K, and $m$dot = 1 kg/sec

Determine the rate of heat added from the heat source, rate of heat removed to the heat sink, power required by the isentropic pump, power produced, by the isentropic turbine, net power produced, and efficiency of the cycle.

To solve this problem by CyclePad, we take the following steps:

1. Build the cycle and its surroundings as shown in Fig. 7.11.
2. Analysis:
   a. Assume that the heat exchangers are isobaric, and turbine and pump are isentropic.
   b. Input heat source fluid = air, $T_5 = 1500$ K, $p_5 = 1$ bar, $T_6 = 650$ K, and $p_6 = 1$ bar; heat sink: fluid = air, $T_7 = 290$ K, $p_7 = 1$ bar, $T_8 = 400$ K, and $p_8 = 1$ bar; steam cycle: fluid = water, $x_2 = 0$, $T_3 = 500$ K, $x_3 = 1$, $T_4 = 420$ K, and $m$dot = 1 kg/sec.
3. Display results:

The answers are: rate of heat added from the heat source = 1827 kW, rate of heat removed to the heat sink = −1535 kW, power required by the isentropic pump = −32.83 kW, power produced by the isentropic turbine = 325.1 kW, net power produced = 292.3 kW, and efficiency of the cycle = 16%. (See Fig. 7.12.)

Example 7.9

An endoreversible (Curzon and Ahlborn) steam cycle operates between a finite heat capacity heat source and a finite heat capacity heat sink. The following information is given:

> Heat source: fluid = air, $T_5 = 1500$ K, $p_5 = 1$ bar, $T_6 = 650$ K, and $p_6 = 1$ bar
>
> Heat sink: fluid = air, $T_7 = 290$ K, $p_7 = 1$ bar, $T_8 = 400$ K, and $p_8 = 1$ bar
>
> Steam cycle: fluid = water, $x_2 = 0$, $x_3 = 1$, $T_4 = 420$ K, and $m$dot = 1 kg/sec

Determine the maximum net power produced and working fluid temperature at the inlet of the turbine with fixed condenser temperature.

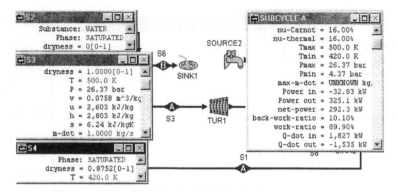

Figure 7.12 Curzon and Ahlborn cycle with finite heat capacity heat source and sink.

To solve this problem by CyclePad, we take the following steps:

1. Build the cycle and its surroundings as shown in Fig. 7.11.
2. Analysis:
   a. Assume that the heat exchangers are isobaric (notice that isobaric is also isothermal in the saturated mixture region), and turbine and pump are isentropic.
   b. Input heat source: fluid = air, $T_5 = 1500$ K, $p_5 = 1$ bar, $T_6 = 650$ K, and $p_6 = 1$ bar; heat sink: fluid = air, $T_7 = 290$ K, $p_7 = 1$ bar, $T_8 = 400$ K, and $p_8 = 1$ bar; steam cycle: fluid = water, $x_2 = 0$, $x_3 = 1$, $T_4 = 420$ K, and $m$dot = 1 kg/sec.
3. Sensitivity analysis: plot net power versus T_3 diagram as shown in Fig. 7.13.

The answers are: maximum net power is about 375.2 kW and T_3 is about 563.7 K.

COMMENT: The partial optimization is only for $\partial(net\ power)/\partial(T_3) = 0$. To have the full optimization, we must let $\partial(net\ power)/\partial(T_4) = 0$ also.

Review Problems 7.5   Curzon and Ahlborn Cycle with Finite Heat Capacity Heat Source and Sink

1. An endoreversible (Curzon and Ahlborn) steam cycle operates between a finite heat capacity heat source and a finite heat capacity heat sink. The following information is given:

Heat source: fluid = air, $T_5 = 1500$ K, $p_5 = 1$ bar, $T_6 = 650$ K, and $p_6 = 1$ bar

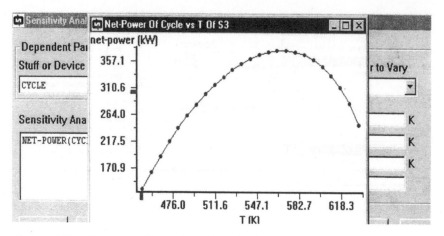

Figure 7.13 Endoreversible cycle with finite heat capacity heat source and sink sensitivity analysis.

Heat sink: fluid = air, $T_7 = 290$ K, $p_7 = 1$ bar, $T_8 = 400$ K, and $p_8 = 1$ bar

Steam cycle: fluid = water, $x_2 = 0$, $T_3 = 450$ K, $x_3 = 1$, $T_4 = 420$ K, and $m$dot = 1 kg/sec

Determine the rate of heat added from the heat source, rate of heat removed to the heat sink, power required by the isentropic pump, power produced by the isentropic turbine, net power produced, and efficiency of the cycle.

ANSWERS: rate of heat added from the heat source = 2026 kW, rate of heat removed to the heat sink = −1891 kW, power required by the isentropic pump = −4.99 kW, power produced by the isentropic turbine = 140.0 kW, net power produced = 135.0 kW, and efficiency of the cycle = 6.67%.

2. An endoreversible (Curzon and Ahlborn) steam cycle operates between a finite heat capacity heat source and a finite heat capacity heat sink. The following information is given:

Heat source: fluid = air, $T_5 = 1500$ K, $p_5 = 1$ bar, $T_6 = 650$ K, and $p_6 = 1$ bar

Heat sink: fluid = air, $T_7 = 290$ K, $p_7 = 1$ bar, $T_8 = 400$ K, and $p_8 = 1$ bar

Steam cycle: fluid = water, $x_2 = 0$, $T_3 = 480$ K, $x_3 = 1$, $T_4 = 410$ K, and $m$dot = 1 kg/sec

Determine the rate of heat added from the heat source, rate of heat removed to the heat sink, power required by the isentropic pump, power produced by the isentropic turbine, net power produced, and efficiency of the cycle.

ANSWERS: rate of heat added from the heat source $= 1913\,\text{kW}$, rate of heat removed to the heat sink $= -1634\,\text{kW}$, power required by the isentropic pump $= -25.32\,\text{kW}$, power produced by the isentropic turbine $= 304.3\,\text{kW}$, net power produced $= 278.9\,\text{kW}$, and efficiency of the cycle $= 14.58\%$.

3. An endoreversible (Curzon and Ahlborn) steam cycle operates between a finite heat capacity heat source and a finite heat capacity heat sink. The following information is given:

Heat source: fluid $=$ air, $T_5 = 1500\,\text{K}$, $p_5 = 1\,\text{bar}$, $T_6 = 650\,\text{K}$, and $p_6 = 1\,\text{bar}$

Heat sink: fluid $=$ air, $T_7 = 290\,\text{K}$, $p_7 = 1\,\text{bar}$, $T_8 = 400\,\text{K}$, and $p_8 = 1\,\text{bar}$

Steam cycle: fluid $=$ water, $x_2 = 0$, $T_4 = 480\,\text{K}$, $x_3 = 1$, $T_4 = 420\,\text{K}$, and $m\text{dot} = 1\,\text{kg/sec}$

Determine the maximum net power produced and working fluid temperature at the inlet of the turbine with fixed condenser temperature.

ANSWERS: maximum net power is about $401.2\,\text{kW}$ and T_3 is about $564.3\,\text{K}$.

7.6 FINITE-TIME RANKINE CYCLE WITH INFINITELY LARGE HEAT RESERVOIRS

The ideal *finite-time Rankine cycle* and its T–s diagram are shown in Figs. 7.14 and 7.15, respectively. The cycle is an endoreversible cycle that consists of two isentropic processes and two isobaric heat-transfer processes. The cycle exchanges heats with its surroundings in the two isobaric external irreversible heat-transfer processes. The heat source and heat sink are infinitely large. Therefore, the temperature of the heat source and heat sink are unchanged during the heat-transfer processes.

Assume that the working fluid flows through the heat engine in a steady-state fashion. The rates of heat rejection and addition of the heat engine are

$$Q\text{dot}_H = U_H A_H \text{LMTD}_H \tag{7.33}$$

$$Q\text{dot}_L = U_L A_L \text{LMTD}_L \tag{7.34}$$

where U_H is the heat-transfer coefficient, A_H is the heat-transfer surface area of the high-temperature side heat exchanger between the heat engine

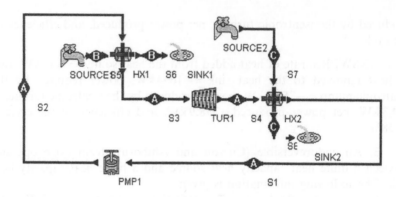

Figure 7.14   Finite-time ideal Rankine cycle with infinitely large heat reservoirs.

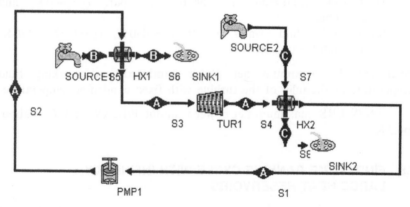

Figure 7.15   Finite-time ideal Rankine cycle T–s diagram.

and the heat source, U_L is the heat-transfer coefficient, and A_L is the heat-transfer surface area of the low-temperature side heat exchanger between the heat engine and the heat sink.

The total heat transfer surface area (A) of the two heat exchangers is assumed to be a constant:

$$A = A_H + A_L \tag{7.35}$$

The power output (P) of the heat engine according to the first law of thermodynamics is

$$P = Q\mathrm{dot}_H - Q\mathrm{dot}_L \tag{7.36}$$

The efficiency (η) of the heat engine is

$$\eta = P/Q\mathrm{dot}_H \tag{7.37}$$

Example 7.10

An endoreversible Rankine steam heat engine with its infinitely large steam heat source and heat sink is shown in Fig. 7.14. The following information is given: $p_1 = 1$ bar, $x_1 = 0$, $m\text{dot}_1 = 1$ kg/sec, $p_3 = 100$ bars, $x_3 = 1$, $p_5 = 200$ bars, $x_5 = 1$, $p_6 = 200$ bars, $x_6 = 0$, $T_5 = 639$ K, $x_5 = 1$, $T_6 = 639$ K, $x_6 = 0$, $p_7 = 0.02$ bar, $x_7 = 0$, $p_8 = 0.02$ bar, and $x_8 = 1$.

Determine the power required by the pump, power produced by the turbine, net power produced by the cycle, rate of heat added by the heat source, rate of heat removed to the heat sink, and cycle efficiency. Optimize the net power produced by the cycle with fixed p_1. Draw the sensitivity diagram of net power versus p_3. Find the maximum net power and p_3 at the maximum net power condition.

To solve this problem by CyclePad, we take the following steps:

1. Build the cycle and its surroundings as shown in Fig. 7.14.

2. Analysis: (a) Assume that the heat exchangers are isobaric, and turbine and pump are isentropic. (b) Input heat source fluid = steam, $T_5 = 639$ K, $x_5 = 1$, $T_6 = 639$ K, and $x_6 = 0$; heat sink: fluid = steam, $p_7 = 0.02$ bar, $x_7 = 0$, $p_7 = 0.02$ bar, and $x_8 = 1$; steam cycle: fluid = water, $x_1 = 0$, $p_1 = 1$ bar, $x_3 = 1$, $p_3 = 200$ bars, and $m\text{dot} = 1$ kg/sec.

3. Display results. The results shown in Fig. 7.16a are: rate of heat added from the heat source = 2297 kW, rate of heat removed to the heat

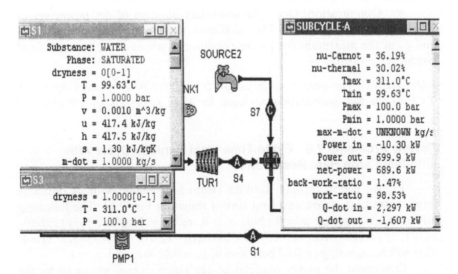

Figure 7.16a Finite-time ideal Rankine cycle with infinitely large heat reservoirs.

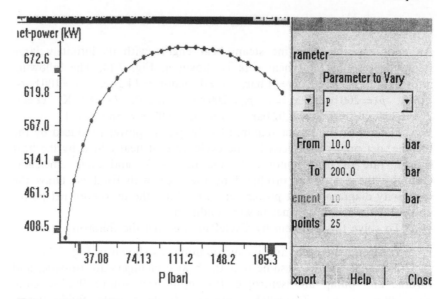

Figure 7.16b   Finite-time ideal Rankine cycle with infinitely large heat reservoirs sensitivity diagram.

sink $= -1507\,kW$, power required by the isentropic pump $= -10.30\,kW$, power produced by the isentropic turbine $= 699.9\,kW$, net power produced $= 689.6\,kW$, and efficiency of the cycle $= 30.02\%$.

   4. Optimization. Draw the sensitivity diagram of net power versus p_3 as shown in Fig. 7.16b. The maximum net power is about $692.4\,kW$, and p_3 at the maximum net power condition is about 116.1 bar with fixed condenser pressure.

COMMENT: The partial optimization is only for $\partial(net\ power)/\partial(p_3) = 0$. To have the full optimization, we must let $\partial(net\ power)/\partial(p_1) = 0$ also.

Review Problems 7.6   Finite Time Ideal Rankine Cycle with Infinitely Large Heat Reservoirs

   1. An endoreversible Rankine steam heat engine with its infinitely large steam heat source and heat sink is shown in Fig. 7.14. The following information is given: $p_1 = 0.1$ bar, $x_1 = 0$, $\dot{m}dot_1 = 1\,kg/sec$, $p_3 = 120$ bars, $x_3 = 1$, $p_5 = 200$ bars, $x_5 = 1$, $p_6 = 200$ bars, $x_6 = 0$, $T_5 = 639\,K$, $x_5 = 1$, $T_6 = 639\,K$, $x_6 = 0$, $p_7 = 0.02$ bar, $x_7 = 0$, $p_8 = 0.02$ bar, and $x_8 = 1$.

   Determine the power required by the pump, power produced by the turbine, net power produced by the cycle, rate of heat added by the heat source, rate of heat removed to the heat sink, and cycle efficiency. Optimize the net power produced by the cycle with fixed p_1. Draw the sensitivity diagram of net power versus p_3. Find the maximum net power and p_3 at the maximum net power condition.

ANSWERS: power required by the pump $= -12.13$ kW, power produced by the turbine $= 948.0$ kW, net power produced by the cycle $= 935.9$ kW, rate of heat added by the heat source $= 2481$ kW, rate of heat removed to the heat sink $= -1545$ kW, and cycle efficiency $= 37.73\%$.

2.   An endoreversible Rankine steam heat engine with its infinitely large steam heat source and heat sink is shown in Fig. 7.14. The following information is given: $p_1 = 0.1$ bar, $x_1 = 0$, $mdot_1 = 1$ kg/sec, $p_3 = 150$ bars, $x_3 = 1$, $p_5 = 200$ bars, $x_5 = 1$, $p_6 = 200$ bars, $x_6 = 0$, $T_5 = 639$ K, $x_5 = 1$, $T_6 = 639$ K, $x_6 = 0$, $p_7 = 0.02$ bar, $x_7 = 0$, $p_8 = 0.02$ bar, and $x_8 = 1$.
Determine the power required by the pump, power produced by the turbine, net power produced by the cycle, rate of heat added by the heat source, rate of heat removed to the heat sink, and cycle efficiency. Optimize the net power produced by the cycle with fixed p_1. Draw the sensitivity diagram of net power versus p_3. Find the maximum net power and p_3 at the maximum net power condition.

ANSWERS: power required by the pump $= -15.5$ kW, power produced by the turbine $= 699.0$ kW, net power produced by the cycle $= 683.5$ kW, rate of heat added by the heat source $= 2177$ kW, rate of heat removed to the heat sink $= -1494$ kW, and cycle efficiency $= 31.40\%$.

7.7   ACTUAL RANKINE CYCLE WITH INFINITELY LARGE HEAT RESERVOIRS

The actual finite-time Rankine cycle is shown in Fig. 7.17. The cycle is an external and internal irreversible cycle that consists of two irreversible internal adiabatic processes (pump and turbine) and two irreversible external isobaric heat-transfer processes. The heat source and heat sink are infinitely large.

Example 7.11

An endoreversible Rankine steam heat engine with its infinitely large steam heat source and heat sink is shown in Fig. 7.17. The following information is given: $\eta_{\text{turbine}} = 85\%$, $\eta_{\text{pump}} = 100\%$, $p_1 = 1$ bar, $x_1 = 0$,

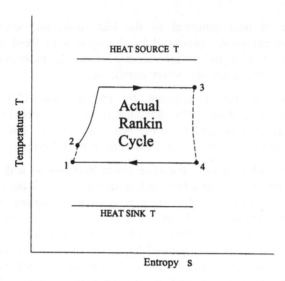

Figure 7.17   Finite-time actual Rankine cycle with infinitely large heat reservoirs.

$mdot_1 = 1$ kg/sec, $p_3 = 100$ bars, $x_3 = 1$, $p_5 = 200$ bars, $x_5 = 1$, $p_6 = 200$ bars, $x_6 = 0$, $p_5 = 639$ K, $x_5 = 1$, $T_6 = 639$ K, $x_6 = 0$, $p_7 = 0.02$ bar, $x_7 = 0$, $p_8 = 0.02$ bar, and $x_8 = 1$.

Determine the power required by the pump, power produced by the turbine, net power produced by the cycle, rate of heat added by the heat source, rate of heat removed to the heat sink, and cycle efficiency. Optimize the net power produced by the cycle with fixed p_1. Draw the sensitivity diagram of net power versus p_3. Find the maximum net power and p_3 at the maximum net power condition.

To solve this problem by CyclePad, we take the following steps:

  1.   Build the cycle and its surroundings as shown in Fig. 7.17.
  2.   Analysis: (a) Assume that the heat exchangers are isobaric, and turbine and pump are isentropic. (b) Input heat source fluid = steam, $T_5 = 639$ K, $x_5 = 1$, $T_6 = 639$ K, and $x_6 = 0$; heat sink: fluid = steam, $p_7 = 0.02$ bar, $x_7 = 0$, $p_7 = 0.02$ bar, $p_8 = 0.02$ bar, and $x_8 = 1$; steam cycle: fluid = water, $x_1 = 0$, $p_1 = 1$ bar, $x_3 = 1$, $p_3 = 200$ bars, and $mdot = 1$ kg/sec.
  3.   Display results. The results shown in Fig. 7.18a are: rate of heat added from the heat source = 2297 kW, rate of heat removed to the heat sink = −1712 kW, power required by the isentropic pump = −10.30 kW, power produced by the isentropic turbine = 594.9 kW, net power produced = 584.6 kW, and efficiency of the cycle = 25.45%.

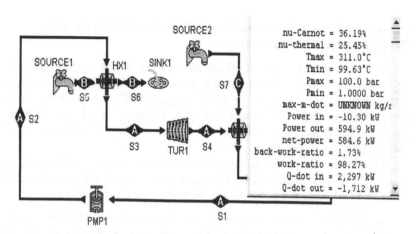

Figure 7.18a Finite-time actual Rankine cycle with infinitely large heat reservoirs.

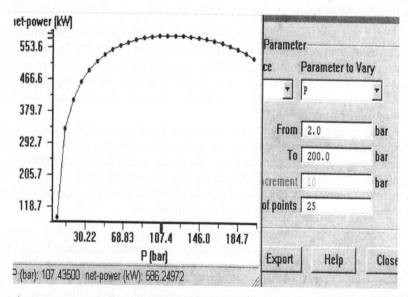

Figure 7.18b Finite-time actual Rankine cycle with infinitely large heat reservoirs sensitivity diagram.

4. Optimization. Draw the sensitivity diagram of net power versus p_3 as shown in Fig. 7.18b. The maximum net power is about 586.2 kW, and p_3 at the maximum net power condition is about 107.4 kPa.

COMMENT: The partial optimization is only for $\partial(\text{net power})/\partial(p_3) = 0$.
To have the full optimization, we must let $\partial(\text{net power})/\partial(p_1) = 0$ also.

Review Problems 7.7 Finite-Time Actual Rankine Cycle with Infinitely Large Heat Reservoirs

1. An endoreversible Rankine steam heat engine with its infinitely large steam heat source and heat sink is shown in Fig. 7.17. The following information is given: $\eta_{turbine} = 85\%$, $\eta_{pump} = 100\%$, $p_1 = 1$ bar, $x_1 = 0$, $mdot_1 = 1$ kg/sec, $p_3 = 150$ bars, $x_3 = 1$, $p_5 = 150$ bars, $x_5 = 1$, $p_6 = 200$ bars, $x_6 = 0$, $p_5 = 639$ K, $x_5 = 1$, $T_6 = 639$ K, $x_6 = 0$, $p_7 = 0.02$ bar, $x_7 = 0$, $p_8 = 0.02$ bar, and $x_8 = 1$.

Determine the power required by the pump, power produced by the turbine, net power produced by the cycle, rate of heat added by the heat source, rate of heat removed to the heat sink, and cycle efficiency.

ANSWERS: power required by the pump $= -15.5$ kW, power produced by the turbine $= 594.2$ kW, net power produced by the cycle $= 578.7$ kW, rate of heat added by the heat source $= 2177$ kW, rate of heat removed to the heat sink $= -1598$ kW, and cycle efficiency $= 26.58\%$.

2. An endoreversible Rankine steam heat engine with its infinitely large steam heat source and heat sink is shown in Fig. 7.17. The following information is given: $\eta_{turbine} = 85\%$, $\eta_{pump} = 100\%$, $p_1 = 0.5$ bar, $x_1 = 0$, $mdot_1 = 1$ kg/sec, $p_3 = 100$ bars, $x_3 = 1$, $p_5 = 150$ bars, $x_5 = 1$, $p_6 = 200$ bars, $x_6 = 0$, $p_5 = 639$ K, $x_5 = 1$, $T_6 = 639$ K, $x_6 = 0$, $p_7 = 0.02$ bar, $x_7 = 0$, $p_8 = 0.02$ bar, and $x_8 = 1$.

Determine the power required by the pump, power produced by the turbine, net power produced by the cycle, rate of heat added by the heat source, rate of heat removed to the heat sink, and cycle efficiency. Optimize the net power produced by the cycle with fixed p_1. Draw the sensitivity diagram of net power versus p_3. Find the maximum net power and p_3 at the maximum net power condition.

ANSWERS: power required by the pump $= -10.27$ kW, power produced by the turbine $= 663.6$ kW, net power produced by the cycle $= 653.3$ kW, rate of heat added by the heat source $= 2374$ kW, rate of heat removed to the heat sink $= -1720$ kW, and cycle efficiency $= 27.52\%$; maximum net power $=$ approximately 653.7 kW and $p_3 = 102.3$ bars at the maximum net power condition.

7.8 IDEAL RANKINE CYCLE WITH FINITE CAPACITY HEAT RESERVOIRS

The ideal *finite-time Rankine cycle* is shown in Fig. 7.19. The cycle is an endoreversible cycle that consists of two isentropic processes and two

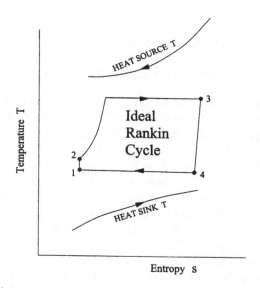

Temperature T

Entropy s

Figure 7.19   Finite-time ideal Rankine cycle with finite heat reservoirs.

isobaric heat-transfer processes. The cycle exchanges heat with its surroundings in the two isobaric external irreversible heat-transfer processes. The heat source and heat sink are not infinitely large. Therefore, the temperature of the heat source and heat sink are unchanged during the heat-transfer processes.

Example 7.12

A finite-time ideal Rankine cycle operates between a finite heat capacity heat source and a finite heat capacity heat sink. The following information is given:

Heat source: fluid = air, $T_5 = 2000°C$, $p_5 = 1$ bar, $T_6 = 800°C$, and $p_6 = 1$ bar

Heat sink: fluid = water, $T_7 = 17°C$, $p_7 = 1$ bar, $T_8 = 30°C$, and $p_8 = 1$ bar

Steam cycle: fluid = water, $x_2 = 0$, $p_3 = 150$ bars, $x_3 = 1$, $p_4 = 1$ bar, and $m$dot $= 1$ kg/sec

Determine the rate of heat added from the heat source, rate of heat removed to the heat sink, power required by the isentropic pump, power produced by the isentropic turbine, net power produced, and efficiency of the cycle.

Optimize the net power produced by the cycle with fixed p_1. Draw the sensitivity diagram of net power versus p_3. Find the maximum net power and p_3 at the maximum net power condition.

To solve this problem by CyclePad, we take the following steps:

1. Build the cycle and its surroundings as shown in Fig. 7.20a.
2. Analysis: (a) Assume that the heat exchangers are isobaric, and turbine and pump are isentropic. (b) Input heat source fluid = air, $T_5 = 2000°C$, $p_5 = 1$ bar, $T_6 = 800°C$, and $p_6 = 1$ bar; heat sink: fluid = water, $T_7 = 17°C$, $p_7 = 1$ bar, $T_8 = 30°C$, and $p_8 = 1$ bar; and steam cycle: fluid = water, $x_2 = 0$, $p_3 = 150$ bars, $x_3 = 1$, $p_4 = 1$ bar, and $mdot = 1$ kg/sec.
3. Display results. The answers are: rate of heat added from the heat source = 2177 kW, rate of heat removed to the heat sink = −1494 kW, power required by the isentropic pump = −15.5 kW, power produced by the isentropic turbine = 699.0 kW, net power produced = 683.5 kW, and efficiency of the cycle = 31.40%. (See Fig. 7.20a.)
4. Optimization. Draw the sensitivity diagram of net power versus p_3 as shown in Fig. 7.20b. The maximum net power is about 695.5 kW, and p_3 at the maximum net power condition is about 103.6 bars.

COMMENT: The partial optimization is only for $\partial(net\ power)/\partial(p_3) = 0$. To have the full optimization, we must let $\partial(net\ power)/\partial(p_1) = 0$ also.

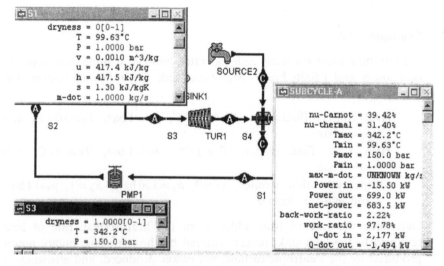

| **Figure 7.20a**   Finite-time ideal Rankine cycle with finite capacity heat reservoirs.

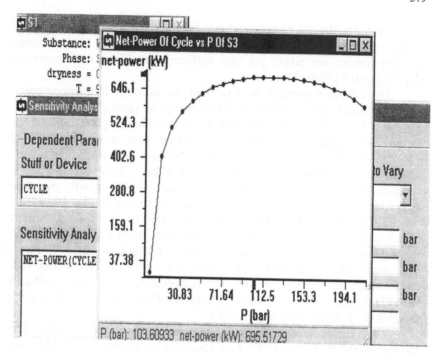

I Figure 7.20b Finite-time ideal Rankine cycle with finite heat reservoirs sensitivity (diagram.

Example 7.13

A finite-time ideal Rankine cycle operates between a finite heat capacity heat source and a finite heat capacity heat sink. The following information is given:

Heat source: fluid = water, $p_5 = 200$ bars, $x_5 = 1$ bar, $x_6 = 0$, and $p_6 = 200$ bars

Heat sink: fluid = water, $x_7 = 0$, $p_7 = 0.02$ bar, $x_8 = 1$, and $p_8 = 0.02$ bar

Steam cycle: fluid = water, $x_1 = 0$, $p_1 = 1$ bar, $x_3 = 1$, $p_3 = 117.6$ bars, and $m$dot $= 1$ kg/sec

Heat exchangers are counter-flow type

Determine the rate of heat added from the heat source, rate of heat removed to the heat sink, power required by the isentropic pump, power produced by the isentropic turbine, net power produced, and efficiency of the cycle.

Optimize the net power produced by the cycle with variable p_3 based on the criterion of (a) net power per unit conductance of heat exchanger, and (b) specific net power per unit surface area of heat exchangers with $U_H = U_H = 0.5 \, kW/[m^2(K)]$.

To solve this problem by CyclePad, we take the following steps:

1. Build the cycle and its surroundings as shown in Fig. 7.19.

2. Analysis: (a) Assume that the heat exchangers are isobaric, and turbine and pump are isentropic. (b) Input heat source fluid = water, $p_5 = 200$ bars, $x_5 = 1$ bar, $x_6 = 0$, and $p_6 = 200$ bars; heat sink: fluid = water, $x_7 = 0$, $p_7 = 0.02$ bar, $x_8 = 1$, and $p_8 = 0.02$ bar; heat exchangers are counter-flow type; and steam cycle: fluid = water, $x_1 = 0$, $p_1 = 1$ bar, $x_3 = 1$, $p_3 = 117.6$ bars, and $m$dot = 1 kg/sec, as shown in Fig. 7.21a.

3. Display results. The answers are: rate of heat added from the heat source $(Q\text{dot}_H) = 2260$ kW, logarithm mean temperature difference of the high-temperature side heat exchanger $(\text{LMTD}_H) = 121.8°C$, rate of heat removed to the heat sink $(Q\text{dot}_L) = -1567$ kW, logarithm mean temperature difference of the low-temperature side heat exchanger $(\text{LMTD}_L) = 82.13°C$, net power produced = 693.0 kW, and efficiency of the cycle = 30.66%. (See Fig. 7.21b.)

The conductances of the heat exchanger are $U_H A_H = 2260/121.8 = 18.56$ kW/K and $U_L A_L = 1567/82.13 = 19.08$ kW/K. The total conductance of the heat exchanger is $18.56 + 19.08 = 37.64$ kW/K. The

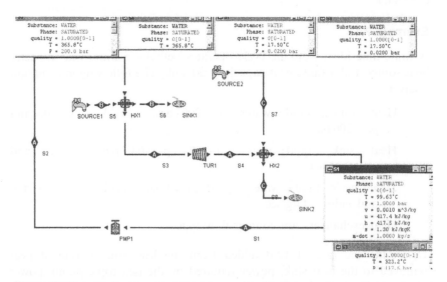

Figure 7.21a Finite-time ideal Rankine cycle with finite heat capacity source and sink input.

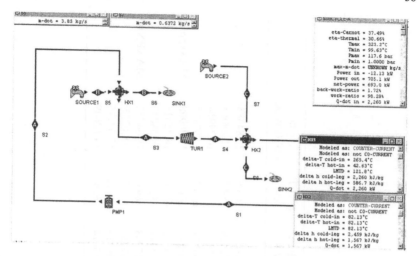

Figure 7.21b Finite-time ideal Rankine cycle with finite heat capacity source and sink output.

net power output per unit conductance of the heat exchanger is $693.0/37.64 = 18.41$ kW/K.

The surface areas of the heat exchanger are $A_H = 18.56/0.5 = 37.12\,m^2$ and $A_L = 19.08/0.5 = 38.16\,m^2$. The total surface area of the heat exchanger is $37.12 + 38.16 = 75.28\,m^2$. The specific power per unit total surface area of the heat exchanger is $693.0/75.28 = 9.206\,m^2$.

Using p_3 as a design parameter, Table 7.2 is prepared. Based on the criterion of (1) net power per unit conductance of heat exchanger, the optimization $p_3 = 120$ bars, and (2) specific net power per unit surface area of heat exchanger, the optimization $p_3 = 80$ bars. (See Fig. 7.21c.)

Example 7.14

A finite-time ideal Rankine OTEC (ocean thermal energy conversion) cycle as shown in Fig. 7.22a operates between a finite heat capacity heat source and a finite heat capacity heat sink. The following information is given:

Heat source: fluid = warm ocean surface water, $T_1 = 26°C$, $p_1 = 101$ kPa, $T_2 = 22°C$, and $p_2 = 101$ kPa
Heat sink: fluid = cold deep ocean water, $T_3 = 5°C$, $p_3 = 101$ kPa, $T_4 = 9°C$, and $p_4 = 101$ kPa
Steam cycle: fluid = ammonia, $x_5 = 0$, $T_5 = 12°C$, $x_7 = 1$, $T_7 = 20°C$, and $m\text{dot} = 1$ kg/sec.
The heat exchangers are counter-flow type, $U_H = 0.4\,kJ/(m^2)K$ and $U_L = 0.4\,kJ/(m^2)K$

Table 7.2 Net Power Per Unit Conductance of Heat Exchanger

p_3 bar	$LMTD_H$ K	$LMTD_L$ K	$QDOT_H$ kW	$QDOT_L$ kW	$U_H A_H$ kW/K	$U_L A_L$ kW/K	$SUM(U_A)$ kW/K	PNET kW	sppnet kW/(kW/K)	
2	254.4	82.13	2277	2171	8.950472	26.4337	35.38417	105.6	2.984385	
10	223.6	82.13	2359	1969	10.55009	23.97419	34.52428	389.8	11.29061	
30	191.2	82.13	2383	1820	12.46339	22.15999	34.62338	562.6	16.24914	
50	170.9	82.13	2371	1741	13.87361	21.1981	35.07171	630.3	17.97175	
80	147.3	82.13	2332	1655	15.83164	20.15098	35.98262	676.8	**18.80908**	MAXsppnet
100	133.5	82.13	2297	1607	17.20599	19.56654	36.77253	689.6	18.75313	
110	126.8	82.13	2277	1584	17.95741	19.2865	37.24391	692.3	18.58827	
120	120.2	82.13	2255	1562	18.7604	19.01863	37.77903	**693**	18.34351	MAXPNET
130	113.6	82.13	2231	1539	19.63908	18.73859	38.37767	691.7	18.0235	
150	99.86	82.13	2177	1494	21.80052	18.19067	39.99119	683.5	17.09126	
180	75.15	82.13	2073	1418	27.58483	17.26531	44.85014	656	14.62649	

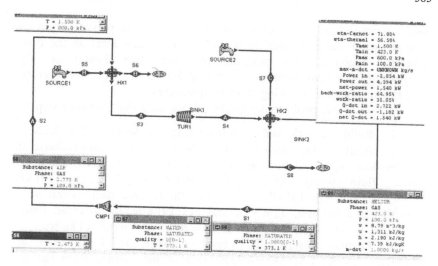

Figure 7.21c Net power per unit conductance of heat exchangers.

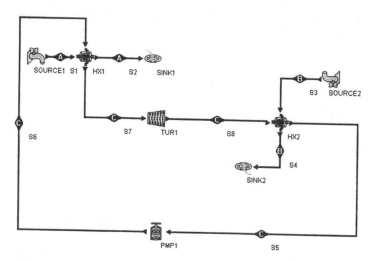

Figure 7.22a Finite-time OTEC cycle.

Determine the rate of heat added from the heat source, rate of heat removed to the heat sink, power required by the isentropic pump, power produced by the isentropic turbine, net power produced, and efficiency of the cycle.

Since the fuel cost of the OTEC is free, the primary cost is that of initial construction. The heat exchangers are the major concern of the initial construction cost. Let us take the specific net power output (net output power per unit total heat exchanger surface area) as the design objective function, and optimize the warm-side (heater or high-temperature side heat exchanger) and cold-side (cooler or low-temperature side heat exchanger) working fluid temperatures.

To solve this problem by CyclePad, we take the following steps:

1. Build the cycle and its surroundings as shown in Fig. 7.22a.

2. Analysis: (a) Assume that the heat exchangers are isobaric and counter-flow type, and turbine and pump are isentropic. (b) Input heat source: fluid = warm ocean surface water, $T_1 = 26°C$, $p_1 = 101$ kPa, $T_2 = 22°C$, and $p_2 = 101$ kPa; heat sink: fluid = cold deep ocean water, $T_3 = 5°C$, $p_3 = 101$ kPa, $T_4 = 9°C$, and $p_4 = 101$ kPa; steam cycle: fluid = ammonia, $x_5 = 0$, $T_5 = 12°C$, $x_7 = 1$, $T_7 = 20°C$, and $mdot = 1$ kg/sec.

3. Display results. The results shown in Fig. 7.22b are: $LMTD_H = 7.56$ K, $LMTD_L = 4.72$ K, rate of heat added from the heat source = 1219 kW, rate of heat removed to the heat sink = −1191 kW, power required by the isentropic pump = −4.38 kW, power produced by the isentropic turbine = 33.20 kW, net power produced = 28.82 kW, and efficiency of the cycle = 2.36%.

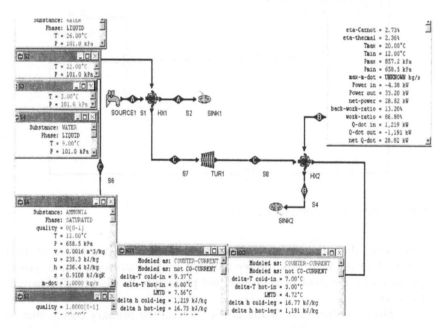

Figure 7.22b   Finite-time OTEC cycle input and output.

4. Calculate $U_H A_H = Q_H/\text{LMTD}_H = 1219/7.56 = 161.2\,\text{kW/K}$, $U_L A_L = Q_L/\text{LMTD}_L = 1191/4.72 = 252.3\,\text{kW/K}$, $U_H A_H + U_L A_L = 161.2 + 252.3 = 413.5\,\text{kW/K}$, and specific net power output $= \dot{W}\text{dot}_\text{net}/(A_H + A_L) = 28.82/(161.2/0.4 + 252.3/0.4) = 0.1742\,\text{kW/m}^2$.

5. To optimize the specific power output of the cycle, we let $\partial(\text{specific power output})/\partial T_7 = 0$ first and then $\partial(\text{specific power output})/\partial T_7 = 0$. It is seen that $\partial(\text{specific power output})/\partial T_7 = 0$ occurs at $T_7 = 24°\text{C}$ as shown in Table 7.3a.

6. Then let $\partial(\text{specific power output})/\partial T_5 = 0$. It is seen that $\partial(\text{specific power output})/\partial T_5 = 0$ occurs at $T_5 = 12°\text{C}$ as shown in Table 7.3b.

7. Let $T_5 = 12°\text{C}$ and $T_7 = 24°\text{C}$; the optimized specific power output of the cycle is $0.1036\,\text{kW/m}^2$. At the maximum optimized specific power output condition, $\text{LMTD}_H = 4.67\,\text{K}$, $\text{LMTD}_H = 4.72\,\text{K}$, rate of heat added from the heat source $= 1220\,\text{kW}$, rate of heat removed to the heat sink $= -1178\,\text{kW}$, power required by the isentropic pump $= -6.58\,\text{kW}$, power produced by the isentropic turbine $= 48.91\,\text{kW}$, net power produced $= 42.33\,\text{kW}$ and efficiency of the cycle $= 3.47\%$, as shown in Fig. 7.22c.

Review Problems 7.8 Finite-Time Ideal Rankine Cycle with Finite Heat Capacity Reservoirs

1. An endoreversible (Curzon and Ahlborn) steam cycle operates between a finite heat capacity heat source and a finite heat capacity heat sink. The following information is given:

Heat source: fluid $=$ air, $T_5 = 2000°\text{C}$, $p_5 = 1\,\text{bar}$, $T_6 = 800°\text{C}$, and $p_6 = 1\,\text{bar}$
Heat sink: fluid $=$ water, $T_7 = 17°\text{C}$, $p_7 = 1\,\text{bar}$, $T_8 = °\text{C}$, and $p_8 = 1\,\text{bar}$
Steam cycle: fluid $=$ water, $x_2 = 0$, $p_3 = 200\,\text{bars}$, $x_3 = 1$, $p_4 = 1\,\text{bar}$, and $m\text{dot} = 1\,\text{kg/sec}$

Determine the rate of heat added from the heat source, rate of heat removed to the heat sink, power required by the isentropic pump, power produced by the isentropic turbine, net power produced, and efficiency of the cycle.

ANSWERS: rate of heat added from the heat source $= 1975\,\text{kW}$, rate of heat removed to the heat sink $= -1353\,\text{kW}$, power required by the isentropic pump $= -20.68\,\text{kW}$, power produced by the isentropic turbine $= 642.7\,\text{kW}$, net power produced $= 622.0\,\text{kW}$, and efficiency of the cycle $= 31.49\%$.

Table 7.3a Specific Power Optimization with Respect to T_7

OTEC	$T_1=26$	$T_2=22$	$T_3=5$	$T_4=9$					Counter Flow HX				
12	12.1	0.001028	0.001285	0.0304	−0.0561	0.4267	0.3706	1217	−1217	11.84	4.72	102.7872	257.839
12	14	0.020189	0.025236	0.611	−1.09	8.53	7.44	1218	−1211	10.88	4.72	111.9485	256.5678
12	16	0.038816	0.04852	1.21	−2.21	16.92	14.71	1219	−1204	9.84	4.72	123.8821	255.0847
12	18	0.055613	0.069516	1.79	−3.28	25.14	21.86	1219	−1197	8.74	4.72	139.4737	253.6017
12	20	0.069685	0.087107	2.36	−4.38	33.2	28.82	1219	−1191	7.56	4.72	161.2434	252.3305
12	21.9	0.07957	0.099463	2.9	−5.42	40.73	35.32	1220	−1184	6.32	4.72	193.038	250.8475
12	22	0.079947	0.099933	2.92	−5.47	41.13	35.66	1220	−1184	6.25	4.72	195.2	250.8475
12	23	0.082671	0.103339	3.2	−6.03	45.02	38.99	1220	−1181	5.51	4.72	221.4156	250.2119
12	24	0.082867	0.103584	3.47	−6.58	48.91	42.33	1220	−1178	4.67	4.72	261.242	249.5763
12	24.5	0.081377	0.101722	3.6	−6.86	50.83	43.97	1220	−1176	4.19	4.72	291.1695	249.1525
12	24.6	0.080922	0.101152	3.63	−6.91	51.21	44.3	1220	−1176	4.09	4.72	298.2885	249.1525
12	24.7	0.080328	0.10041	3.66	−6.97	51.59	44.62	1220	−1175	3.98	4.72	306.5327	248.9407
12	24.8	0.079672	0.09959	3.68	−7.02	51.97	44.95	1220	−1175	3.87	4.72	315.2455	248.9407

Table 7.3b Specific Power Optimization with Respect to T_5

OTEC	$T_1=26$	$T_2=22$	$T_3=5$	$T_4=9$					Counter Flow HX				
12	12.1	0.001028	0.001285	0.0304	−0.0561	0.4267	0.3706	1217	−1217	11.84	4.72	102.7872	257.839
12	14	0.020189	0.025236	0.611	−1.09	8.53	7.44	1218	−1211	10.88	4.72	111.9485	256.5678
12	16	0.038816	0.04852	1.21	−2.21	16.92	14.71	1219	−1204	9.84	4.72	123.8821	255.0847
12	18	0.055613	0.069516	1.79	−3.28	25.14	21.86	1219	−1197	8.74	4.72	139.4737	253.6017
12	20	0.069685	0.087107	2.36	−4.38	33.2	28.82	1219	−1191	7.56	4.72	161.2434	252.3305
12	21.9	0.07957	0.099463	2.9	−5.42	40.73	35.32	1220	−1184	6.32	4.72	193.038	250.8475
12	22	0.079947	0.099933	2.92	−5.47	41.13	35.66	1220	−1184	6.25	4.72	195.2	250.8475
12	23	0.082671	0.103339	3.2	−6.03	45.02	38.99	1220	−1181	5.51	4.72	221.4156	250.2119
12	**24**	**0.082867**	**0.103584**	**3.47**	**−6.58**	**48.91**	**42.33**	**1220**	**−1178**	**4.67**	**4.72**	**261.242**	**249.5763**
12	24.5	0.081377	0.101722	3.6	−6.86	50.83	43.97	1220	−1176	4.19	4.72	291.1695	249.1525
12	24.6	0.080922	0.101152	3.63	−6.91	51.21	44.3	1220	−1176	4.09	4.72	298.2885	249.1525
12	24.7	0.080328	0.10041	3.66	−6.97	51.59	44.62	1220	−1175	3.98	4.72	306.5327	248.9407
12	24.8	0.079672	0.09959	3.68	−7.02	51.97	44.95	1220	−1175	3.87	4.72	315.2455	248.9407
9.1	24	0.040529	0.050661	4.32	−7.8	61.09	53.29	1232	−1179	5.52	1.08	223.1884	1091.667
9.5	24	0.059103	0.073878	4.21	−7.64	59.4	51.76	1231	−1179	5.4	1.82	227.963	647.8022
10	24	0.0705	0.088125	4.06	−7.44	57.28	49.84	1228	−1179	5.26	2.49	233.4601	473.494
11	24	0.080891	0.101114	3.77	−7.01	53.11	46.1	1224	−1178	4.97	3.64	246.2777	323.6264
12	**24**	**0.082867**	**0.103584**	**3.47**	**−6.58**	**48.91**	**42.33**	**1220**	**−1178**	**4.67**	**4.72**	**261.242**	**249.5763**
15	24	0.066312	0.082889	2.6	−5.14	36.5	31.36	1207	−1176	3.74	7.83	322.7273	150.1916
18	24	0.037198	0.046497	1.72	−3.59	24.18	20.6	1195	−1174	2.68	10.88	445.8955	107.9044
20	24	0.018341	0.022927	1.15	−2.45	16.08	13.62	1186	−1173	1.82	12.9	651.6484	90.93023
21	24	0.010039	0.012549	0.8623	−1.86	12.05	10.19	1182	−1172	1.27	13.9	930.7087	84.31655
21.5	24	0.005931	0.007414	0.7177	−1.57	10.03	8.47	1180	−1171	0.8762	14.41	1346.724	81.26301

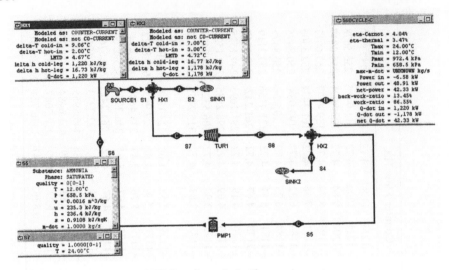

Figure 7.22c   Finite-time OTEC cycle optimization.

2. An endoreversible (Curzon and Ahlborn) steam cycle operates between a finite heat capacity heat source and a finite heat capacity heat sink. The following information is given:

Heat source: fluid = air, $T_5 = 2000°C$, $p_5 = 1\,bar$, $T_6 = 800°C$, and $p_6 = 1\,bar$
Heat sink: fluid = water, $T_7 = 17°C$, $p_7 = 1\,bar$, $T_8 = °C$, and $p_8 = 1\,bar$
Steam cycle: fluid = water, $x_2 = 0$, $p_3 = 200\,bars$, $x_3 = 1$, $p_4 = 0.2\,bar$, and $mdot = 1\,kg/sec$

Determine the rate of heat added from the heat source, rate of heat removed to the heat sink, power required by the isentropic pump, power produced by the isentropic turbine, net power produced, and efficiency of the cycle.

ANSWERS: rate of heat added from the heat source = 2141 kW, rate of heat removed to the heat sink = −1366 kW, power required by the isentropic pump = −20.25 kW, power produced by the isentropic turbine = 795.5 kW, net power produced = 775.2 kW, and efficiency of the cycle = 36.20%.

3. An endoreversible (Curzon and Ahlborn) steam cycle operates between a finite heat capacity heat source and a finite heat capacity heat sink. The following information is given:

Heat source: fluid = air, $T_5 = 2000°C$, $p_5 = 1\,bar$, $T_6 = 800°C$, and $p_6 = 1\,bar$

Heat sink: fluid = water, $T_7 = 17°C$, $p_7 = 1$ bar, $T_8 = °C$, and $p_8 = 1$ bar
Steam cycle: fluid = water, $x_2 = 0$, $p_3 = 200$ bars, $x_3 = 1$, $p_4 = 0.2$ bar, and $m$dot = 1 kg/sec

Optimize the net power produced by the cycle with fixed p_4. Draw the sensitivity diagram of net power versus p_3. Find the maximum net power and p_3 at the maximum net power condition.

ANSWERS: The maximum net power is about 691.7 kW, and p_3 at the maximum net power condition is about 103.6 bars.

4. A finite-time ideal Rankine OTEC cycle as shown in Fig. 7.22a operates between a finite heat capacity heat source and a finite heat capacity heat sink. The following information is given:

Heat source: fluid = warm ocean surface water, $T_1 = 26°C$, $p_1 = 101$ kPa, $T_2 = 20°C$, and $p_2 = 101$ kPa
Heat sink: fluid = cold deep ocean water, $T_3 = 5°C$, $p_3 = 101$ kPa, $T_4 = 10°C$, and $p_4 = 101$ kPa
Steam cycle: fluid = ammonia, $x_5 = 0$, $T_5 = 12°C$, $x_7 = 1$, $T_7 = 20°C$, and $m$dot = 1 kg/sec
The heat exchangers are counter-flow type, $U_H = 0.4$ kJ/(m²)K and $U_L = 0.4$ kJ/(m²)K

Determine the rate of heat added from the heat source, rate of heat removed to the heat sink, power required by the isentropic pump, power produced by the isentropic turbine, net power produced, and efficiency of the cycle.

Since the fuel cost of the OTEC is free, the primary cost is that of initial construction. The heat exchangers are the major concern of the initial construction cost. Let us take the specific net power output (net output power per unit total heat exchanger surface area) as the design objective function, and optimize the warm-side (heater or high-temperature side heat exchanger) and cold-side (cooler or low-temperature side heat exchanger) working fluid temperatures.

5. A finite time ideal Rankine OTEC cycle as shown in Fig. 7.22a operates between a finite heat capacity heat source and a finite heat capacity heat sink. The following information is given:

Heat source: fluid = warm ocean surface water, $T_1 = 26°C$, $p_1 = 101$ kPa, $T_2 = 20°C$, and $p_2 = 101$ kPa
Heat sink: fluid = cold deep ocean water, $T_3 = 5°C$, $p_3 = 101$ kPa, $T_4 = 10°C$, and $p_4 = 101$ kPa
Steam cycle: fluid = ammonia, $x_5 = 0$, $T_5 = 12°C$, $x_7 = 1$, $T_7 = 20°C$, and $m$dot = 1 kg/sec

The heat exchangers are co-current (parallel) flow type, $U_H = 0.4\,\text{kJ}/(\text{m}^2)\text{K}$ and $U_L = 0.4\,\text{kJ}/(\text{m}^2)\text{K}$

Determine the rate of heat added from the heat source, rate of heat removed to the heat sink, power required by the isentropic pump, power produced by the isentropic turbine, net power produced, and efficiency of the cycle.

Since the fuel cost of the OTEC is free, the primary cost is that of initial construction. The heat exchanger is the major concern of the initial construction cost. Let us take the specific net power output (net output power per unit total heat exchanger surface area) as the design objective function, and optimize the warm-side (heater or high-temperature side heat exchanger) and cold-side (cooler or low-temperature side heat exchanger) working fluid temperatures.

7.9 ACTUAL RANKINE CYCLE WITH FINITE CAPACITY HEAT RESERVOIRS

The actual finite time Rankine cycle is shown in Fig. 7.23. The cycle is an actual Rankine cycle that consists of two adiabatic processes and two isobaric heat-transfer processes. The cycle exchanges heat with its surroundings in the two isobaric external irreversible heat-transfer processes.

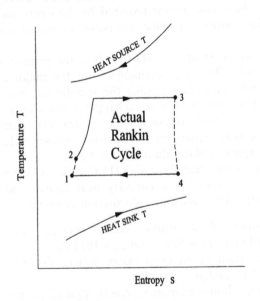

Figure 7.23 Finite-time actual Rankine cycle with finite heat reservoirs.

The heat source and heat sink are not infinitely large. Therefore, the temperature of the heat source and heat sink change during the heat-transfer processes.

Example 7.15

A finite-time actual Rankine cycle operates between a finite heat capacity heat source and a finite heat capacity heat sink. The following information is given:

> Heat source: fluid = air, $T_5 = 2000°C$, $p_5 = 1$ bar, $T_6 = 800°C$, and $p_6 = 1$ bar
>
> Heat sink: fluid = water, $T_7 = 17°C$, $p_7 = 1$ bar, $T_8 = 30°C$, and $p_8 = 1$ bar
>
> Steam cycle: fluid = water, $x_2 = 0$, $p_3 = 200$ bars, $x_3 = 1$, $p_4 = 0.2$ bar, and $mdot = 1$ kg/sec, $\eta_{turbine} = 85\%$

Determine the rate of heat added from the heat source, rate of heat removed to the heat sink, power required by the isentropic pump, power produced by the isentropic turbine, net power produced, and efficiency of the cycle.

Optimize the net power produced by the cycle with fixed p_1. Draw the sensitivity diagram of net power versus p_3. Find the maximum net power and p_3 at the maximum net power condition.

To solve this problem by CyclePad, we take the following steps:

1. Build the cycle and its surroundings as shown in Fig. 7.23.

2. Analysis: (a) Assume that the heat exchangers are isobaric, and turbine and pump are isentropic. (b) Input heat source fluid = air, $T_5 = 2000°C$, $p_5 = 1$ bar, $T_6 = 800°C$, and $p_6 = 1$ bar; heat sink: fluid = water, $T_7 = 17°C$, $p_7 = 1$ bar, $T_8 = 30°C$, and $p_8 = 1$ bar; and steam cycle: fluid = water, $x_2 = 0$, $p_3 = 200$ bars, $x_3 = 1$, $p_4 = 0.2$ bar, and $mdot = 1$ kg/ sec, $\eta_{turbine} = 85\%$.

3. Display results. The answers are: rate of heat added from the heat source = 2141 kW, rate of heat removed to the heat sink = −1486 kW, power required by the isentropic pump = −20.25 kW, power produced by the turbine = 676.1 kW, net power produced = 655.9 kW, and efficiency of the cycle = 30.63%. (See Fig. 7.24a.)

4. Optimization. Draw the sensitivity diagram of net power versus p_3 as shown in Fig. 7.24b. The maximum net power is about 733.8 kW, and p_3 at the maximum net power condition is about 90.69 bars.

COMMENT: The partial optimization is only for $\partial(net\ power)/\partial(p_3) = 0$. To have the full optimization, we must let $\partial(net\ power)/\partial(p_1) = 0$ also.

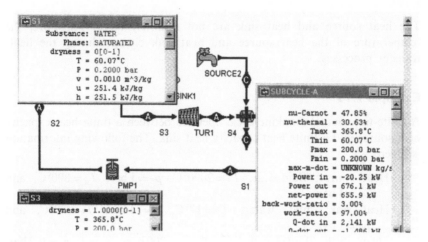

Figure 7.24a   Finite-time actual Rankine cycle with finite capacity heat reservoirs.

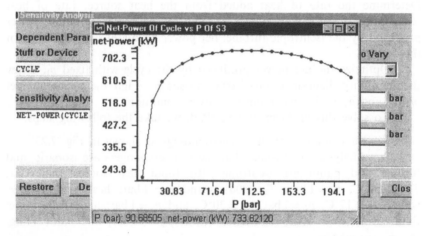

Figure 7.24b   Finite-time actual Rankine cycle with finite heat reservoirs sensitivity diagram.

Example 7.16

A finite time actual Rankine cycle operates between a finite heat capacity heat source and a finite heat capacity heat sink. The following information is given:

Heat source: fluid = air, $T_5 = 2000°C$, $p_5 = 1$ bar, $T_6 = 800°C$, and $p_6 = 1$ bar

Heat sink: fluid = water, $T_7 = 17°C$, $p_7 = 1$ bar, $T_8 = 30°C$, and $p_8 = 1$ bar

Steam cycle: fluid = water, $x_2 = 0$, $p_3 = 150$ bars, $T_8 = 400°C$ (superheated vapor), $p_4 = 0.1$ bar, $m$dot = 1 kg/sec, and $\eta_{turbine} = 85\%$

Determine the rate of heat added from the heat source, rate of heat removed to the heat sink, power required by the isentropic pump, power produced by the isentropic turbine, net power produced, and efficiency of the cycle.

Optimize the net power produced by the cycle with fixed p_1. Draw the sensitivity diagram of net power versus p_3. Find the maximum net power and p_3 at the maximum net power condition.

To solve this problem by CyclePad, we take the following steps:

1. Build the cycle and its surroundings as shown in Fig. 7.25a.
2. Analysis: (a) Assume that the heat exchangers are isobaric, and turbine and pump are isentropic. (b) Input heat source fluid = air, $T_5 = 2000°C$, $p_5 = 1$ bar, $T_6 = 800°C$, and $p_6 = 1$ bar; heat sink: fluid = water, $T_7 = 17°C$, $p_7 = 1$ bar, $T_8 = 30°C$, and $p_8 = 1$ bar; and steam cycle: fluid = water, $x_2 = 0$, $p_3 = 150$ bars, $T_8 = 400°C$ (superheated vapor), $p_4 = 0.1$ bar, and $m$dot = 1 kg/sec, and $\eta_{turbine} = 85\%$.
3. Display results. The answers are: rate of heat added from the heat source = 2768 kW, rate of heat removed to the heat sink = −1836 kW, power required by the pump = −15.14 kW, power produced by the turbine = 947.3 kW, net power produced = 932.1 kW, and efficiency of the cycle = 33.68%. (See Fig. 7.25a.)

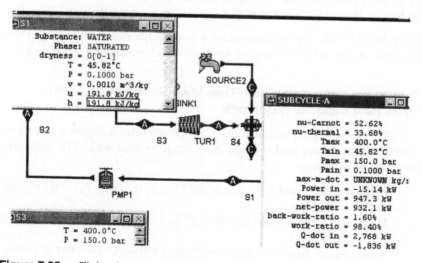

Figure 7.25a Finite-time actual Rankine cycle with finite capacity heat reservoirs.

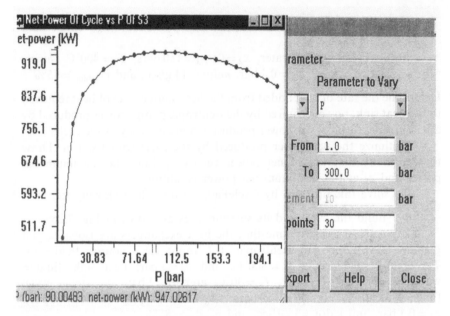

Figure 7.25b   Finite-time actual Rankine cycle with finite heat reservoirs sensitivity diagram.

4.   Optimization. Draw the sensitivity diagram of net power versus p_3 as shown in Fig. 7.25b. The maximum net power is about 947.0 kW, and p_3 at the maximum net power condition is about 90.0 bars.

COMMENT: The partial optimization is only for $\partial(net\ power)/\partial(p_3) = 0$. To have the full optimization, we must let $\partial(net\ power)/\partial(p_1) = 0$ also.

Review Problems 7.9   Finite-Time Ideal Rankine Cycle with Finite Heat Capacity Reservoirs

1.   A finite-time actual Rankine cycle operates between a finite heat capacity heat source and a finite heat capacity heat sink. The following information is given:

Heat source: fluid = air, $T_5 = 2000°C$, $p_5 = 1$ bar, $T_6 = 800°C$, and $p_6 = 1$ bar
Heat sink: fluid = water, $T_7 = 17°C$, $p_7 = 1$ bar, $T_8 = 30°C$, and $p_8 = 1$ bar
Steam cycle: fluid = water, $x_2 = 0$, $p_3 = 120$ bars, $T_8 = 400°C$ (superheated vapor), $p_4 = 0.1$ bar, $\dot{m} = 1$ kg/sec, and $\eta_{turbine} = 85\%$

Determine the rate of heat added from the heat source, rate of heat removed to the heat sink, power required by the isentropic pump, power produced by the turbine, net power produced, and efficiency of the cycle.

ANSWERS: rate of heat added from the heat source $= 2847\,\text{kW}$, rate of heat removed to the heat sink $= -1900\,\text{kW}$, power required by the pump $= -12.13\,\text{kW}$, power produced by the turbine $= 959.2\,\text{kW}$, net power produced $= 947.1\,\text{kW}$, and efficiency of the cycle $= 33.27\%$.

2. A finite-time actual Rankine cycle operates between a finite heat capacity heat source and a finite heat capacity heat sink. The following information is given:

Heat source: fluid $=$ air, $T_5 = 2000°\text{C}$, $p_5 = 1\,\text{bar}$, $T_6 = 800°\text{C}$, and $p_6 = 1\,\text{bar}$

Heat sink: fluid $=$ water, $T_7 = 17°\text{C}$, $p_7 = 1\,\text{bar}$, $T_8 = 30°\text{C}$, and $p_8 = 1\,\text{bar}$

Steam cycle: fluid $=$ water, $x_2 = 0$, $p_3 = 120\,\text{bars}$, $T_8 = 400°\text{C}$ (superheated vapor), $p_4 = 0.5\,\text{bar}$, $m\text{dot} = 1\,\text{kg/sec}$, and $\eta_{\text{turbine}} = 85\%$

Determine the rate of heat added from the heat source, rate of heat removed to the heat sink, power required by the isentropic pump, power produced by the turbine, net power produced, and efficiency of the cycle.

ANSWERS: rate of heat added from the heat source $= 2698\,\text{kW}$, rate of heat removed to the heat sink $= -1908\,\text{kW}$, power required by the pump $= -12.32\,\text{kW}$, power produced by the turbine $= 802.2\,\text{kW}$, net power produced $= 789.9\,\text{kW}$, and efficiency of the cycle $= 29.28\%$.

3. A finite-time actual Rankine cycle operates between a finite heat capacity heat source and a finite heat capacity heat sink. The following information is given:

Heat source: fluid $=$ air, $T_5 = 2000°\text{C}$, $p_5 = 1\,\text{bar}$, $T_6 = 800°\text{C}$, and $p_6 = 1\,\text{bar}$

Heat sink: fluid $=$ water, $T_7 = 17°\text{C}$, $p_7 = 1\,\text{bar}$, $T_8 = 30°\text{C}$, and $p_8 = 1\,\text{bar}$

Steam cycle: fluid $=$ water, $x_2 = 0$, $p_3 = 120\,\text{bars}$, $T_8 = 400°\text{C}$ (superheated vapor), $p_4 = 0.5\,\text{bar}$, $m\text{dot} = 1\,\text{kg/sec}$, and $\eta_{\text{turbine}} = 85\%$

Optimize the net power produced by the cycle with fixed p_1. Draw the sensitivity diagram of net power versus p_3. Find the maximum net power and p_3 at the maximum net power condition.

ANSWERS: The maximum net power is about $789.4\,\text{kW}$, and p_3 at the maximum net power condition is about $100.9\,\text{bars}$.

4.   A finite-time actual Rankine OTEC cycle as shown in Fig. 7.25a operates between a finite heat capacity heat source and a finite heat capacity heat sink. The following information is given:

Pump efficiency $= 85\%$ and turbine efficiency $= 85\%$

Heat   source: fluid $=$ warm   ocean   surface   water, $T_1 = 26°C$, $p_1 = 101$ kPa, $T_2 = 20°C$, and $p_2 = 101$ kPa

Heat sink: fluid $=$ cold deep ocean water, $T_3 = 5°C$, $p_3 = 101$ kPa, $T_4 = 10°C$, and $p_4 = 101$ kPa

Steam cycle: fluid $=$ ammonia, $x_5 = 0$, $T_5 = 12°C$, $x_7 = 1$, $T_7 = 20°C$, and $m$dot $= 1$ kg/sec

The heat exchangers are counter-flow type, $U_H = 0.4$ kJ/(m^2)K and $U_L = 0.4$ kJ/(m^2)K

Determine the rate of heat added from the heat source, rate of heat removed to the heat sink, power required by the isentropic pump, power produced by the isentropic turbine, net power produced, and efficiency of the cycle.

Since the fuel cost of the OTEC is free, the primary cost is that of the initial construction cost. The heat exchangers are the major concern of the initial construction cost. Let us take the specific net power output (net output power per unit total heat exchanger surface area) as the design objective function, and optimize the warm-side (heater or high-temperature side heat exchanger) and cold-side (cooler or low-temperature side heat exchanger) working fluid temperatures.

4.   A finite-time actual Rankine OTEC cycle as shown in Fig. 7.25a operates between a finite heat capacity heat source and a finite heat capacity heat sink. The following information is given:

Pump efficiency $= 85\%$ and turbine efficiency $= 85\%$

Heat   source: fluid $=$ warm   ocean   surface   water, $T_1 = 26°C$, $p_1 = 101$ kPa, $T_2 = 20°C$, and $p_2 = 101$ kPa

Heat sink: fluid $=$ cold deep ocean water, $T_3 = 5°C$, $p_3 = 101$ kPa, $T_4 = 10°C$, and $p_4 = 101$ kPa

Steam cycle: fluid $=$ ammonia, $x_5 = 0$, $T_5 = 12°C$, $x_7 = 1$, $T_7 = 20°C$, and $m$dot $= 1$ kg/sec

The heat exchangers are co-current (parallel) flow type, $U_H = 0.4$ kJ/(m^2)K and $U_L = 0.4$ kJ/(m^2)K

Determine the rate of heat added from the heat source, rate of heat removed to the heat sink, power required by the isentropic pump, power produced by the isentropic turbine, net power produced and efficiency of the cycle.

Since the fuel cost of the OTEC is free, the primary cost is that of the initial construction. The heat exchanger is the major concern of the initial construction cost. Let us take the specific net power output (net output power per unit total heat exchanger surface area) as the design objective function, optimize the warm-side (heater or high-temperature side heat exchanger) and cold-side (cooler or low-temperature side heat exchanger) working fluid temperatures.

7.10 FINITE-TIME BRAYTON CYCLE

The schematic and T–s diagrams of the ideal *finite-time Brayton cycle* are shown in Fig. 7.26 and 7.27. The cycle is an endoreversible cycle that consists of two isentropic processes and two isobaric heat-transfer processes. The cycle exchanges heat with its surroundings in the two isobaric external irreversible heat-transfer processes. By taking into account the rates of heat transfer associated with the cycle, the upper bound of the power output of the cycle can be found as illustrated in Example 7.17.

Example 7.17

A finite-time ideal Brayton cycle operates between a heat source and a heat sink. The following information is given:

Heat source: fluid = air, $T_5 = 2773$ K, $p_5 = 100$ kPa, $T_6 = 2473$ K, and $p_6 = 100$ kPa

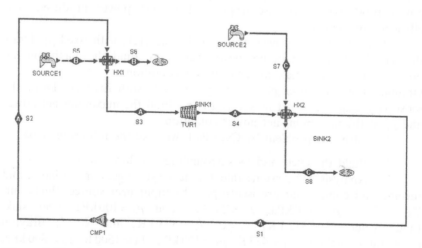

Figure 7.26 Schematic diagram of the ideal finite-time Brayton cycle.

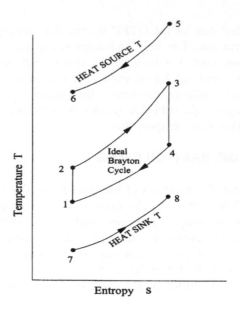

Figure 7.27 T–s diagrams of the ideal finite-time Brayton cycle.

Heat sink: fluid = water, $T_7 = 373.1$ K, $x_7 = 0$, $T_8 = 373.1$ K, and $x_8 = 1$
Brayton cycle: fluid = helium, $T_1 = 423$ K, $p_1 = 100$ kPa, $T_3 = 1500$ K, $p_3 = 800$ kPa, and $m$dot = 1 kg/sec

Determine the rate of heat added from the heat source, rate of heat removed to the heat sink, power required by the isentropic compressor, power produced by the isentropic turbine, net power produced, and efficiency of the cycle.

Optimize the net power produced by the cycle with fixed p_1. Draw the sensitivity diagram of net power versus p_3. Find the maximum net power and p_3 at the maximum net power condition.

Optimize the net power produced by the cycle with fixed p_3. Draw the sensitivity diagram of net power versus p_1. Find the maximum net power and p_1 at the maximum net power condition.

To solve this problem by CyclePad, we take the following steps:

1.   Build the cycle and its surroundings as shown in Fig. 7.26.

2.   Analysis: (a) Assume that the heat exchangers are isobaric, and turbine and compressor are isentropic. (b) Input heat source: fluid = air, $T_5 = 2773$ K, $p_5 = 100$ kPa, $T_6 = 2473$ K, and $p_6 = 100$ kPa; heat sink: fluid = water, $T_7 = 373.1$ K, $x_7 = 0$, $T_8 = 373.1$ K, and $x_8 = 1$; Brayton cycle: fluid = helium, $T_1 = 423$ K, $p_1 = 100$ kPa, $T_3 = 1500$ K, $p_3 = 800$ kPa, and $m$dot = 1 kg/sec.

3. Display results. The answers are: rate of heat added from the heat source = 2722 kW, rate of heat removed to the heat sink = −1182 kW, power required by the compressor = −2854 kW, power produced by the turbine = 4394 kW, net power produced = 1540 kW, and efficiency of the cycle = 56.58%. (See Fig. 7.28a.)

4. Optimization. Draw the sensitivity diagram of net power versus p_3 as shown in Fig. 7.28b. The maximum net power is about 1697 kW, and p_3 at the maximum net power condition is about 491.6 kPa.

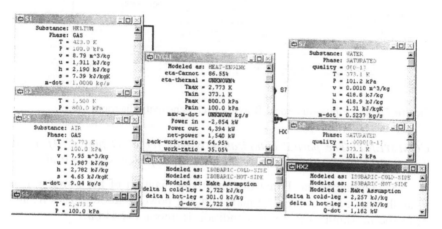

Figure 7.28a   Finite-time ideal Brayton cycle.

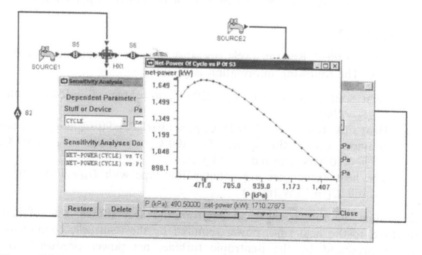

Figure 7.28b   Finite-time ideal Brayton cycle sensitivity diagram.

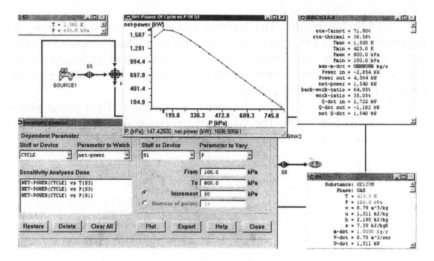

Figure 7.28c   Finite-time ideal Brayton cycle sensitivity diagram.

COMMENT: The partial optimization is only for $\partial(net\ power)/\partial(p_3) = 0$. To have the full optimization, we must let $\partial(net\ power)/\partial(p_1) = 0$ also.

5.   Optimization. Draw the sensitivity diagram of net power versus p_1 as shown in Fig. 7.28c. The maximum net power is about 1699 kW, and p_3 at the maximum net power condition is about 147.4 kPa.

COMMENT: The partial optimization is only for $\partial(net\ power)/\partial(p_3) = 0$. To have the full optimization, we must let $\partial(net\ power)/\partial(p_1) = 0$ also.

Example 7.18

A finite-time ideal Brayton cycle operates between a heat source and a heat sink. The following information is given:

> Heat source: fluid = air, $T_4 = 2500°C$, $p_4 = 1$ bar, $T_5 = 1500°C$, and $p_5 = 1$ bar
> Heat sink: fluid = air, $T_7 = 15°C$, $p_7 = 1$ bar, $T_8 = 90°C$, and $p_8 = 1$ bar
> Brayton cycle: fluid = air, $T_1 = 100°C$, $p_1 = 1$ bar, $T_3 = 1200°C$, $p_3 = 10$ bars, and $mdot = 1$ kg/sec
> The heat exchangers are counter-flow type with $U_H = 1$ kW/(m^{2}K) and $U_L = 1$ kW/(m^{2}K)

Determine the rate of heat added from the heat source, rate of heat removed to the heat sink, power required by the isentropic compressor, power produced by the isentropic turbine, net power produced, and efficiency of the cycle.

(a) Optimize the cycle based on cycle efficiency with respect to p_3, (b) optimize the cycle based on net power with respect to p_3, (c) optimize the cycle based on net power per unit conductance of heat exchanger with respect to p_3, and (d) optimize the cycle based on net power per unit surface of heat exchanger with respect to p_3.

To solve this problem by CyclePad, we take the following steps:

1. Build the cycle and its surroundings as shown in Fig. 7.26.
2. Analysis: (a) Assume that the heat exchangers are isobaric and counter-flow type, and turbine and compressor are isentropic. (b) Input heat source: fluid = air, $T_4 = 2500°C$, $p_4 = 1$ bar, $T_5 = 1500°C$, and $p_5 = 1$ bar; heat sink: fluid = air, $T_7 = 15°C$, $p_7 = 1$ bar, $T_8 = 90°C$, and $p_8 = 1$ bar; and Brayton cycle: fluid = air, $T_1 = 100°C$, $p_1 = 1$ bar, $T_3 = 1200°C$, $p_3 = 10$ bars, and $m$dot = 1 kg/sec, as shown in Fig. 7.29a.
3. Display results. The answers are: rate of heat added from the heat source = 755.3 kW, rate of heat removed to the heat sink = −391.2 kW, net power produced = 364.1 kW, $LMTD_H = 1172$ K, $LMTD_L = 203.3$ K, and efficiency of the cycle = 48.21% as shown in Fig. 7.29b.

Change $p_3 = 2, 3, 4, 5, 6, 7, 8, 9.$ 11, 12, 14, and 15 bars. The results are as shown in Table 7.4 and we have: (a) optimize the cycle based on cycle efficiency with respect to p_3, $\eta_{max} = 52.95\%$ at $p_3 = 14$ bars; (b) optimize the cycle based on net power with respect to p_3, net power$_{max} = 364.7$ kW at

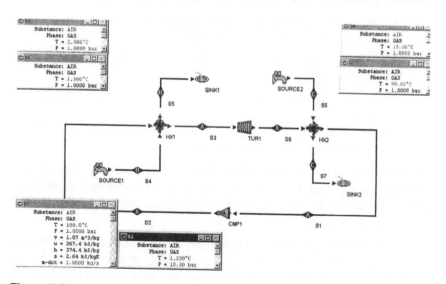

Figure 7.29a Finite-time Brayton cycle input.

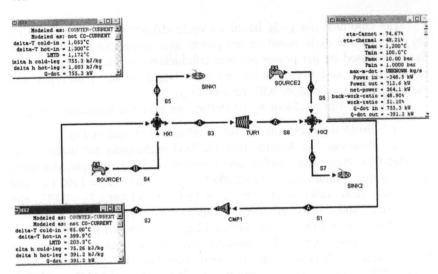

Figure 7.29b Finite-time Brayton cycle output.

$p_3 = 11$ bars; and (c) optimize the cycle based on net power per unit conductance of heat exchangers with respect to p_3, and net power per unit conductance of heat exchangers$_{max} = 151.8 \, kW/(kW/K)$ at $p_3 = 14$ bars.

Review Problems 7.10   Finite-Time Ideal Brayton Cycle

1. A finite-time ideal Brayton cycle operates between a heat source and a heat sink. The following information is given:

Heat source: fluid = air, $T_5 = 2773 \, K$, $p_5 = 100 \, kPa$, $T_6 = 2473 \, K$, and $p_6 = 100 \, kPa$

Heat sink: fluid = water, $T_7 = 373.1 \, K$, $x_7 = 0$, $T_8 = 373.1 \, K$, and $x_8 = 1$

Brayton cycle: fluid = helium, $T_1 = 423 \, K$, $p_1 = 100 \, kPa$, $T_3 = 1800 \, K$, $p_3 = 800 \, kPa$, and $m$dot = 1 kg/sec

Determine the rate of heat added from the heat source, rate of heat removed to the heat sink, power required by the isentropic compressor, power produced by the isentropic turbine, net power produced, and efficiency of the cycle.

ANSWERS: rate of heat added from the heat source = 4275 kW, rate of heat removed to the heat sink = −1856 kW, power required by the compressor = −2854 kW, power produced by the turbine = 5272 kW, net power produced = 2419 kW, and efficiency of the cycle = 56.58%.

Table 4 Finite-Time Brayton Cycle Optimization

$p3$ bar	CY EFF %	$QDOT_H$ kW	$QDOT_L$ kW	$LMTD_H$ K	$LMTD_L$ K	$U_H A_H$ kW/K	$U_L A_L$ kW/K	$SUM(U_A)$ kW/K	PNET kW	PNET/U_A kW/kW/K	A_H m²	A_L m²	PNET/A kW/m²
2	17.97	1022	838.2	1309	331	0.780749	2.532326	3.313075	183.6	55.4168	0.013793	5.064653	36.1528
3	26.94	965.7	705.5	1281	295.3	0.753864	2.389096	3.14296	260.2	82.7882	0.009182	4.778192	54.35131
4	32.7	921.8	620.3	1259	271.6	0.732168	2.283873	3.016042	301.5	99.96546	0.007575	4.567747	65.89701
5	36.86	885.2	558.9	1240	254	0.713871	2.200394	2.914265	326.3	111.9665	0.006743	4.400787	74.03238
6	40.07	853.4	511.5	1224	240.1	0.697222	2.130362	2.827585	341.9	120.9159	0.006231	4.260725	80.12739
7	42.65	825.3	473.3	1209	228.7	0.68263	2.069523	2.752154	352	127.8998	0.005879	4.139047	84.92311
8	44.8	799.9	441.6	1196	219	0.668813	2.016438	2.685251	358.3	133.4326	0.005628	4.032877	88.72096
9	46.62	776.7	414.6	1183	210.7	0.656551	1.967727	2.624278	362.1	137.9808	0.005434	3.935453	91.88286
10	48.21	755.3	391.2	1172	203.3	0.644454	1.92425	2.568704	364.1	141.7446	0.005285	3.8485	94.47855
11	49.6	735.5	370.6	1161	196.8	0.633506	1.88313	2.516636	364.7	144.9157	0.005164	3.76626	96.70088
12	50.83	716.6	352.3	1151	190.9	0.622589	1.845469	2.468058	364.3	147.6059	0.005066	3.690938	98.56593
14	52.95	682.3	321	1132	180.6	0.602739	1.777409	2.380147	361.3	151.7973	0.004919	3.554817	101.4963
15	51.95	699	335.9	1142	185.5	0.612084	1.810782	2.422866	363.1	149.8639	0.004987	3.621563	100.1227

2.   A finite-time ideal Brayton cycle operates between a heat source and a heat sink. The following information is given:

Heat source: fluid = air, $T_5 = 2773$ K, $p_5 = 100$ kPa, $T_6 = 2473$ K, and $p_6 = 100$ kPa

Heat sink: fluid = water, $T_7 = 373.1$ K, $x_7 = 0$, $T_8 = 373.1$ K, and $x_8 = 1$

Brayton cycle: fluid = helium, $T_1 = 423$ K, $p_1 = 100$ kPa, $T_3 = 1800$ K, $p_3 = 800$ kPa, and $m$dot $= 1$ kg/sec

Optimize the net power produced by the cycle with fixed p_3. Draw the sensitivity diagram of net power versus p_1. Find the maximum net power and p_1 at the maximum net power condition.

ANSWERS: The maximum net power is about 2457 kW, and p_1 at the maximum net power condition is about 145.2 kPa.

3.   A finite-time ideal Brayton cycle operates between a heat source and a heat sink. The following information is given:

Heat source: fluid = air, $T_5 = 2773$ K, $p_5 = 100$ kPa, $T_6 = 2473$ K, and $p_6 = 100$ kPa

Heat sink: fluid = water, $T_7 = 373.1$ K, $x_7 = 0$, $T_8 = 373.1$ K, and $x_8 = 1$

Brayton cycle: fluid = helium, $T_1 = 423$ K, $p_1 = 100$ kPa, $T_3 = 1800$ K, $p_3 = 800$ kPa, and $m$dot $= 1$ kg/sec

Determine the rate of heat added from the heat source, rate of heat removed to the heat sink, power required by the isentropic compressor, power produced by the isentropic turbine, net power produced, and efficiency of the cycle.

ANSWERS: rate of heat added from the heat source = 3803 kW, rate of heat removed to the heat sink = -1510 kW, power required by the compressor = -3326 kW, power produced by the turbine = 5619 kW, net power produced = 2293 kW, and efficiency of the cycle = 60.30%.

4.   A finite-time ideal Brayton cycle operates between a heat source and a heat sink. The following information is given:

Heat source: fluid = air, $T_5 = 2773$ K, $p_5 = 100$ kPa, $T_6 = 2473$ K, and $p_6 = 100$ kPa

Heat sink: fluid = water, $T_7 = 373.1$ K, $x_7 = 0$, $T_8 = 373.1$ K, and $x_8 = 1$

Brayton cycle: fluid = helium, $T_1 = 423$ K, $p_1 = 100$ kPa, $T_3 = 1800$ K, $p_3 = 1$ MPa, and $m$dot $= 1$ kg/sec

Optimize the net power produced by the cycle with fixed p_1. Draw the sensitivity diagram of net power versus p_3. Find the maximum net power and p_3 at the maximum net power condition.

ANSWERS: The maximum net power is about 2460 kW, and p_3 at the maximum net power condition is about 608.6 kPa.

5. A finite-time ideal Brayton cycle operates between a heat source and a heat sink. The following information is given:

Heat source: fluid = air, $T_5 = 2773$ K, $p_5 = 100$ kPa, $T_6 = 2473$ K, and $p_6 = 100$ kPa

Heat sink: fluid = water, $T_7 = 373.1$ K, $x_7 = 0$, $T_8 = 373.1$ K, and $x_8 = 1$

Brayton cycle: fluid = air, $T_1 = 423$ K, $p_1 = 100$ kPa, $T_3 = 1500$ K, $p_3 = 800$ kPa, and $m$dot $= 1$ kg/sec

Determine the rate of heat added from the heat source, rate of heat removed to the heat sink, power required by the isentropic compressor, power produced by the isentropic turbine, net power produced, and efficiency of the cycle.

Optimize the net power produced by the cycle with fixed p_1. Draw the sensitivity diagram of net power versus p_3. Find the maximum net power and p_3 at the maximum net power condition.

ANSWERS: rate of heat added from the heat source $= 736.3$ kW, rate of heat removed to the heat sink $= -406.4$ kW, power required by the compressor $= -344.0$ kW, power produced by the turbine $= 674.2$ kW, net power produced $= 329.8$ kW, and efficiency of the cycle $= 44.80\%$.

The maximum net power is about 330.7 kW, and p_3 at the maximum net power condition is about 836.8 kPa.

7.11 ACTUAL BRAYTON FINITE TIME CYCLE

The actual finite-time Brayton cycle as shown in Fig. 7.30 consists of two adiabatic processes and two isobaric heat-transfer processes. The cycle exchanges heat with its surroundings in the two isobaric external irreversible heat-transfer processes. By taking into account the rates of heat transfer associated with the cycle, the upper bound of the power output of the cycle can be found as illustrated in Example 7.19.

Example 7.19

A finite time actual Brayton cycle operates between a heat source and a heat sink. The following information is given:

Heat source: fluid = air, $T_5 = 2773$ K, $p_5 = 100$ kPa, $T_6 = 2473$ K, and $p_6 = 100$ kPa

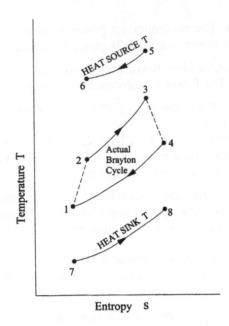

Figure 7.30   Actual finite-time Brayton cycle.

Heat sink: fluid = water, $T_7 = 373.1$ K, $x_7 = 0$, $T_8 = 373.1$ K, and $x_8 = 1$
Brayton   cycle: fluid = air, $\eta_{compressor} = 85\%$, $\eta_{turbine} = 85\%$,
$T_1 = 423$ K, $p_1 = 100$ kPa, $T_3 = 1500$ K, $p_3 = 800$ kPa,   and
$m$dot $= 1$ kg/sec

Determine the rate of heat added from the heat source, rate of heat removed to the heat sink, power required by the isentropic compressor, power produced by the isentropic turbine, net power produced, and efficiency of the cycle.

Optimize the net power produced by the cycle with fixed p_1. Draw the sensitivity diagram of net power versus p_3. Find the maximum net power and p_3 at the maximum net power condition.

Optimize the net power produced by the cycle with fixed p_3. Draw the sensitivity diagram of net power versus p_1. Find the maximum net power and p_3 at the maximum net power condition.

To solve this problem by CyclePad, we take the following steps:

    1.   Build the cycle and its surroundings as shown in Fig. 7.30.

    2.   Analysis: (a) Assume that the heat exchangers are isobaric, and turbine and compressor are isentropic. (b) Input heat source: fluid = air, $T_5 = 2773$ K, $p_5 = 100$ kPa, $T_6 = 2473$ K, and $p_6 = 100$ kPa; heat

sink: fluid = water, $T_7 = 373.1$ K, $x_7 = 0$, $T_8 = 373.1$ K, and $x_8 = 1$; Brayton cycle: fluid = air, $\eta_{compressor} = 85\%$, $\eta_{turbine} = 85\%$, $T_1 = 423$ K, $p_1 = 100$ kPa, $T_3 = 1500$ K, $p_3 = 800$ kPa, and $m$dot = 1 kg/sec.

3.   Display results. The answers are: rate of heat added from the heat source = 675.5 kW, rate of heat removed to the heat sink = −507.6 kW, power required by the compressor = −405.2 kW, power produced by the turbine = 573.1 kW, net power produced = 167.9 kW, and efficiency of the cycle = 24.86%. (See Fig. 7.31a.)

4.   Optimization. Draw the sensitivity diagram of net power versus p_3 as shown in Fig. 7.31b. The maximum net power is about 179.8 kW, and p_3 at the maximum net power condition is about 544.3 kPa.

COMMENT: The partial optimization is only for $\partial(net\ power)/\partial(p_3) = 0$. To have the full optimization, we must let $\partial(net\ power)/\partial(p_1) = 0$ also.

5.   Optimization. Draw the sensitivity diagram of net power versus p_3 as shown in Fig. 7.31c. The maximum net power is about 179.8 kW, and p_1 at the maximum net power condition is about 147.4 kPa.

COMMENT: The partial optimization is only for $\partial(net\ power)/\partial(p_{31}) = 0$. To have the full optimization, we must let $\partial(net\ power)/\partial(p_3) = 0$ also.

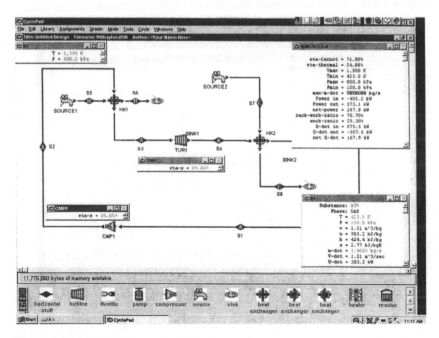

Figure 7.31a   Finite-time actual Brayton cycle.

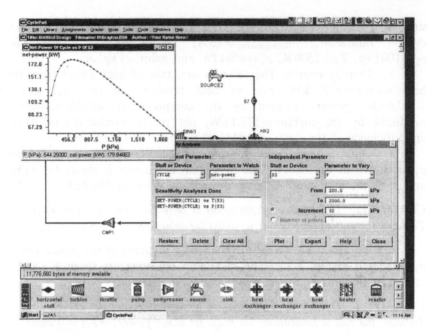

Figure 7.31b Finite-time actual Brayton cycle sensitivity diagram.

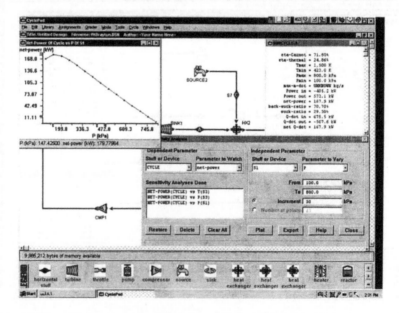

Figure 7.31c Finite-time actual Brayton cycle sensitivity diagram.

Review Problems 7.11   Finite-Time Actual Brayton Cycle

1.   A finite-time actual Brayton cycle operates between a heat source and a heat sink. The following information is given:

Heat source: fluid = air, $T_5 = 2773$ K, $p_5 = 100$ kPa, $T_6 = 2473$ K, and $p_6 = 100$ kPa

Heat sink: fluid = water, $T_7 = 373.1$ K, $x_7 = 0$, $T_8 = 373.1$ K, and $x_8 = 1$

Brayton cycle: fluid = air, $\eta_{compressor} = 80\%$, $\eta_{turbine} = 80\%$, $T_1 = 423$ K, $p_1 = 100$ kPa, $T_3 = 1500$ K, $p_3 = 800$ kPa, and $m$dot $= 1$ kg/sec

Determine the rate of heat added from the heat source, rate of heat removed to the heat sink, power required by the isentropic compressor, power produced by the isentropic turbine, net power produced, and efficiency of the cycle.

ANSWERS: rate of heat added from the heat source = 650.2 kW, rate of heat removed to the heat sink = −541.3 kW, power required by the compressor = −430.5 kW, power produced by the turbine = 539.4 kW, net power produced = 108.9 kW, and efficiency of the cycle = 16.74%.

2.   A finite-time actual Brayton cycle operates between a heat source and a heat sink. The following information is given:

Heat source: fluid = air, $T_5 = 2773$ K, $p_5 = 100$ kPa, $T_6 = 2473$ K, and $p_6 = 100$ kPa

Heat sink: fluid = water, $T_7 = 373.1$ K, $x_7 = 0$, $T_8 = 373.1$ K, and $x_8 = 1$

Brayton cycle: fluid = air, $\eta_{compressor} = 80\%$, $\eta_{turbine} = 80\%$, $T_1 = 423$ K, $p_1 = 100$ kPa, $T_3 = 1500$ K, $p_3 = 800$ kPa, and $m$dot $= 1$ kg/sec

Optimize the net power produced by the cycle with fixed p_1. Draw the sensitivity diagram of net power versus p_3. Find the maximum net power and p_3 at the maximum net power condition.

ANSWERS: The maximum net power is about 135.0 kW, and p_3 at the maximum net power condition is about 200.0 kPa.

3.   A finite-time actual Brayton cycle operates between a heat source and a heat sink. The following information is given:

Heat source: fluid = air, $T_5 = 2773$ K, $p_5 = 100$ kPa, $T_6 = 2473$ K, and $p_6 = 100$ kPa

Heat sink: fluid = water, $T_7 = 373.1$ K, $x_7 = 0$, $T_8 = 373.1$ K, and $x_8 = 1$

Brayton cycle: fluid = air, $\eta_{compressor} = 90\%$, $\eta_{turbine} = 90\%$, $T_1 = 423$ K, $p_1 = 100$ kPa, $T_3 = 1500$ K, $p_3 = 800$ kPa,   and $m$dot $= 1$ kg/sec

Determine the rate of heat added from the heat source, rate of heat removed to the heat sink, power required by the isentropic compressor, power produced by the isentropic turbine, net power produced, and efficiency of the cycle.

Optimize the net power produced by the cycle with fixed p_1. Draw the sensitivity diagram of net power versus p_3. Find the maximum net power and p_3 at the maximum net power condition.

ANSWERS: rate of heat added from the heat source $= 698.0$ kW, rate of heat removed to the heat sink $= -473.9$ kW, power required by the compressor $= -382.7$ kW, power produced by the turbine $= 606.8$ kW, net power produced $= 224.1$ kW, and efficiency of the cycle $= 32.11\%$.
The maximum net power is about 227.0 kW, and p_3 at the maximum net power condition is about 608.6 kPa.

4. A finite-time actual Brayton cycle operates between a heat source and a heat sink. The following information is given:

Heat source: fluid $=$ air, $T_5 = 2773$ K, $p_5 = 100$ kPa, $T_6 = 2473$ K, and $p_6 = 100$ kPa
Heat sink: fluid $=$ water, $T_7 = 373.1$ K, $x_7 = 0$, $T_8 = 373.1$ K, and $x_8 = 1$
Brayton cycle: fluid $=$ helium, $\eta_{compressor} = 90\%$, $\eta_{turbine} = 90\%$, $T_1 = 423$ K, $p_1 = 100$ kPa, $T_3 = 1500$ K, $p_3 = 800$ kPa, and $m$dot $= 1$ kg/sec

Determine the rate of heat added from the heat source, rate of heat removed to the heat sink, power required by the isentropic compressor, power produced by the isentropic turbine, net power produced, and efficiency of the cycle.

Optimize the net power produced by the cycle with fixed p_1. Draw the sensitivity diagram of net power versus p_3. Find the maximum net power and p_3 at the maximum net power condition.

ANSWERS: rate of heat added from the heat source $= 2405$ kW, rate of heat removed to the heat sink $= -1621$ kW, power required by the compressor $= -3171$ kW, power produced by the turbine $= 3954$ kW, net power produced $= 783.6$ kW, and efficiency of the cycle $= 32.59\%$.
The maximum net power is about 1158 kW, and p_3 at the maximum net power condition is about 357.1 kPa.

5. A finite-time actual Brayton cycle operates between a heat source and a heat sink. The following information is given:

Heat source: fluid $=$ air, $T_5 = 2773$ K, $p_5 = 100$ kPa, $T_6 = 2473$ K, and $p_6 = 100$ kPa

Heat sink: fluid = water, $T_7 = 373.1$ K, $x_7 = 0$, $T_8 = 373.1$ K, and $x_8 = 1$
Brayton cycle: fluid = helium, $\eta_{compressor} = 90\%$, $\eta_{turbine} = 90\%$,
$T_1 = 423$ K, $p_1 = 100$ kPa, $T_3 = 1500$ K, $p_3 = 800$ kPa, and
$mdot = 1$ kg/sec

Optimize the net power produced by the cycle with fixed p_3. Draw the sensitivity diagram of net power versus p_1. Find the maximum net power and p_3 at the maximum net power condition.

ANSWERS: The maximum net power is about 1163 kW, and p_1 at the maximum net power condition is about 197.5 kPa.

7.12 OTHER FINITE TIME CYCLES

Finite-time thermodynamics is one of the newest and most challenging areas in thermodynamics. A book entitled *Recent Advances in Finite Time Thermodynamics* (editors: Chih Wu, Lingen Chen, and Jincan Chen, Nova Science Publishers, Inc., New York, USA, 1999, ISBN 1-56072-644-4) provides results from research, which continues at an impressive rate. The book contains many academic and industrial papers that are relevant to current problems and practice. The numerous contributions from the international thermodynamic community are indicative of the continuing global interest in finite-time thermodynamics.

The readers should find the papers listed in the Bibliography informative and useful for analysis and design of various finite-time thermodynamic cycles. It is hoped that these papers will provide interest and encouragement for further study in the area of finite-time thermodynamics.

7.13 SUMMARY

Maximum efficiency and maximum coefficient of performance are not necessarily the primary concern in the design of a real cycle. Net power output and specific net power output in a heat engine, cooling load, and specific cooling load in a refrigerator, and heating load and specific heating load in a heat pump are probably more important in the industrial design of thermodynamic cycles. A different criteria of real cycle performance is provided by finite-time thermodynamics. The basic finite-time thermodynamic cycles are the Carnot, Brayton, and Rankine cycles. Literature concerning other finite-time thermodynamic cycles are also provided in the Bibliography.

7.14 BIBLIOGRAPHY

7.14.1 Carnot Cycle

Wu, C., Power optimization of a finite-time Carnot heat engine. *Energy: The International Journal*, **13**(9), 681–687, 1988.

Wu, C. and Kiang, R.L., Finite time thermodynamic analysis of a Carnot engine with internal irreversibility. *Energy: The International Journal*, **17**(12), 1173–1178, 1992.

Wu, C., Specific heating load of an endo-reversible Carnot heat pump. *International Journal of Ambient Energy*, **14**(1), 25–28, 1993.

Wu, C., Maximum obtainable specific cooling load of a Carnot refrigerator. *Energy Conversion and Management*, **36**(1), 7–10, 1995.

Chen, L. and Wu, C., Performance of an endo-reversible Carnot refrigerator. *Energy Conversion and Management*, **37**(10), 1509–1512, 1996.

Chen, L. and Wu, C., The influence of heat transfer law on the endo-reversible Carnot refrigerator. *Journal of the Institute of Energy*, **69**(480), 96–100, 1996.

Wu, C., General performance characteristics of a finite-speed Carnot refrigerator. *Applied Thermal Engineering*, **16**(4), 299–304, 1996.

Chen, L. and Wu, C., Optimal performance of an endo-reversible Carnot heat pump. *Energy Conversion and Management*, **38**(14), 1439–1444, 1997.

Chen, L. and Wu, C., Effect of heat transfer law on the performance of a generalized Carnot heat pump. *Journal of the Institute of Energy*, **72**(2), 64–68, 1999.

Chen, L. and Wu, C., Effect of heat transfer law on the performance of generalized irreversible Carnot refrigerator. *Journal of Engineering Thermophysics*, **20**(2), 10–13, 1999.

7.14.2 Brayton Cycle

Wu, C., Work and power optimization of a finite-time Brayton cycle. *International Journal of Ambient Energy*, **11**(3), 129–136, 1990.

Wu, C., Power optimization of an endo-reversible Brayton gas heat engine. *Energy Conversion and Management*, **31**(6), 561–565, 1991.

Wu, C. and Kiang, R.L., Power performance of a non-isentropic Brayton cycle. *ASME Journal of Engineering for Gas Turbines and Power*, **113**, 501–504, 1991.

Wu, C., Performance of a regenerative Brayton heat engine. *Energy: The International Journal*, **21**(2), 71–76, 1996.

Chen, L., Sun, F., and Wu, C., Performance analysis for a real closed regenerated Brayton cycle via methods of finite-time thermodynamics. *International Journal of Ambient Energy*, **20**(2), 95–104, 1999.

Chen, L., Sun, F., and Wu, C., Performance analysis of a closed regenerated Brayton heat pump with internal irreversibility. *International Journal of Energy Research*, **23**, 1039–1050, 1999.

7.14.3 Rankine Cycle

Wu, C., Power optimization of a finite-time Rankine heat engine. *International Journal of Heat and Fluid Flow*, **10**(2), 134–138, 1989.

Wu, C., Intelligent computer aided design on optimization of specific power of finite-time Rankine cycle using CyclePad. *Journal of Computer Application in Engineering Education*, **6**(1), 9–13, 1998.

Other finite-time thermodynamic cycle literature including Atkinson, combined and cascaded, Diesel, dual, Ericsson, Otto, Rallis, and Stirling cycles are provided in the following:

7.14.4 Atkinson Cycle

Chen, L., Sun, F., and Wu, C., Efficiency of an Atkinson engine at maximum power density. *Energy Conversion and Management*, **39**(3/4), 337–342, 1998.

7.14.5 Combined and Cascaded Cycle

Wu, C., Power performance of a cascade Endo-reversible cycle. *Energy Conversion and Management*, vol. 30, no. 3, pp. 261–266, 1990.

Wu, C., Karpouzian, G., and Kiang, R.L., The optimal power performance of an endo-reversible Combined cycle. *Journal of the Institute of Energy*, vol. **65**(462), pp. 41–45, 1992.

Chen, J. and Wu, C., Maximum specific power output of a two-stage endo-reversible Combined cycle. *Energy: The International Journal*, **20**(4), 305–309, 1995.

Wu, C., Maximum obtainable power of a Carnot combined power plant. *Heat Recovery Systems and CHP*, **15**(4), 351–355, 1995.

Wu, C., Performance of a cascade endo-reversible heat pump system. *The Institute of Energy Journal*, **68**(476), 137–141, 1995.

Wu, C., Finite-time thermodynamic analysis of a two-stage combined heat pump system. *International Journal of Ambient Energy*, **16**(4), 205–208, 1995.

Chen, J. and Wu, C., General performance characteristics of a N-stage endo-reversible combined power cycle system at maximum specific power output. *Energy Conversion and Management*, **37**(9), 1401–1406, 1996.

Chen, L., Sun, F., and Wu, C., The equivalent cycles of an *n*-stage irreversible combined refrigeration system. *International Journal of Ambient Energy*, **18**(4), 197–204, 1997.

Wu, C., Intelligent computer aided analysis of a Rankine/Rankine combined cycle. *International Journal of Energy, Environment and Economics*, **7**(2), 239–244, 1998.

Chen, L., Sun, F., and Wu, C., A generalized model of a combined refrigeration cycle and its performance. *International Journal of Thermal Sciences (Revue Generale de Thermique)*, **38**(8), 712–718, 1999.

Chen, J. and Wu, C., Thermoeconomic analysis on the performance characteristics of a multi-stage irreversible combined heat pump system. *ASME Journal of Energy Resources Technology*, **122**(4), 212–216, 2000.

7.14.6 Diesel Cycle

Wu, C. and Blank, D.A., The effect of combustion on a power optimized endo-reversible Diesel cycle. *Energy Conversion and Management*, **34**(6), 493–498, 1993.

Chen, L. and Wu, C., Heat transfer effect on the net work and/or power versus efficiency characteristics for the air standard Diesel cycles. *Energy: The International Journal*, **21**(12), 1201–1205, 1996.

7.14.7 Dual Cycle

Wu, C. and Blank, D.A., The effect of combustion on a power optimized endo-reversible Dual cycle. *International Journal of Power and Energy Systems*, **14**(3), 98–103, 1994.

Chen, L., Sun, F., and Wu, C., Finite thermodynamics performance of a Dual cycle. *International Journal of Energy Research*, **23**(9), 765–772, 1999.

7.14.8 Ericsson Cycle

Blank, D.A. and Wu, C., Performance potential of a terrestrial solar-radiant Ericsson power cycle from finite-time thermodynamics. *International Power and Energy Systems*, **15**(2), 78–84, 1995.

Blank, D.A. and Wu, C., Power limit of an endo-reversible Ericsson cycle with regeneration. *Energy Conversion and Management*, **37**(1), 59–66, 1996.

Chen, L., Sun, F., and Wu, C., Cooling and heating rate limits of a reversed reciprocating Ericsson cycle at steady state. *Proceedings of the Institute of Mechanical Engineers, Part A, Journal of Power and Energy*, **214**, 75–85, 2000.

7.14.9 Otto Cycle

Wu, C. and Blank, D.A., The effect of combustion on a work optimized endo-reversible Otto cycle. *Journal of the Institute of Energy*, **65**(463), 86–89, 1992.

Chen, L., Sun, F., and Wu, C., Heat transfer effects on the net work output and efficiency characteristic for an air-standard Otto cycles. *Energy Conversion and Management*, **39**(7), 643–648, 1998.

7.14.10 Rallis Cycle

Wu, C., Analysis of an endo-reversible Rallis cooler. *Energy Conversion and Management*, **35**(1), 79–85, 1994.

7.14.11 Stirling Cycle

Wu, C., Analysis of an endo-reversible Stirling cooler. *Energy Conversion and Management*, **34**(12), 1249–1253, 1993.

Blank, D.A. and Wu, C., Power optimization of an endo-reversible Stirling cycle with regeneration. *Energy: The International Journal*, **19**(1), 125–133, 1994.

Blank, D.A. and Wu, C., Power optimization of an extra-terrestrial, solar radiant Stirling heat engine. *Energy: The International Journal*, **20**(6), 523–530, 1995.

Chen, L., Wu, C., and Sun, F., Optimum performance of irreversible Stirling engine with imperfect regeneration. *Energy Conversion and Management*, **39**(8), 727–732, 1998.

Chen, L., Wu, C., and Sun, F., Optimal performance of an irreversible Stirling cryocooler. *International Journal of Ambient Energy*, **20**(1), 39–44, 1999.

Chen, L., Wu, C., and Sun, F., Performance characteristic of an endo-reversible Stirling refrigerator. *International Journal of Power and Energy Systems*, **19**(1), 79–82, 1999.

Appendix

q	Specific heat transfer, kJ/kg
Q	Heat transfer, kJ
Q_H	Heat transfer with high-temperature thermal reservoir, kJ
Q_L	Heat transfer with low-temperature thermal reservoir, kJ
Q_{dot}	Heat transfer rate, kW
r	Compression ratio
r_c	Cut-off ratio
r_p	Pressure ratio
R	Gas constant, kJ/[kg(K)]
R_u	Universal gas constant, kJ/[kmol(K)]
s	Specific entropy, kJ/[kg(K)]
S	Entropy, kJ/K
s_{gen}	Specific entropy generation, kJ/[kg(K)]
S_{gen}	Entropy generation, kJ/K
t	Time, s
T	Temperature, K
T_H	Temperature of high-temperature thermal reservoir, K
T_L	Temperature of low-temperature thermal reservoir, K
T_o	Temperature of surroundings, K
u	Specific internal energy, kJ/kg
U	Internal energy, kJ
v	Specific volume, m^3/kg
V	Volume, m^3
V	Velocity, m/s
$Vdot$	Rate of volumetric flow, m^3/s
w	Specific work, kJ/kg
W	Work, kJ
$Wdot$	Power, kW
W_{in}	Work input, kJ
W_{irr}	Irreversible work, kJ
W_{out}	Work output, kJ
W_{rev}	Reversible work, kJ
x	Quality
z	Elevation, m

Greek letters

β	Coefficient of performance
β_R	Coefficient of performance of a refrigerator
β_{HP}	Coefficient of performance of a heat pump
Δ	Finite change in a quantity
δ	Differential change of a path function
η	Heat engine efficiency

η_{Car}	Carnot heat engine efficiency
λ	co-generation ratio
ρ	Density, kg/m^3

Subscripts

abs	Absolute
act	Actual
atm	Atmospheric
e	Exit section
f	Saturated liquid
fg	Difference between saturated vapor and saturated liquid
g	Saturated vapor
gen	Generation
H	High temperature
i	Inlet section
int	Internally
irrev	Irreversible
L	Low temperature
rev	Reversible
s	Isentropic
surr	Surroundings
sys	System
1	Initial state or inlet state
2	Final state or exit state

Index

T - #0024 - 111024 - C0 - 229/152/25 - PB - 9780367394912 - Gloss Lamination